Airborne Pulsed Doppler Radar

Second Edition

For a complete listing of the *Artech House Radar Library,*
turn to the back of this book.

Airborne Pulsed Doppler Radar

Second Edition

Guy Morris
Linda Harkness

Editors

Artech House
Boston • London

Library of Congress Cataloging-in-Publication Data
Airborne pulsed doppler radar/Guy Morris, Linda Harkness, editors.—2nd ed.
    p.   cm.
Includes bibliographical references and index.
ISBN 0-89006-867-4 (alk. paper)
    1. Doppler radar. 2. Pulse compression radar. 3. Radar in aeronautics. I. Morris, G. V.
(Guy V.) , 1935-   . II. Harkness, L. (Linda), 1956-   .
TK6592.D6A37   1996
621.3848'5—dc20   96-42989
                                          CIP

British Library Cataloguing in Publication Data
Airborne pulsed doppler radar. – 2nd ed.
    1. Doppler radar 2. Radar in aeronautics   I. Morris, Guy II. Harkness, Linda
    621.3'8485

    ISBN 0-89006-867-4

Cover design by Lucia Collela

© 1996 ARTECH HOUSE, INC.
685 Canton Street
Norwood, MA 02062

International Standard Book Number: 0-89006-867-4
Library of Congress Catalog Card Number: 96-42989

10 9 8 7 6 5 4 3 2 1

Contents

Preface

The second edition of *Airborne Pulsed Doppler Radar* has been updated to include the technologies that have appeared in user systems since the original publication, or that will appear in the next five years. Examples of these significant revisions include:

1. A chapter on electronic counter-countermeasures;
2. A chapter on phased-array antennas;
3. Discussions of stretch and stepped-frequency high-range-resolution waveforms.

This book evolved from the lecture notes of a continuing education course presented by the Georgia Institute of Technology. The course faculty are the authors and their colleagues at the Georgia Tech Research Institute.

Chapters 1 and 2 present the basic principles of pulse-doppler radar without resorting to a heavily mathematical treatment.

Chapters 3 to 6 describe the high-, medium-, and low-pulse repetition frequency (PRF) modes and presents the advantages and disadvantages of each.

Chapters 7 to 13 explain the major signal-processing functions of doppler filtering, pulse compression, tracking, synthetic aperture, selection of medium PRFs, and resolving range ambiguities. This part also includes a discussion of phased-array antennas to multimode airborne radar.

Chapters 14 to 17 show how to predict the performance of a pulse-doppler radar in the presence of noise and clutter.

Chapter 18 contains a discussion about electronic protection (formerly electronic counter-countermeasures).

CHAPTER 1

Principles of Pulse-Doppler Radar

Guy V. Morris

The term *doppler radar* refers to any radar that is capable of measuring the frequency shift between the transmitted frequency and the frequency of reflections received from objects. Doppler radars are most often used to discriminate between the return from a desired *target* and that from undesired objects, usually *ground clutter*. This chapter presents the basic principles of pulse doppler without resorting to an excessively mathematical treatment. The equations required by the radar practitioner will be treated in some detail in subsequent chapters. The principles will be explained by starting with the simplest doppler radar, a stationary radar employing continuous-wave (CW) transmission and reception, and then moving step by step to the more complex pulsed waveforms. The topics to be discussed include:

1. Types of pulse-doppler radar;
2. Definition of terms;
3. Spectra and waveforms of stationary CW radar;
4. Spectra and waveforms of stationary pulsed radar.

1.1 TYPES OF DOPPLER RADAR

Figure 1.1 depicts the types of doppler radar, all of which differ in performance. The optimum choice is dependent on the operational requirements of the radar. Radars are called upon to operate over such a wide range of conditions that no single type is universally acceptable. Many modern radars are *multimode;* that is, they can function as two or more of the types shown in Figure 1.1.

    CW radars, as the name implies, transmit and receive a continuous waveform. Figure 1.2 shows a simplified block diagram. In high-power CW radar, separate transmit and receive antennas, with adequate isolation between them, are required

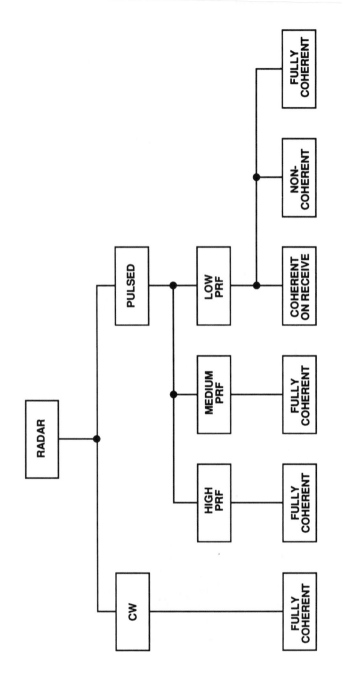

Figure 1.1 Types of radar.

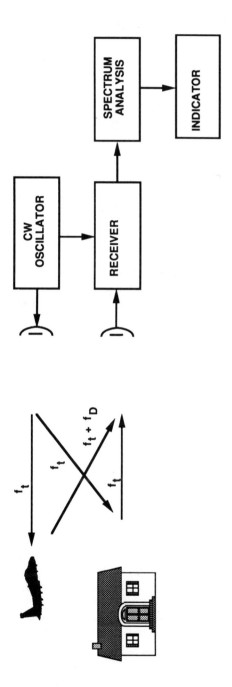

Figure 1.2 CW radar block diagram.

to prevent receiver desensitization. CW radars are not used for airborne surveillance because of the constraints associated with achieving an acceptable antenna installation. Spectrum analysis of the received signal provides discrimination between ground return and moving targets, and permits unambiguous measurement of target velocities.

Low-PRF radars are defined as those radars with a *pulse repetition frequency* (PRF) sufficiently low that range is measured unambiguously. The transmitted pulse travels to and from the range of maximum interest during the interpulse period before the transmission of the next pulse. The unambiguous range R_u is calculated using

$$R_u = \frac{c}{2\text{PRF}} \tag{1.1}$$

where c is the speed of light.

The dividing line between low and medium PRF is not a specific number but depends somewhat on the application. A typical airborne radar might have a PRF of 1,000 Hz that provides an unambiguous range of 150 km (80 nmi). Low-PRF doppler radars often do not unambiguously measure the velocity, that is, the doppler shift.

High-PRF radars are defined as radars with a PRF sufficiently high that all target velocities of interest are unambiguously measured. The maximum doppler shift that can be unambiguously measured is given by

$$f_{d_{\max}} = \text{PRF} = \frac{2fv_{\max}}{c} \tag{1.2}$$

where:

f = transmitter frequency;
$v_{\max}$ = maximum target velocity.

A typical X-band (9-GHz) radar operating in a Mach 2 aircraft might have a PRF of 250 kHz to ensure detection and unambiguous velocity measurement of a Mach 2 target. From (1.1) we can see that the corresponding unambiguous range is only 600m (2,000 ft). Therefore, a target at a range of 150 km (80 nmi) would be highly ambiguous in range. The radar platform velocity must also be taken into account in setting the PRF, as explained in Chapter 5.

Medium-PRF radars are defined as radars with a PRF that produces ambiguities in both range and doppler. It might seem that medium PRF combines the worst features of both high- and low-PRF radars. However, we will show in Chapter 6 that medium PRF is often the best choice of waveform for airborne radar.

1.2 DEFINITIONS

Several terms commonly used in radar technology have already been introduced without definition. It is appropriate to pause here and define some of these terms as they apply to radar, since they do not always have precisely the same meanings in other disciplines.

1.2.1 Doppler Shift

Doppler shift is the shift in frequency between the transmitted radio frequency (RF) carrier and the echoes reflected from moving objects. Such frequency shifts are named after Christian Johann Doppler (1803–1853), who first predicted the phenomenon. The frequency f_1 received by an observer on the ground in Figure 1.3, predicted by the theory of relativity, is given by

$$f_1 = f \frac{c + v}{\sqrt{c^2 - v^2}} \tag{1.3}$$

where:

f = transmitted frequency;
v = component of aircraft velocity in direction of observer;
c = speed of light.

Defining the doppler shift as $f_d = f_1 - f$ and using the fact that the aircraft speed is very small compared to the speed of light, (1.3) reduces to

$$f_d = \frac{fv}{c} \tag{1.4}$$

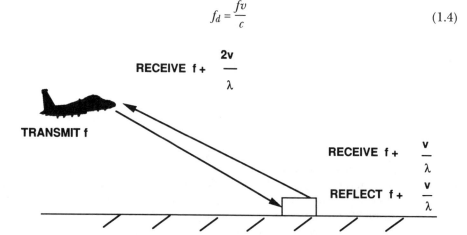

Figure 1.3 Observed doppler shift.

The reflection from the ground may be viewed as a reradiation at frequency f_1. Equation (1.3) or the very accurate approximation (1.4) can be applied to calculate the doppler shift observed in the aircraft. The total doppler shift observed is twice the *one-way* shift and is given by the familiar forms

$$f_d = \frac{2fv}{c} = \frac{2v}{\lambda} \tag{1.5}$$

where:

$f\lambda \quad = \quad c;$
$\lambda \quad = \quad$ wavelength.

The doppler shift can be derived in a second way that is perhaps more instructive for radar applications. Consider a target at a range R. The round-trip distance is $2R$, and the total phase difference between the transmitted and received wave is given by

$$\phi = -2\pi\left(\frac{2R}{\lambda}\right) \tag{1.6}$$

where the negative sign indicates a phase delay. Using the definition of frequency,

$$f = \frac{1}{2\pi}\left(\frac{d\phi}{dt}\right) \tag{1.7}$$

the change in frequency (i.e., the doppler shift), seen at the radar, resulting from a target with changing range, is given by

$$f_d = -\frac{2}{\lambda}\left(\frac{dR}{dt}\right) = \frac{2v}{\lambda} \tag{1.8}$$

1.2.2 Translation to Zero Intermediate Frequency

Equation (1.8) indicates that the doppler shift will be positive, that is, at a higher frequency if the target is approaching (dR/dt negative) and negative if the target is receding. The ability to discriminate between closing (approaching) and opening (receding) targets is often a valuable attribute in airborne applications. Most radars do not perform the frequency comparison between the transmitted wave and the received wave directly at RF, but down-convert the received signal to a convenient intermediate frequency (IF). The most prevalent method of providing the spectral analysis function in modern radars is a *fast Fourier transform* (FFT) digital signal processor. The received spectrum is translated so that zero doppler shift corresponds to zero frequency (baseband). Another choice, common in airborne radars, is to translate the frequency corresponding to main-lobe clutter to zero frequency. This will be discussed in Chapter 2. The discrimination between closing and opening targets can be retained by using the down-conversion method illustrated in Figure 1.4. Figure 1.5 shows a comparison of the *in-phase* (I) and *quadrature* (Q) components of the video for an opening and closing target with the same magnitude of velocity. Note the phase reversal of the Q component. An I/Q pair will be referred to as one complex sample. A spectrum analysis of a series of complex samples is performed using a complex FFT. Opening and closing targets will occupy distinctly different positions in the resulting spectrum.

Some simple radars are *single-channel* and extract only one component, such as the I component. The series of single or "real" samples are analyzed by a "real" FFT. Opening targets and closing targets cannot be differentiated in the resulting spectrum.

1.2.3 Doppler Ambiguities and Blind Speeds in Pulsed Radars

Equation (1.8) suggests that any magnitude of target speed can be measured. However, the familiar Nyquist criterion states that the minimum sampling frequency needed to capture correctly the frequency content of a signal is equal to twice the signal bandwidth. For a pulsed radar, the sampling frequency is the PRF. (A CW radar that samples and uses digital spectrum analysis is subject to the same sampling considerations.) The simple single-channel radar described earlier extracts one sample each pulse repetition interval (PRI) and thus can measure the doppler shift over an unambiguous frequency interval equal to PRF/2. The radar that extracts two samples per PRI (i.e., an I and a Q sample) has an effective sampling rate of twice the PRF and provides an unambiguous frequency interval of PRF.

Equation (1.6) shows the total phase delay between the transmitted and receive waveforms. The phase change between pulses (i.e., samples) is

$$\Delta \phi = 2\pi \left(\frac{2\Delta R}{\lambda} \right) \qquad (1.9)$$

where ΔR is the range change between pulses.

If the phase change between pulses is less than 2π, the doppler frequency can be measured unambiguously. If the phase change is equal to 2π, then the doppler frequency equals the PRF. Obviously, a shift of 2π cannot be distinguished from a shift of any integral multiple of 2π, including zero. The speeds that cause the doppler shift to be an integral multiple of 2π have been historically called *blind speeds*, because the presence of a large ground return at zero frequently prevents

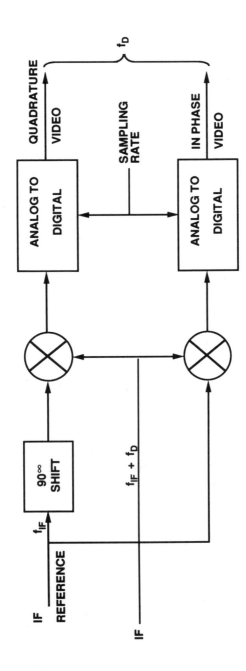

Figure 1.4 Translation to zero IF.

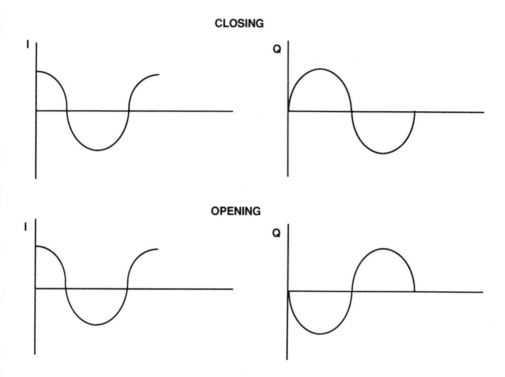

Figure 1.5 Waveforms of opening and closing targets.

the detection of the target of interest. If the phase change is greater than 2π, the targets will generally be detectable, but the observed doppler shift will not correctly represent the target speed. The observed doppler frequency will be incorrect by an integral multiple of the PRF. Multiple PRFs are used to eliminate blind speeds and to resolve ambiguous target speed measurements.

The doppler shift *observed* by a pulsed radar is more correctly shown by modifying (1.8) to read

$$f_d = \left(\frac{2v}{\lambda}\right) \text{ modulo PRF} \tag{1.10}$$

1.2.4 Range Ambiguities and Blind Ranges in Pulsed Radars

Obviously, when the same antenna is used for high-power transmitting and receiving, the radar receiver must be turned off during transmission. The receiver off time must be long enough for the transmitter energy to decay to levels that do not harm the receiver and is therefore longer than the 3-dB transmitter pulse width.

The result is a blind range. The blind range can be substantial in radars that use pulse compression.

Targets at ranges greater than the unambiguous range R_u and not at a blind range may be detected. The observed apparent range R_a will be incorrect by an integral multiple of R_u and is given by

$$R_a = \left(\frac{c\tau_t}{2}\right) \text{ modulo } R_u \tag{1.11}$$

where τ_t is the propagation time to the target.

1.2.5 Fast Fourier Transform

Many modern radars perform the spectrum analysis function in a digital signal processor using an FFT. An FFT is used because it is an efficient digital method of spectral analysis that requires fewer multiplications than other Fourier transform algorithms. Several different software and hardware architectures have been developed to implement an FFT, each having advantages and disadvantages.

The input to the FFT is a sequence of 2^m time samples, where m is an integer. The output is 2^m complex numbers (i.e., having I and Q components) representing the frequency spectrum. The output is equivalent to a bank of uniformly spaced filters covering the frequency region from zero to PRF as shown in Figure 1.6. The filter spacing is PRF/2^m. If the input consists of complex samples, then the entire frequency region from zero to PRF is unambiguous. If the input samples are real, then the unambiguous region is PRF/2. The upper half of the filters contains the complex conjugates of the lower and has no additional information. For this reason, some FFT algorithms operating on real data output only half of the filters.

Since the number of doppler filters depends only on the number of time samples, if the PRF is varied to eliminate blind speeds or resolve doppler ambiguities, the bandwidth and frequency spacing automatically adjust. Thus, the target return may move from one filter to another as the PRF is changed, even if the doppler frequency is unambiguous.

1.2.6 Coherence

Coherent transmitted pulses have a phase continuity from pulse to pulse as if they were gated portions of a continuous RF signal, as shown in Figure 1.7. In fact, coherent pulses are most often generated as a low-level continuous RF signal and are then processed by one or more stages of pulsed amplification, as shown in Figure 1.8. By contrast, pulse trains that are generated by pulsed oscillators (e.g., magnetrons) have random starting phases compared to a CW reference and are referred to as *noncoherent*. (Methods have been developed for low-PRF radars that

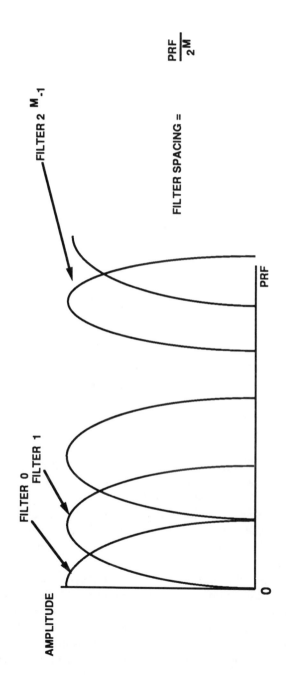

Figure 1.6 Fast Fourier transform.

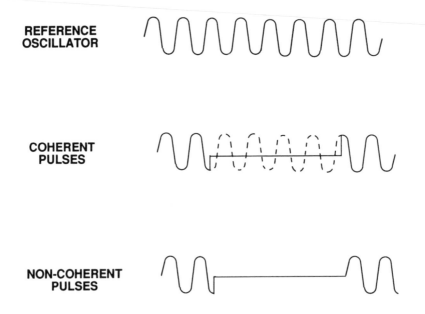

REFERENCE OSCILLATOR

COHERENT PULSES

NON-COHERENT PULSES

Figure 1.7 Coherence.

use a noncoherent transmitter to produce a coherent pulse train within the receiver. These techniques are aptly named *coherent on receive*.) The term MOPA (master oscillator power amplifier) is used frequently in the literature to describe a radar transmitter consisting of a low-level CW oscillator and pulsed power amplifier. The term *fully coherent* will be used occasionally to describe a radar with a MOPA transmitter.

A primary benefit of transmitting coherent pulses is that a sequence of return pulses can be subjected to doppler analysis. The doppler analysis bandwidths are small compared to the IF bandwidth; therefore, the process provides a signal-to-noise improvement. However, the most important benefit is the ability to differentiate between relatively small differences in velocity. A more rigorous and scientific treatment is presented in Chapter 7, but a heuristic example might be helpful at this point. Filters, including doppler filters and FFTs, are resonant circuits. The output is maximum when the input frequency matches the natural resonant frequency of the filter. Consider pushing a child on a swing. The swing is a form of pendulum that is a tuned circuit. If properly timed (i.e., phased), a series of gentle pushes will result in a large amplitude of the swing. Conversely, ill timed pushes may decrease the amplitude. The analogy of the swing is not unlike the action of a series of coherent pulses upon whatever resonant circuit is used to perform the doppler analysis.

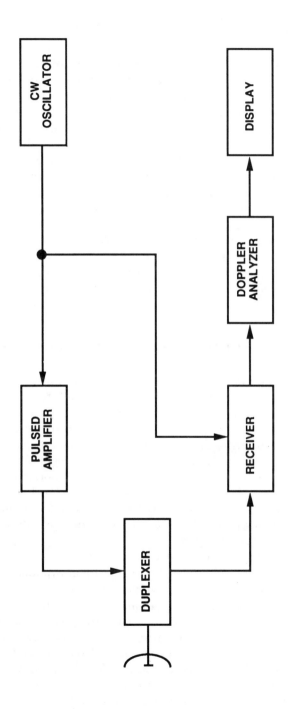

Figure 1.8 Fully coherent pulse-doppler radar.

1.3 SPECTRA AND WAVEFORMS OF STATIONARY CW RADAR

The idealized time waveforms of a stationary CW radar are shown in Figure 1.9. The return from a fixed target is a delayed replica of the transmitted wave at the same frequency as the transmitter. If it were possible to view the moving-target return by itself, it might appear as shown. The moving-target return has been purposely drawn much smaller than the fixed return, since this is the usual situation in doppler radar applications. The moving-target return is shifted in frequency relative to the transmitter. However, if we are viewing the microwave carrier or an IF that is high compared to the doppler shift (e.g., 30-MHz IF and 10-kHz doppler) with our hypothetical oscilloscope, the difference in frequency would be imperceptible. The total return signal is the sum of the returns from the fixed (clutter) and moving targets. Because of the low signal-to-clutter ratio, the total signal resembles the fixed target return with what appears to be a slight distortion.

Subjecting the received signal to spectral analysis yields the spectrum shown in Figure 1.10. Moving targets far smaller than the fixed target, such as −60 dB (amplitude ratio of 0.001), can be detected on the spectral presentation, whereas they cannot be on the time waveform presentation.

1.4 WAVEFORMS AND SPECTRA OF STATIONARY HIGH-PRF RADAR

The high-PRF transmitter waveform is a series of coherent pulses of an RF carrier, as shown in Figure 1.11. The typical transmitter duty factors are from 0.3 to 0.5, but lower values may be used. The receiver is gated off during the time the transmitter is on and vice versa, thus permitting the radar to operate from a single antenna. In fact, the development of high-PRF radar was motivated by the desire to have the advantages of the high average power and doppler detection performance of a CW radar without paying the penalty for a dual antenna installation in an aircraft. The high-PRF radar has much in common with the CW radar, such as unambiguous doppler measurement, the need to use a secondary modulation to measure range, and the methods of signal processing. A descriptive, but never widely used, name for the high-PRF radar was *interrupted CW*.

The spectrum of the transmitter waveform is also shown in Figure 1.11. In addition to the central spectral line at the carrier, there are other spectral lines above and below the carrier at integral multiples of the PRF. The envelope of the transmitter spectrum is dependent on the pulse shape (e.g., $(\sin x/x)^2$ for the rectangular pulse illustrated). The spectrum of a moving target is a shifted replica of the transmitter spectrum and also contains lines spaced at the PRF. The PRF is chosen sufficiently high so that each spectral line of the moving-target return is adjacent to its corresponding transmitter spectral line. Each received spectral line can be readily associated with the transmitter spectral line. This is simply another way of stating that the PRF is chosen high enough to provide unambiguous

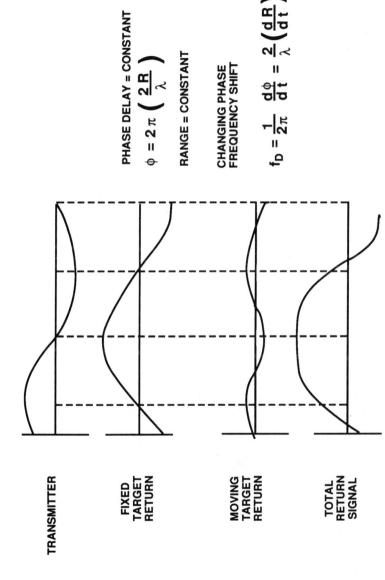

PHASE DELAY = CONSTANT

$$\phi = 2\pi \left(\frac{2R}{\lambda} \right)$$

RANGE = CONSTANT

CHANGING PHASE
FREQUENCY SHIFT

$$f_D = \frac{1}{2\pi} \frac{d\phi}{dt} = \frac{2}{\lambda} \left(\frac{dR}{dt} \right)$$

TRANSMITTER

FIXED
TARGET
RETURN

MOVING
TARGET
RETURN

TOTAL
RETURN
SIGNAL

Figure 1.9 CW time waveforms.

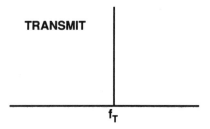

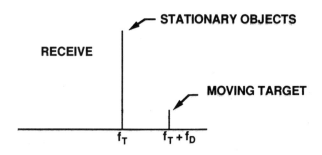

Figure 1.10 CW spectra.

measurement of the doppler frequency. Typical PRFs are 250 kHz for an X-band fire control radar and up to 1 MHz for a K-band missile seeker. In Figure 1.11, the received pulse is illustrated arriving at the most fortuitous time. The entire pulse is received while the transmitter is off. As the target moves, the total round-trip propagation time changes. Part or all of the target return may arrive when the receiver is turned off, resulting in partial or total *eclipsing*. The eclipsed zones for a given PRF may be visualized, as shown in Figure 1.12, as a set of concentric spherical shells that travel with the aircraft and through which the target flies. The distance between shells is the unambiguous range R_u, which is given by

$$R_u = \frac{c}{2\text{PRF}} \tag{1.12}$$

The time that a target spends in an eclipsed state depends on the relative closing rate. High-PRF radars use multiple PRFs to reduce the effects of these blind or eclipsed ranges by preventing long periods of target signal loss. The multiple PRFs do not eliminate the blind doppler region around zero velocity, and by definition there are no higher blind speeds in high-PRF radar.

- **DUTY FACTOR 0.3 TO 0.5**

- **RECEIVED PULSE CAN BE ECLIPSED BY TRANSMIT**

- **TYPICAL PRF's - 250 KHz   X-BAND FIRE CONTROL**
  1 MHz   35 GHz SEEKER

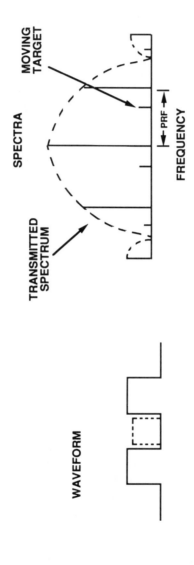

Figure 1.11 High-PRF waveform and spectra.

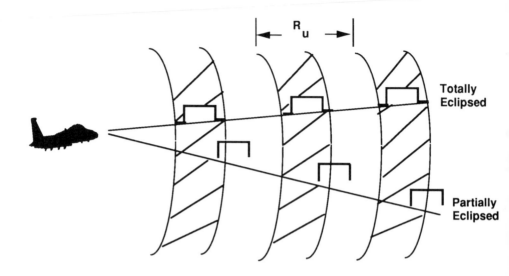

Figure 1.12 Eclipse zones.

1.5 SPECTRA AND WAVEFORMS OF STATIONARY LOW-PRF RADAR

The low-PRF radar transmits a pulse and then turns on its receiver for a time that is equal to the round-trip propagation time to the greatest range of interest. The transmitter duty factors are typically on the order of 0.001. The remainder of the time is available for reception. Figure 1.13 is a representation of the time waveform if an oscilloscope were connected to either the I or Q component of the video. Note that the return video can be in any phase relationship to the transmitter reference: in-phase, quadrature, or 180 degrees out of phase. The video is referred to as *bipolar*. When the returns from many pulses are superimposed on the oscilloscope, the returns from the fixed targets will overlay. If a fixed target and a moving target are at the same range—the difference in propagation times is less than a pulse width—the received video is the vector sum of the two returns. On subsequent pulses, when the target moves even a fraction of the transmitter wavelength, the vector sum of the fixed and moving-target returns changes. As the target moves, the composite return cycles through relative maxima and minima, causing the superimposed sweeps to not overlay at the target range. The appearance of a moving target in this A-scope (range versus amplitude) presentation is called a *butterfly* by some practitioners.

In the idealized illustration of Figure 1.13, the fixed target interference was not much larger than the moving target. As a result, the moving target was readily visible on the oscilloscope. In a more typical case, the target might be on the order of receiver noise and 30 dB less than the fixed target clutter, and therefore not visible. Increased detectability can be provided by doppler signal processing. During

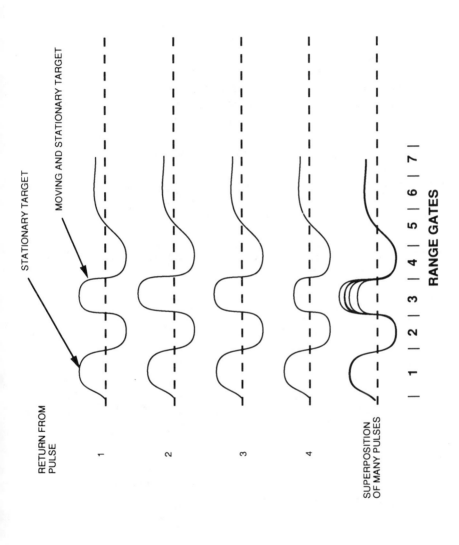

Figure 1.13 Fully coherent low-PRF time waveform.

each interpulse period, the received signal is sampled at many times (ranges), the usual time between samples being the transmitter pulse width. Each of these sample times is called a *range cell* or *range bin*. A collection of samples, taken at the same range cell on consecutive transmitter pulses, is processed by some form of doppler analyzer. A low-PRF radar has the potential for having a large number of range gates if contiguous coverage is provided from zero to maximum range. Each range gate requires doppler processing as shown in Figure 1.14. Sometimes the processing is simply a filter that rejects zero doppler shift (fixed targets), known as *moving-target indication* (MTI). Other radars perform a spectral analysis for each range gate. Figure 1.15 depicts a map of range versus doppler frequency versus amplitude that results from performing spectral analysis at each range gate. Fixed targets are shown at various ranges along the zero frequency axis; moving targets are shown with nonzero frequencies. Remember that this presentation depicts the *apparent* doppler shift. The doppler frequency measured by low-PRF radars is generally highly ambiguous. The true doppler may be the apparent plus some integral multiple of the PRF. Moving targets could also appear to have zero doppler, due to the waveform ambiguity, and could be obscured by the fixed targets.

1.6 SPECTRA AND WAVEFORMS OF STATIONARY MEDIUM-PRF RADAR

The time waveforms and spectra in medium-PRF radar are similar to those previously shown for low-PRF radar. Medium-PRF radar is a compromise between high- and low-PRF, having some of the desirable features of each. The transmitter duty factor is often on the order of 0.01. Since the interpulse period is shorter than that of low-PRF, there are generally fewer range gates. The PRF is high enough for there to be range ambiguities, but the number is small compared to that of high-PRF radar. There are also doppler ambiguities, but fewer than with low-PRF. The range-doppler map of a medium-PRF radar resembles the one for low-PRF radar shown in Figure 1.15 except for having fewer range gates.

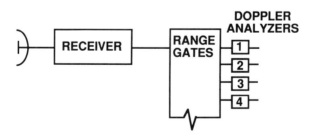

Figure 1.14 Range-gated doppler processing.

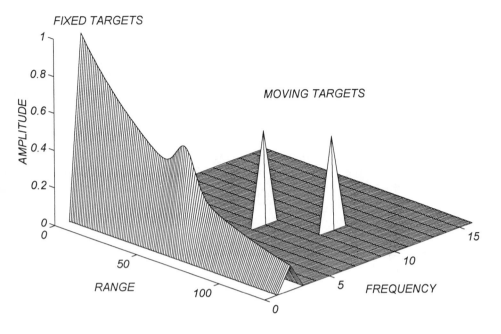

Figure 1.15 Low-PRF clutter map.

The main advantage of the medium-PRF waveform is the capability it provides to a fast-moving airborne radar for detecting targets with low relative velocities. The advantages of the medium-PRF waveform have not been apparent in the preceding discussion of stationary radars, but will be fully explained, beginning in Chapter 2.

Effects of Platform Motion on Clutter

Guy V. Morris

The returns from stationary objects are all at the transmitter frequency, regardless of range or angle from the stationary radar described in Chapter 1. When the radar is in motion, then objects that are stationary with respect to the ground are in motion with respect to the radar. As in Chapter 1, the effect of platform motion will first be described for a CW radar, and then the effects of the pulsed waveform will be described.

The apparent velocity of the object is proportional to the component of the velocity of the radar in the direction of the object. The doppler frequency of the object or clutter patch is given by

$$f_d = \frac{2v_a}{\lambda} \cos \phi \tag{2.1}$$

where:

v_a = aircraft velocity;
ϕ = angle between the velocity vector and the line of sight (LOS);
λ = transmitter wavelength.

The geometry is illustrated in Figure 2.1.

Note that the angle ϕ is the total cone angle between the velocity vector and the LOS. Normally directions are specified using a coordinate system with two orthogonal angles; for example, azimuth a and elevation e referenced to the body of the aircraft, or azimuth a_s and elevation e_s referenced to an inertial reference frame centered on the aircraft. The vertical plane of the aircraft-referenced and the inertial-referenced (sometimes referred to as *space-stabilized* or *roll-and-pitch-stabilized*) coordinate systems is shown in Figure 2.2. Even in level flight, the aircraft-referenced coordinate system will generally differ from the inertial-referenced

---

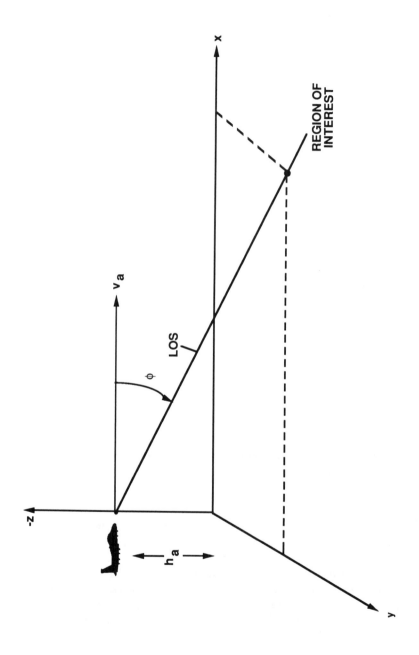

Figure 2.1 Aircraft to terrain geometry.

X, Z ARE AIRCRAFT REFERENCED

X$_S$, Z$_S$ ARE INERTIAL REFERENCED

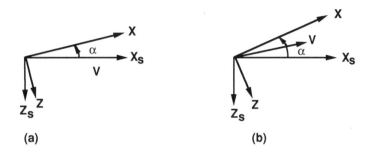

(a)   **(b)**

Figure 2.2 Coordinate reference systems: (a) level flight and (b) climb.

coordinates due to the angle of attack α and crosswind. For the simple case of level flight, angle ϕ is related to the space-stabilized azimuth and elevation angles by

$$\cos \phi = \cos a_s \cdot \cos e_s \tag{2.2}$$

Equation (2.1) shows that the loci of points having a constant apparent doppler due to the platform motion is a cone of revolution about the velocity vector with a half-angle ϕ. For illustration, assume that the aircraft is in level flight above a plane earth; that is, the velocity vector is parallel to the plane. The intersection of the cone and the plane is a hyperbola. All clutter located along each hyperbola will produce returns at the same doppler frequency, and thus the hyperbolas are given the name *isodops*. One such hyperbola is illustrated in Figure 2.3. Note that the range to the closest point of the isodop is a function of the aircraft altitude h_a.

2.1 CW RADAR CLUTTER SPECTRUM

Figure 2.4 shows an idealized CW clutter spectrum. The clutter from all ranges on a single isodop is collapsed into a single doppler resolution cell. In the limiting case, a clutter patch on the radar horizon directly in front of the aircraft has an apparent closing rate equal to the aircraft velocity v_a; that is, the angle ϕ is approximately zero. Similarly, a clutter patch on the radar horizon directly behind the aircraft will appear to be opening at the aircraft velocity, that is, at $-v_a$. In theory, all apparent velocities between v_a and $-v_a$ can be present. The strength of each spectral component depends on the effective radar cross section of the ground and the radar antenna gain at each of the LOS angles. Two large peaks are shown in the spectrum. The one at zero velocity is called the *altitude line*. Relating to

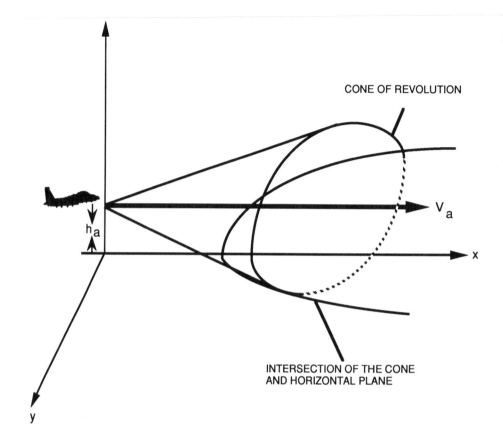

Figure 2.3 Contours of constant doppler.

Figure 2.3, the altitude line is the straight-line isodop corresponding to $\phi = 90$ degrees. All clutter patches along the isodop contribute to it, but the return is dominated by the patches directly under the aircraft that have a nearly normal incidence angle. In actual radar systems, the component at zero velocity may be the result of direct leakage between the transmitter and the receiver. The second large peak shown in the spectrum corresponds to the pointing angle of the main beam of a high-gain antenna; this is called *main-lobe clutter*. The center and the width of main-lobe clutter depend on the antenna pointing angle. Therefore, the spectral position of main-lobe clutter can change rapidly in scanning radar systems.

Figure 2.4 also relates the regions of the clutter spectrum to the relative heading of the target. The radar is closing on all targets that have a velocity on the right side of zero. The distance from zero to the peak of main-lobe clutter is the radar contribution to the total closing rate. The target component extends from the peak of main-lobe clutter to the target return. If the total closing rate is greater

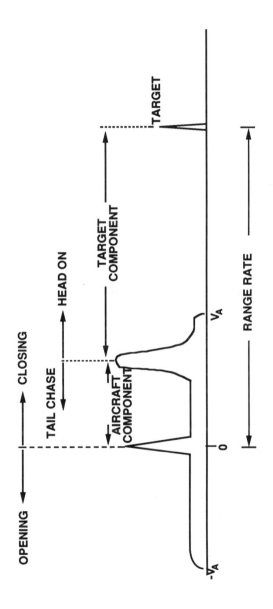

Figure 2.4 CW radar clutter spectrum.

than the radar contribution, then the target is moving toward the radar. Such encounters are often referred to as *nose-on* or *head-on*. If the total closing rate is less than the radar contribution but greater than zero, then the radar is overtaking the target from the rear. The clutter-free region above v_a provides the greatest detection sensitivity. Even though the clutter returns from many ranges are collapsed into the same doppler cell in a CW radar, it is instructive to examine the zones on the ground that contribute to each doppler cell. Obviously, the shortest slant range from which a ground return can be received is the point located directly below the aircraft, whose range is the altitude of the aircraft h_a. The patch of ground at long range, directly in front of the aircraft at the radar horizon, exhibits the maximum doppler shift. For our plane earth illustration, the slant range R is given by

$$R = \frac{h_a}{\sin e} \tag{2.3}$$

where:

h_a = aircraft altitude;
e = elevation angle.

Combining (2.1) through (2.3), we can express the boundary of the side-lobe clutter region f_{de} as a function of the slant range R:

$$f_{de} = \frac{2v_a}{\lambda}\sqrt{1 - \left(\frac{h_a}{R}\right)^2} \tag{2.4}$$

Figure 2.5 depicts the contributions from the various ranges to the total CW clutter spectrum. Equation (2.4) is plotted as the boundary of the side-lobe clutter region. Note that the main-lobe clutter results only from those ranges illuminated by the antenna main beam.

2.2 HIGH-PRF RADAR CLUTTER SPECTRUM

The PRF is chosen sufficiently high to ensure that the doppler region of interest is unambiguously covered. For a typical airborne radar operating at 9.5 GHz, the resulting PRFs are on the order of 250 to 310 kHz. The unambiguous range for a PRF of 250 kHz is 600m (2,000 ft). The spatial position corresponding to the uneclipsed receiver on time may be visualized as a series of concentric spheres centered on the radar and spaced at 600m, as shown in Figure 2.6. The clutter return comes from all ranges at which the concentric spheres intersect the ground. The target return must compete with this composite clutter, some of which comes from ranges far less than the target range and is generally of far greater signal strength.

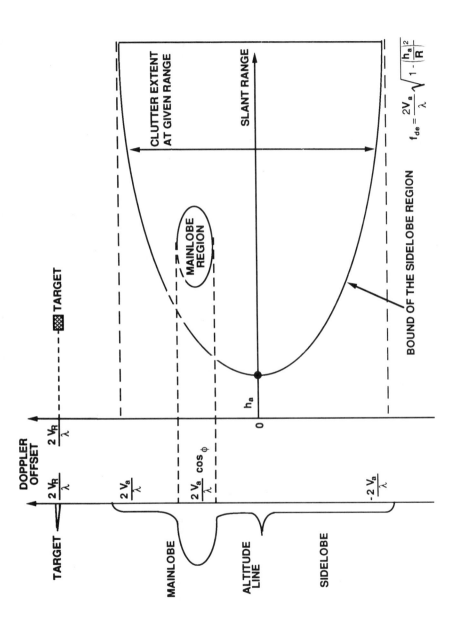

Figure 2.5 Range contributions to clutter spectra for CW waveform.

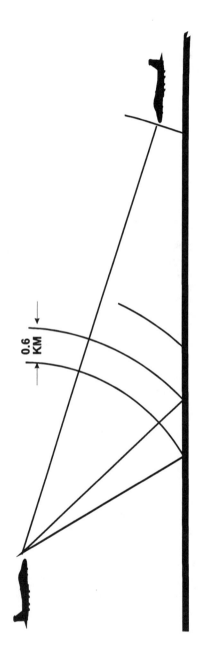

Figure 2.6 High-PRF clutter ranges.

The shape of the high-PRF clutter spectrum is essentially the same as the CW spectrum, except that it is replicated at each PRF transmitted spectral line, as shown in Figure 2.7. A conservative estimate of the minimum PRF, PRF_{min}, necessary to ensure that high velocity head-on targets appear in the clutter-free region is given by

$$PRF_{min} = \frac{2}{\lambda}(2v_{amax} + v_{tmax}) \qquad (2.5)$$

where:

$v_{amax} =$ maximum aircraft velocity;
$v_{tmax} =$ maximum target velocity.

Equation (2.5) ensures that clutter from the back lobes corresponding to velocities of $-v_{amax}$ associated with the spectrum component at $(f_0 + PRF)$ does not encroach on the region reserved for high-velocity targets. Some radar designers use v_{amax} instead of $2v_{amax}$ in (2.5) if the antenna back lobes are judged to be low enough for back-lobe clutter to be negligible.

2.3 LOW-PRF RADAR CLUTTER SPECTRUM

The PRF is chosen low enough to ensure that the range region of interest is unambiguously covered. The return from the target at range R_t has to compete only with clutter from the same range and from ranges equal to R_t plus integral multiples of the unambiguous range as shown in Figure 2.8. Thus, the clutter-to-target ratio will be significantly less than the high-PRF waveform. The low-PRF waveform results in a small unambiguous doppler region. For example, an unambiguous range of 75 km (40 nmi) requires a PRF of 2 kHz or less. For a transmit frequency of 9.5 GHz, the unambiguous velocity is only 32 m/s (104 ft/s), which is less than the aircraft velocity for most airborne radars. In contrast to high-PRF waveforms, there is no clutter-free doppler region; side-lobe clutter occupies all range-doppler cells for which the range is greater than the aircraft altitude. In addition, main-lobe clutter is present at those ranges corresponding to the footprint of the antenna main lobe on the ground. The center frequency of main-lobe clutter, f_{mlc}, is given by

$$f_{mlc} = \left(\frac{2v_a}{\lambda} \cos \phi_{ant}\right) \text{ modulo PRF} \qquad (2.6)$$

where ϕ_{ant} is the angle between the antenna main beam and the velocity vector.

The apparent frequency of main-lobe clutter can be at any frequency between zero and PRF. The apparent frequency sweeps rapidly through the spectral region as the antenna scans. Therefore, most airborne radar implementations perform a

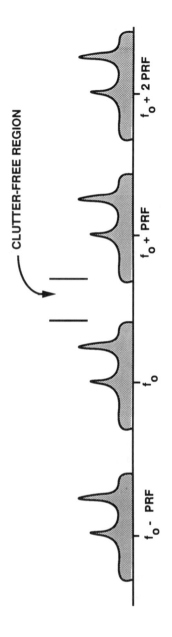

Figure 2.7 High-PRF clutter spectrum. PRF must be sufficiently high to provide desired clutter-free region.

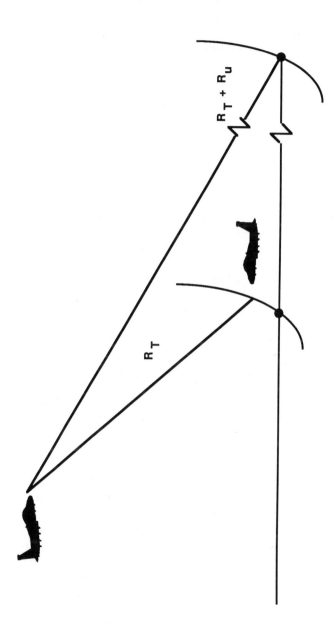

Figure 2.8 Low-PRF clutter ranges.

frequency translation that is a function of the antenna scan angle at some point in the signal-processing chain, so that main-lobe clutter always appears at a constant known frequency, usually zero.

The spectral width of main-lobe clutter is also a function of the antenna scan angle. The spectral width, δf_{mlc}, can be approximated using

$$\delta f_{mlc} = \left(\frac{2v_a}{\lambda} \sin \phi_{ant}\right) \delta\phi \tag{2.7}$$

where $\delta\phi$ is the antenna beamwidth measured between the first nulls on either side of the main lobe in radians (≤ 0.3), and v_a is in the same length units as λ (e.g., meters per second and meters, respectively). The null-to-null beamwidth can be approximated as $\delta\phi = 2.5 \cdot$ (3-dB beamwidth). Continuing our 9.5-GHz example and assuming a scan angle of 60 degrees, a 3-dB beamwidth of 2.5 degrees, and an aircraft velocity of 300 m/s (984 ft/s), the main-lobe clutter width is 1,800 Hz. Therefore, 90% of the 2-kHz spectral region would be occupied by main-lobe clutter.

2.4 MEDIUM-PRF RADAR CLUTTER SPECTRUM

Medium-PRF is a compromise between the classic high-PRF and low-PRF waveforms. Both range and doppler ambiguities exist within the range-doppler region of interest. Many of the general comments made about the low-PRF spectrum apply to the medium-PRF spectrum; for example, side-lobe clutter fills the entire spectral region in each range cell. Figure 2.9 shows the clutter ranges that compete with the target return. A typical value of a medium PRF for a 9.5-GHz radar is on the order of 10 kHz. Equations (2.6) and (2.7) apply. The main-lobe clutter spectral width of 1,800 Hz that was calculated in our example in Section 2.3 occupies less than 20% of the unambiguous spectral width.

Equation (2.7) provides good insights into the rationale for using medium-PRF radar. The main-lobe clutter spectral width is proportional to the aircraft velocity and antenna scan angle and inversely proportional to the antenna beamwidth. The antenna beamwidth is in turn proportional to the transmitter wavelength and inversely proportional to the physical antenna size. The only methods for reducing the spectral width of main-lobe clutter that are permitted by the laws of physics are reducing the aircraft velocity, reducing the antenna scan angle, or increasing the antenna size. All three methods are unacceptable in many applications. Increasing the transmitter frequency reduces the antenna beamwidth but increases the doppler scale factor (the conversion factor between meters per second and Hertz). Thus, the main-lobe clutter width is independent of transmitter frequency. For high-speed aircraft, the available antenna size often results in an unacceptably small, perhaps zero, spectral region clear of main-lobe clutter. The only remaining alternative is to increase the PRF and thereby increase the relative portion of the doppler spectrum that is free.

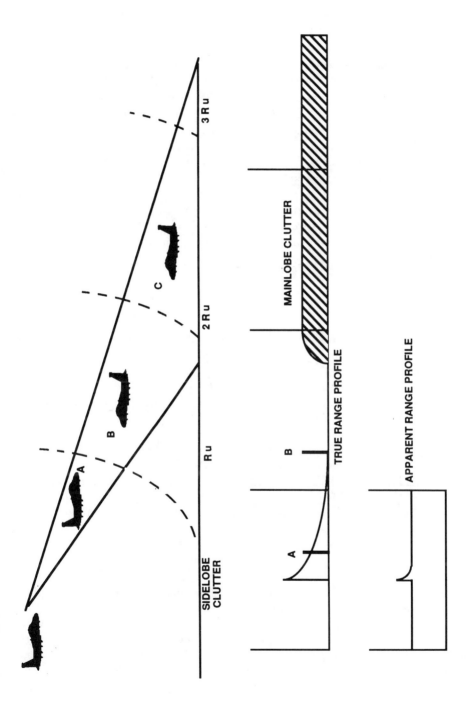

Figure 2.9 Medium-PRF clutter ranges.

2.5 CALCULATION OF CLUTTER POWER

Calculation of clutter power will be discussed in detail in Chapter 15. However, some overall principles will be summarized here. An estimate of the total clutter power that will be present in a single range cell is needed during system design to establish the system gains and dynamic range requirements. The total power in a range cell is the sum of all the returns in the constant range circle shown in Figure 2.10. Main-lobe clutter dominates, and, therefore, relatively straightforward estimates can be made using the radar cross section of the main beam ground patch in the radar range equation, especially for low-PRF radar, for which essentially only one ground patch is to be considered. For high-and medium-PRF radars, the process is tedious and is best implemented via a computer program.

Estimating the clutter spectra at each range cell is far more difficult but is necessary to design and predict the performance of signal-processing and detection

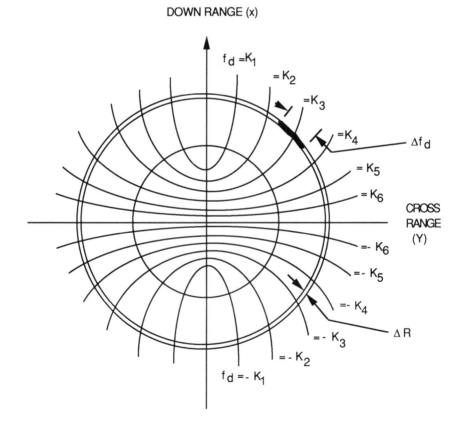

Figure 2.10 Locus of points for constant range and doppler.

circuits. Within each range cell, the power in each doppler cell results from the return corresponding to the intersection between the isorange circle and each isodoppler hyperbola in Figure 2.10. In general, the physical sizes of each of these ground patches that produce returns in the same range-doppler cell vary with angle. Figure 2.11 illustrates the ambiguities produced by the medium-PRF waveform. For a given range, returns are not received from a single isorange circle but also from isorange circles located integral multiples of the unambiguous range away. Similarly, an isorange circle may cross a multitude of isodops separated in frequency by the PRF. The shaded areas of Figure 2.11 contribute to the same range-doppler cell. It should be readily apparent that a computer program is very helpful in estimating the clutter power in each range-doppler cell.

In all of our preceding examples, we assumed that the aircraft was in level flight. If the aircraft is diving, then not all the intersections between the isodoppler cones of revolution and the ground plane are hyperbolas, as shown in Figure 2.12. Obviously, the diving geometry further complicates the estimation of the clutter power in each range-doppler cell.

A computer program that is simple and small enough to operate on an IBM™ compatible personal computer is included in Appendix A. The program calculates an estimate of the normalized power from each range-doppler cell using

$$\frac{P_c(f_{da}, R_a)}{P_t} = \frac{\lambda^2 G(\phi)^2 \sigma_c}{(4\pi)^3 R^4} \text{ modulo PRF, modulo } R_u \qquad (2.8)$$

where:

$G(\phi)$	=	antenna gain;
R	=	actual slant range to ground;
R_u	=	unambiguous range;
R_a	=	apparent (ambiguous) range;
σ_c	=	radar cross section of clutter;
f_{da}	=	apparent (ambiguous) doppler.

Figures 2.13 and 2.14 are examples generated with the program. Figure 2.13 depicts an atypically large antenna and a slow platform. The main-lobe clutter is a narrow peak that is completely contained within a single unambiguous range interval. Figure 2.14 was generated using more representative X-band and medium-PRF parameters. Due to the shallow elevation depression angle, main-lobe clutter is seen throughout the range interval.

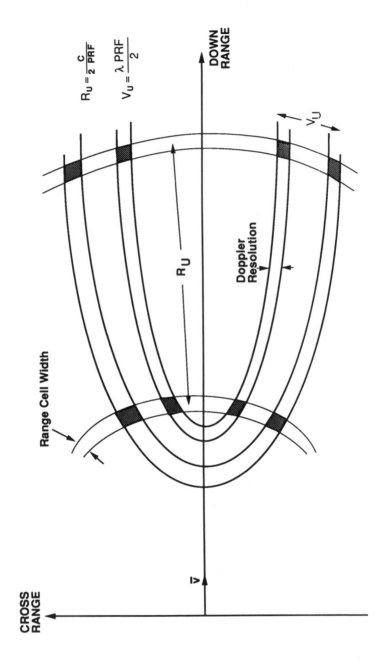

Figure 2.11 Range-doppler cell ambiguities in level flight. (Shaded areas contribute to same range/doppler cell.)

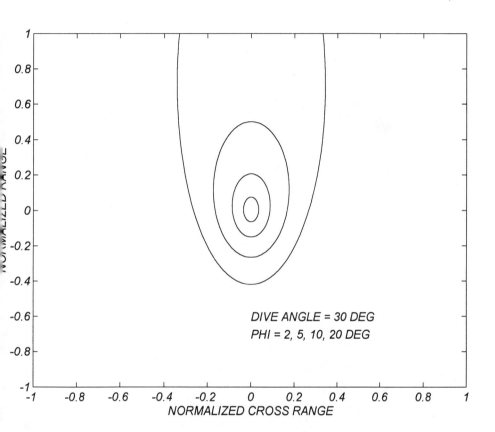

Figure 2.12 Range-doppler cell ambiguities in a dive.

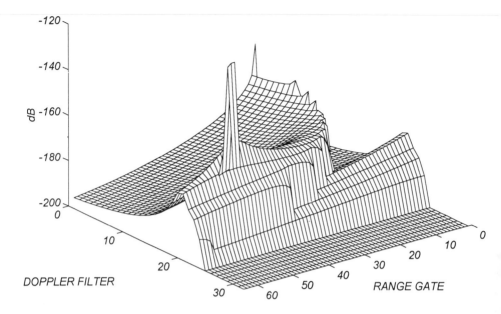

Figure 2.13 Range-doppler clutter map using a narrow-beamwidth, slow platform.

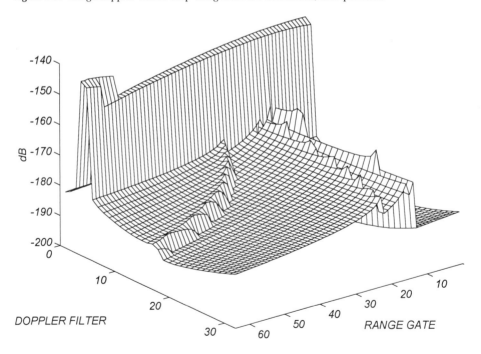

Figure 2.14 Range-doppler clutter map using typical medium-PRF parameters.

CHAPTER 3

Spectral Characteristics of a Pulsed Waveform

Linda L. Harkness

Some fundamental concepts about the spectral characteristics of a pulsed waveform and coherent radar detection are presented in this chapter (after Holm [1]). The chapter begins with a brief discussion of the Fourier transform and the spectra of stationary and moving targets, because these concepts are necessary for a complete understanding of doppler processing.

The signal received by a pulsed radar is a time sequence of pulses for which the amplitude and phase are measured. Doppler processing techniques are based on measuring the spectral (frequency) content of this signal. The frequency content of this *time-domain* signal is obtained by taking its Fourier transform, thus turning it into a *frequency-domain* signal, or spectrum of the time-domain signal. Figure 3.1 shows a plot of amplitude versus time of an arbitrary time-domain signal and the resulting plot of the amplitude versus frequency of the frequency-domain signal obtained by taking the Fourier transform of the time-domain signal. If the time-domain signal is described by the complex function $f(t)$, the Fourier transform of $f(t)$ is given by

$$F(\omega) = \int_{-\infty}^{\infty} f(t) e^{-j\omega t} dt \qquad (3.1)$$

The time-domain signal $f(t)$ can be regained by taking the inverse Fourier transform of $F(\omega)$, which is

$$f(t) = \frac{1}{2\pi} \int_{-\infty}^{\infty} F(\omega) e^{j\omega t} d\omega \qquad (3.2)$$

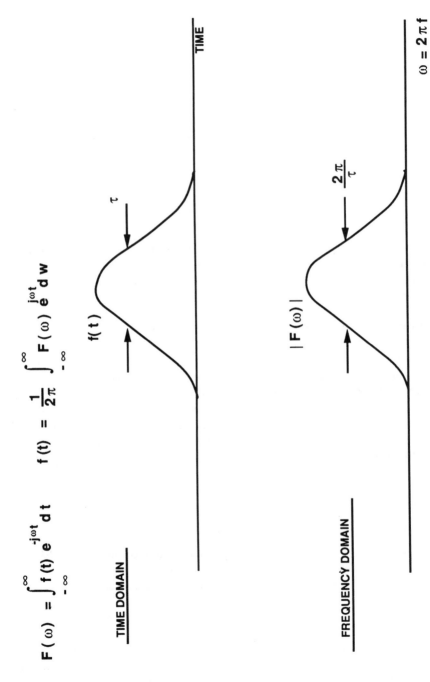

$$F(\omega) = \int_{-\infty}^{\infty} f(t)\, e^{-j\omega t}\, dt \qquad f(t) = \frac{1}{2\pi} \int_{-\infty}^{\infty} F(\omega)\, e^{j\omega t}\, dw$$

Figure 3.1 The Fourier transform [1].

Time-domain pulsed radar signals can be described in terms of four time scales: wave period, pulse width, PRI, and dwell time. The frequency-domain pulsed radar signal can be described in terms of four corresponding frequency scales: wave frequency, pulse bandwidth, PRF, and spectral line bandwidth. A fundamental property of Fourier analysis is that time and frequency scales have an inverse relationship to each other: the larger the wave period, the smaller the wave frequency; the larger the pulse width is, the smaller the pulse bandwidth, and so on. This relationship and its resulting consequences are crucial to the understanding of doppler-processing techniques. The pulsed waveform will now be described considering these four time and frequency scales.

3.1 CONTINUOUS-WAVE SIGNAL: ONE TIME SCALE

The simplest radar waveform is the infinite CW signal of frequency f_0. Only one time scale is required in the description of this waveform: wave period t_0. Figure 3.2 shows the time-domain plot of this waveform as a function of time and the resulting frequency-domain amplitude versus frequency plot. The frequency-domain plot consists of two impulse, or delta (Dirac), functions: one at f_0 and the other at $-f_0$.

3.2 SINGLE PULSE OF A CARRIER: TWO TIME SCALES

Figure 3.3 shows a time-domain plot of a single rectangular pulse of pulse width τ and its corresponding frequency-domain plot. The frequency-domain plot has a $\sin(x)/x$ form centered at zero frequency with a bandwidth of approximately $1/\tau$. Notice that the nulls occur at integer multiples of $1/\tau$. The two waveforms given in Figures 3.2 and 3.3 are combined in Figure 3.4. Two time scales, t_0 and τ, are required in the description of this waveform. The plot of the amplitude of the Fourier transform of this waveform has the form of two $\sin(x)/x$ curves of bandwidth $1/\tau$ centered at $\pm f_0$.

3.3 INFINITE PULSE TRAIN: THREE TIME SCALES

The time-domain and frequency-domain plots of an infinite sequence of pulses are shown in Figure 3.5. Here, the pulses are separated in time by the T_r, the PRI. The resulting frequency-domain plot shows that the $\sin(x)/x$ curves have been resolved into PRF spectral lines within the two $\sin(x)/x$ envelopes. The spectral lines have zero bandwidth and are separated by $f_r = 1/T_r$, that is, the PRF.

3.4 FINITE PULSE TRAIN: FOUR TIME SCALES

Finally, to obtain the spectral characteristics of a realistic pulsed waveform, the infinite series of pulses is truncated into a finite series of duration T. The resulting

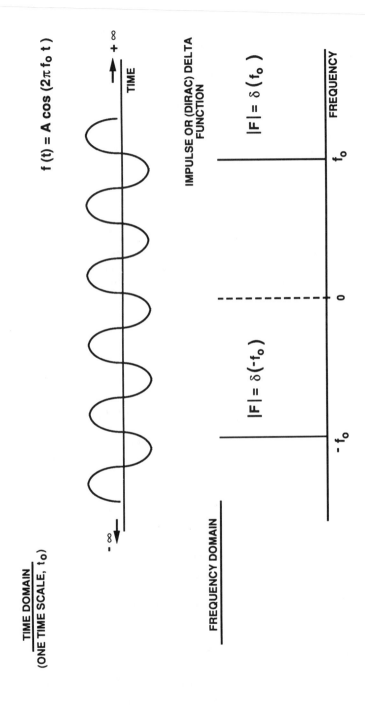

Figure 3.2 Infinite CW signal of frequency f_0 [1].

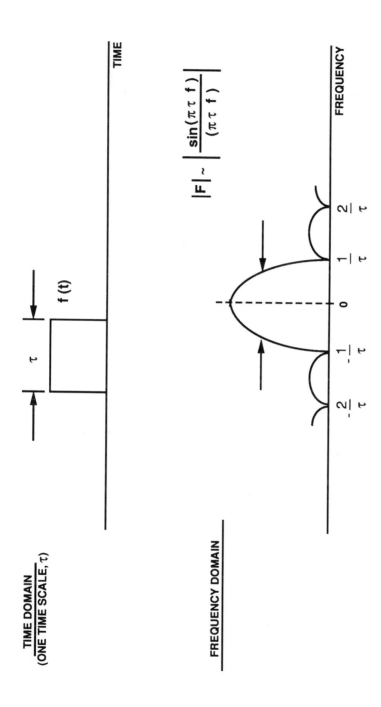

Figure 3.3 Pulse envelope of pulse width τ [1].

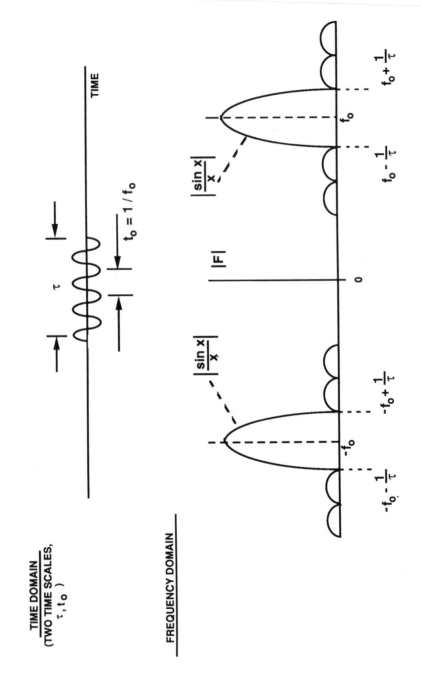

Figure 3.4 Single-pulse (τ) signal of frequency f_0 [1].

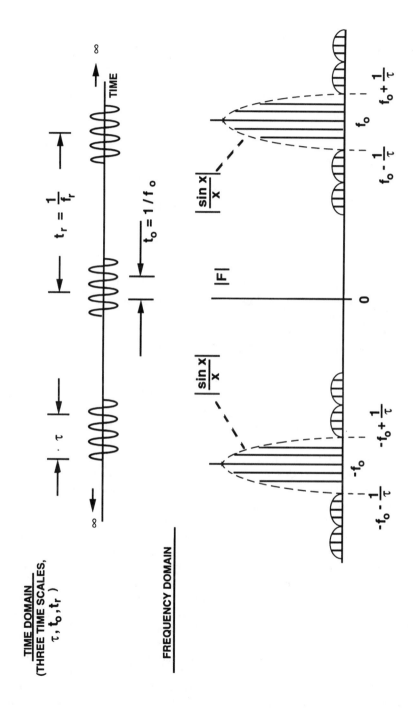

Figure 3.5 Infinite pulse (τ) train signal of PRF f_r and frequency f_0 [1].

time-domain and frequency-domain plots are shown in Figure 3.6. In the frequency-domain plot, the PRF spectral lines now have a nonzero bandwidth given by $1/T$. Thus, four frequency scales are required in the description of a pulsed waveform: bandwidth of spectral lines $(1/T)$, spacing of spectral lines $(\text{PRF} = 1/T_r)$, bandwidth of $\sin(x)/x$ envelopes $(1/\tau)$, and center frequencies of $\sin(x)/x$ envelopes $(\pm f_0 = \pm 1/t_0)$.

If the pulsed radar return is from a cell that contains both a stationary target (clutter) and a moving target, then the return will consist of a superposition of signals. The signal from the clutter will be at the transmitted frequency, and the signal from the moving target will be at the transmitted frequency plus the doppler frequency shift.

The doppler effect is the shift in frequency of a wave transmitted, reflected, or received by a moving object. In airborne radars this motion may be generated by either the radar platform, the target, or both. The doppler effect allows a radar to measure the corresponding target doppler frequency shift, which is related to the velocity of the detected target. The doppler frequency shift, produced by a simple ground target, is illustrated in Figure 3.7. In this illustration, the radar remains stationary while the target closes in range with a velocity v_t. The wavelength (i.e., separation in range between two transmitted wavefronts) is represented by λ. At t_0, wavefront 1 hits the target and is reflected back toward the radar. During time period T, the target travels a distance $\lambda - v_t T$ before wavefront 2 hits it. Since both wavefronts are traveling the same speed, wavefront 1 travels the same distance $(\lambda - v_t T)$ back toward the radar. If the speed of the target is known, the distance between the reflected wavefronts at time T can be calculated. As shown, the wavelength has been compressed by $2v_t T$. If the radar had also been moving, the wavelength compression would have included the velocity of the radar.

In general, the doppler frequency shift f_d has the magnitude $f_d = 2v/\lambda$, where v is the component of the moving target's velocity along the line of sight between the radar and the target, and λ is the wavelength of the transmitted wave. The doppler frequency shift is positive for approaching targets and negative for receding targets. Doppler frequency shifts for various radar frequency bands and target speeds are shown in Table 3.1. The resulting spectrum of the received signal from a moving target plus stationary clutter is shown in Figure 3.8.

The basic moving-target discrimination approach inherent in doppler processing techniques is now obvious. In the spectrum of the received signal, the moving and stationary targets are separated. Thus, by applying the appropriate filtering, the stationary target can be removed from the spectrum, leaving only the signature of the moving target. Before this can occur, however, the received signal must be detected. Coherent radar detection of a pulsed waveform is discussed next.

3.5 COHERENT DETECTION OF A PULSED WAVEFORM

Coherent radar systems extract not only amplitude, but also phase information from the signal reflected from the target. Such coherency is usually required because

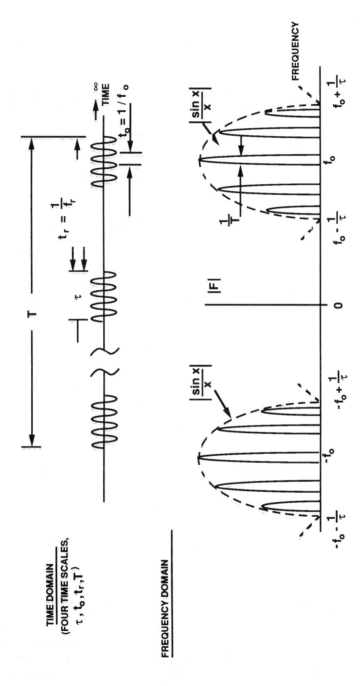

Figure 3.6 Finite (T) pulse (τ) train signal of PRF f_r and frequency f_0 [1].

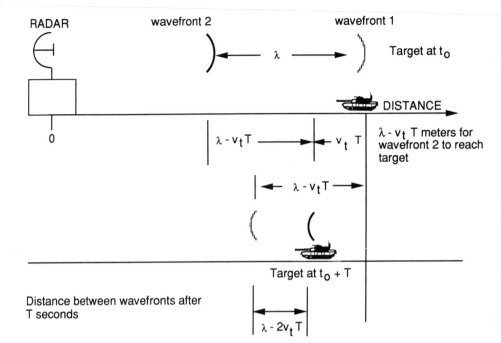

Figure 3.7 The doppler effect.

Table 3.1
Doppler Frequency Shifts (Hz) for Various Radar Frequency Bands and
Target Speeds [1]

	Radial Target Speed		
Radar Frequency Band	1 m/s	1 kn	1 mi/h
L (1 GHz)	6.67	3.43	2.98
S (3 GHz)	20.0	10.3	8.94
C (5 GHz)	33.3	17.1	14.9
X (10 GHz)	66.7	34.3	29.8
Ku (16 GHz)	107	54.9	47.7
Ka (35 GHz)	233	120	104
W (95 GHz)	633	326	283

the bandwidth of a single pulse is usually several orders of magnitude greater than the expected doppler frequency shift, $1/\tau \gg f_d$. Thus, the doppler frequency shift is lost in the single-pulse spectrum. To extract the doppler frequency shift, the returns from many pulses over an observation time T must be frequency-analyzed

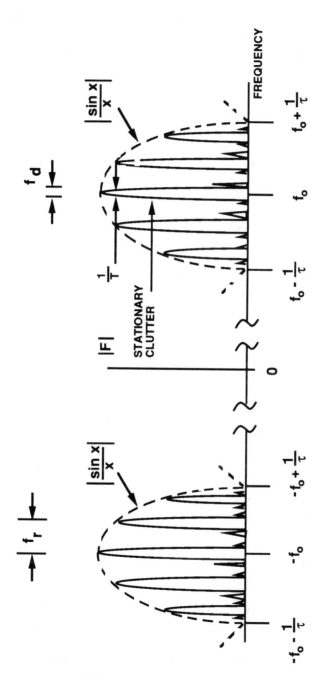

Figure 3.8 Spectrum of the received signal from a moving target plus stationary clutter [1].

so that the single-pulse spectrum will separate into individual PRF lines with bandwidths approximately given by $1/T$. Now, as was shown in Figure 3.8, the doppler frequency shift of the moving object becomes apparent and separates out from any stationary object return. For this process to work, however, a deterministic phase relationship must be maintained from pulse to pulse and measured over the observation time T. Radars capable of maintaining and measuring such a deterministic phase are called *coherent*.

3.5.1 Single-Channel Detection

A block diagram of a typical single-channel coherent radar is shown in Figure 3.9. This transmitter configuration is often referred to as a MOPA configuration. Here, the transmitted signal is obtained by mixing an RF signal from a stable local oscillator (STALO) with an IF, f_{IF}, signal (usually on the order of 30 to 60 MHz) from a coherent oscillator (COHO), which is also very stable. The resulting CW signal at frequency f_0 is then pulse-modulated to produce the pulsed waveform. The reflected signal from a single stationary specular scatterer has the spectrum shown in Figure 3.6. The first step in the detection process is to down-convert this signal to IF by mixing it with the STALO signal. In one mixing technique, the signal is simply added to the STALO output and the fluctuation in the amplitude of the sum is extracted. The spectrum of the resulting signal is the same as that shown in Figure 3.8, except now the $\sin(x)/x$ envelopes are centered at plus and minus the IF; that is, this down-conversion has the effect of sliding the center positions of the two $\sin(x)/x$ envelopes from $\pm f_0$ toward each other to $\pm f_{IF}$. The resulting IF spectrum is shown in Figure 3.10 for a signal reflected from stationary clutter and an approaching target. Amplification is now done at IF to enhance the sensitivity of the receiver.

A subtle aspect of frequency translation in a single-channel system is the creation of image frequencies. The image frequency results from using a single channel and extracting only amplitude modulation. Two signals whose frequencies are an equal amount above and below the STALO output will produce the same amplitude modulation. As a result, the spectrum of a signal processed by a single-channel radar will display this ambiguity in frequency by creating an image of the true target frequency on the opposite side of the IF, as shown by the two $\sin(x)/x$ curves in Figure 3.10.

The signal is down-converted a second time to baseband (video) by mixing it with the COHO signal. The $\sin(x)/x$ envelopes now overlap each other, since both are now centered at zero frequency. This is shown in Figure 3.11 for the same IF spectrum of clutter and an approaching target shown in Figure 3.10. After down-conversion to video, the image frequency and the moving-target spectral lines appear on either side of the stationary-target spectral lines. That is, the image and the actual target frequency are indistinguishable and the sense of the direction of

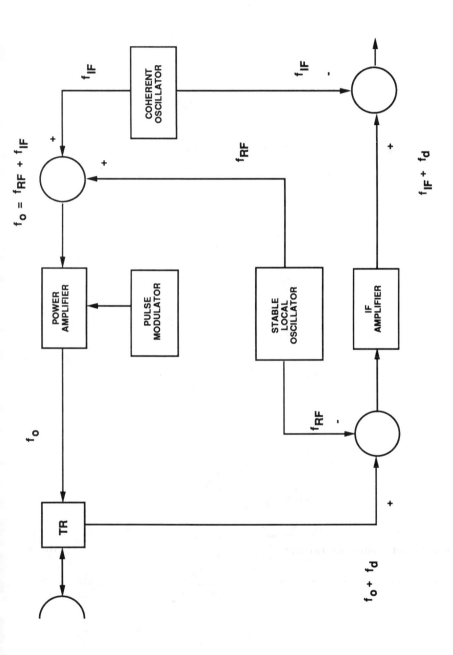

Figure 3.9 Block diagram of a simple coherent radar [1].

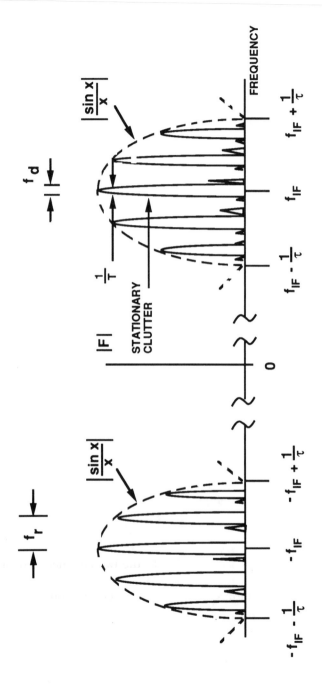

Figure 3.10 IF spectrum of the received signal from a target approaching the radar plus stationary clutter [1].

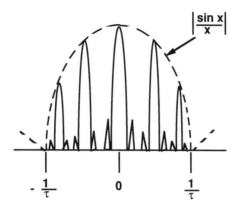

Figure 3.11 Video spectrum of the received signal from a target approaching the radar plus stationary clutter [1].

motion of the moving target is lost, since a target receding from the radar at the same rate would produce exactly the same video spectrum. This ambiguity can be avoided by employing an I/Q detector as shown in the block diagram of Figure 3.12.

3.5.2 I/Q Detection

In an I/Q detector, one or two synchronous detectors are employed to divide the IF signal into two channels: one channel (Q channel) is phase-shifted 90 degrees with respect to the other channel (I channel). The synchronous detector, shown in Figure 3.13, compares a doppler-shifted input signal with a reference signal and produces an output (I channel) whose amplitude is proportional to the amplitude (A) of the input signal times the cosine of the phase (ϕ) of the input signal relative to the reference signal. If the phase of the reference signal is shifted 90 degrees, an output (Q channel) is produced that is proportional to A times the sine of ϕ. The use of synchronous detectors allows the radar to distinguish the direction of motion of the detected signal. If the output of the two channels is depicted as a rotating phasor, shown in Figure 3.14, a positive doppler frequency will cause the target phasor to rotate counterclockwise, while a negative doppler frequency will cause the target's phasor to rotate clockwise. A two-channel (I/Q) radar system is represented by visualizing the first detector (channel) as the projection of the phasor onto the x-axis, and the second detector (channel) as the projection onto the y-axis. If the second detector's output (y) lags behind the first detector's output (x), the phasor is rotating counterclockwise: the target's doppler frequency is positive. If y leads x, the phasor is rotating clockwise: the target's doppler frequency is

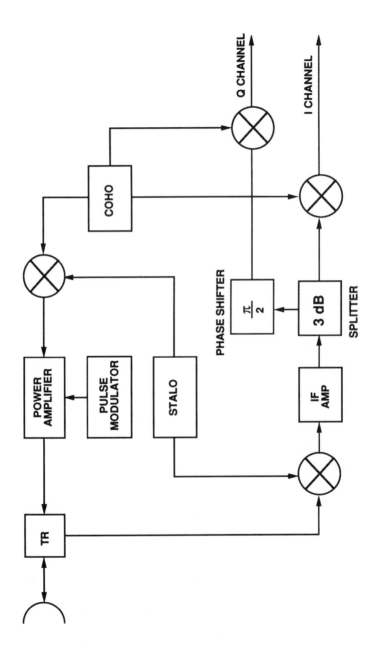

Figure 3.12 Block diagram of a coherent pulsed radar with a synchronous (I/Q) detector [1].

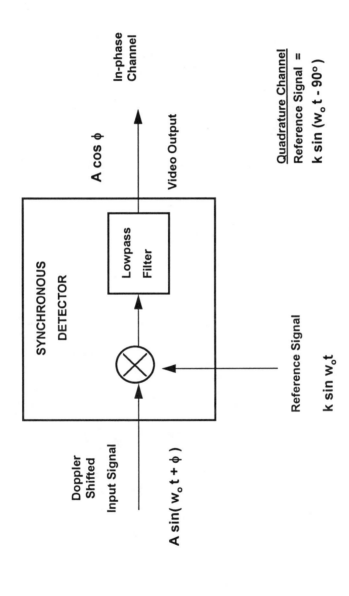

Figure 3.13 Block diagram of synchronous detector.

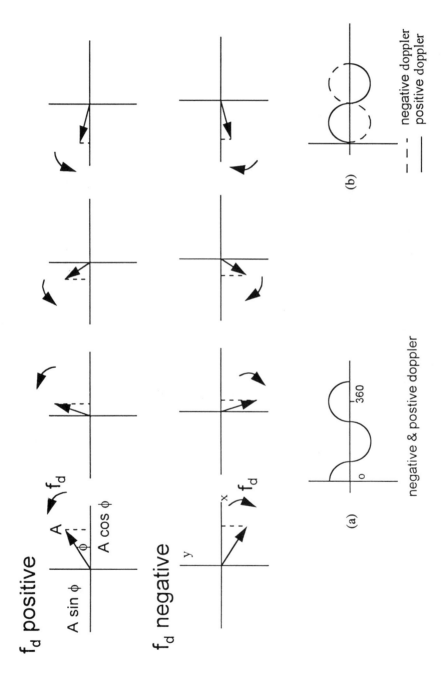

Figure 3.14 Projection of the phasor representation of a positive and negative doppler frequency (a) on the *x*-axis and (b) on the *y*-axis.

negative. By taking advantage of a two-channel system employing synchronous detectors, a receiver's mixer stage can be designed to reject images.

If the power spectra of the I-channel signal and the Q-channel signal in a receiver employing I/Q detection and the power spectrum of the signal from a single-channel receiver were obtained, all three would conceptually appear as shown in Figure 3.15. The Fourier transform of the complex I/Q channel signal differentiates the rate and direction of the target phasor's rotation, thus rejecting the image frequency displayed in the single-channel system spectrum.

The final translation of the single-channel spectrum to zero frequency results in the video spectrum of Figure 3.16(a), which is the same result as that shown in Figure 3.11. However, the down-conversion of the complex I/Q spectrum effectively results in the mirror image of the negative portion of the spectrum being added to the positive portion, causing the target returns to appear at the same frequency and be added coherently as shown in Figure 3.16(b). The spectral lines of the moving target now appear only on the right-hand side of the clutter spectral lines, indicating an approaching target.

The retention of the ability to discriminate between approaching and receding targets can be shown as follows. Consider the I/Q components of the signal after the final down-conversion.

$$I_s = \cos \omega_d t$$

$$Q_s = -\sin \omega_d t$$

where ω_d is the doppler (radian) frequency. The I/Q outputs of the single doppler filter, I_0 and Q_0, with a center frequency ω_r, are given by

$$I_0 = \int_0^T \text{Re}[(I_s + jQ_s)(I_r + jQ_r)] \, dt \tag{3.3}$$

$$Q_0 = \int_0^T \text{Im}[(I_s + jQ_s)(I_r + jQ_r)] \, dt \tag{3.4}$$

where the reference signals I_r and Q_r are given by

$$I_r = \cos \omega_r t$$

$$Q_r = -\sin \omega_r t$$

$$T = 2\pi n/\omega_r$$

and n is an integer.

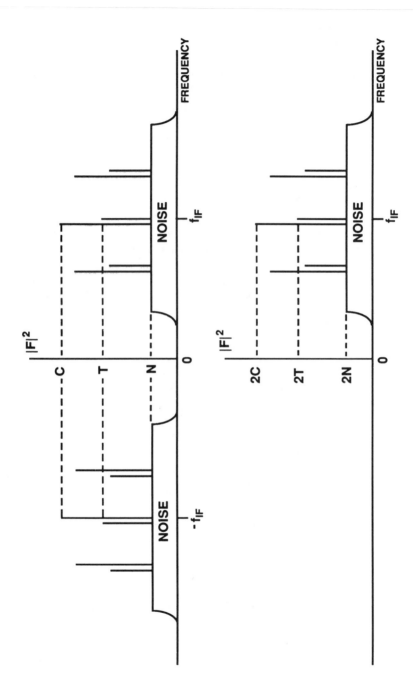

Figure 3.15 IF power spectrum of a coherent radar with (a) a single-channel receiver and (b) an I/Q channel receiver [1].

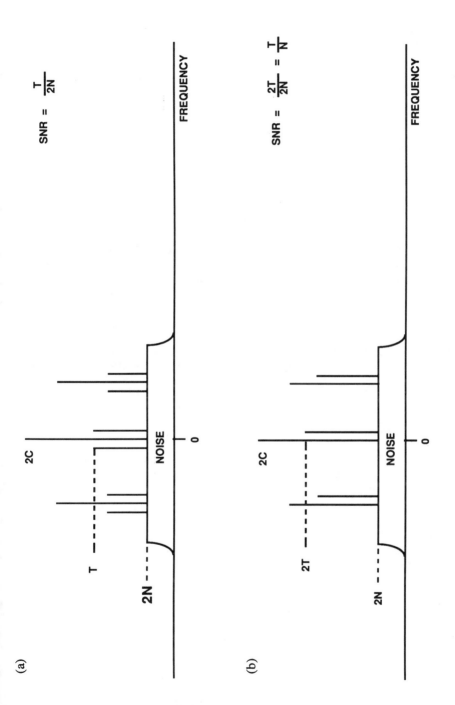

Figure 3.16 Video power spectrum of a coherent radar with (a) a single-channel receiver and (b) an I/Q channel receiver [1].

Equations (3.3) and (3.4) can be written as

$$I_o = \int_0^T \cos(\omega_d - \omega_r)t \, dt$$
$$= T \text{ for } \omega_d = \omega_r$$
$$= 0 \text{ for } \omega_d = -\omega_r \tag{3.5}$$

$$Q_o = \int_0^T -\sin(\omega_d + \omega_r)t \, dt$$
$$= 0 \text{ for } \omega_d = \omega_r$$
$$= 0 \text{ for } \omega_d = -\omega_r \tag{3.6}$$

Another benefit of I/Q detection over single-channel detection is a 3-dB gain in target-to-noise ratio at the output of the doppler spectral analysis function, as shown in Figure 3.16.

Reference

[1] Holm, W. A., "MMW Radar Signal Processing Techniques," Chap. 6 in *Principles and Applications of Millimeter-Wave Radar*, N. C. Currie and C. E. Brown, eds., Norwood, MA: Artech House, 1987.

CHAPTER 4

Low-PRF Mode

Linda L. Harkness

An airborne multimode radar may have a number of low-PRF modes. Table 4.1 shows typical radar modes and the system functions to which they contribute. Although the functions are very diverse, all depend upon the unambiguous range

Table 4.1
Low-PRF Modes

Radar Mode	Typical System Function
Coherent Modes:	
Doppler	Airborne moving-target detection
	Ground moving-target detection
Doppler beam sharpening	Improved resolution ground map for navigation
Synthetic aperture	Stationary-target detection
Noncoherent Modes:	
Ground map	Navigation
Terrain avoidance	Covert navigation
Air-to-air ranging	Short-range gun and missile attack
Air-to-ground ranging	Bomb delivery
Terrain following	Covert navigation

measurement provided by the low-PRF waveform. All the modes will be discussed briefly for completeness except for the synthetic aperture mapping modes, which will be explained in Chapter 10, but the emphasis is on the coherent or doppler modes. This chapter will also cover some of the basics such as antenna scan patterns that are common to the medium- and high-PRF modes. The benefits of the low-PRF mode are summarized in Table 4.2.

4.1 ANTENNA SCAN PATTERNS

In the search phase of the mission, the antenna beam is scanned in azimuth. A variety of azimuth scan pattern selections are usually provided, such as ±20 degrees, ±45 degrees, and ±60 degrees, as shown in Figure 4.1. Similarly, a section of elevation scan rasters is available. Common choices are 1-, 2-, and 4-bar rasters, although a 3-bar raster is sometimes used as shown in Figure 4.2. The separation of the elevation bars is selected to provide an acceptable compromise between maximizing the total elevation coverage and loss of gain and detection sensitivity between elevation bars. The bar separation is usually chosen to be less than the 3-dB beamwidth of the antenna pattern.

Figure 4.3 is a functional diagram of a roll-and-pitch-stabilized scan generator. The operator selects the azimuth scan center, azimuth scan width, elevation scan center, and number of elevation bars. The commanded antenna azimuth angle θ_s and the commanded elevation angle ϵ_s are generated in a space-stabilized, or more accurately a roll-and-pitch-stabilized, coordinate system. The $\sec\epsilon$ compensation is used in some systems to prevent the azimuth scan width from becoming narrower

Table 4.2
Low-PRF Mode Characteristics

Benefits	Limitations
Precise range measurement	Low probability of detection or high false alarm rate in look-down missions
Ground maps	High peak power or pulse compression ratio
Simple method to achieve long unambiguous range	Highly ambiguous doppler
Range gating rejects side-lobe clutter	
Simple data processing	

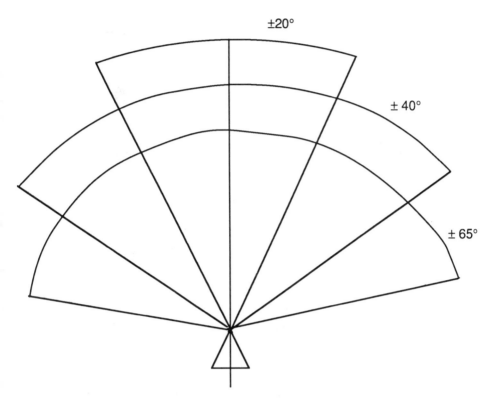

Figure 4.1 Typical azimuth scan patterns.

as the elevation angle increases, in the same way that the distance between longitude lines on the earth decreases with increasing latitude. The commanded azimuth and elevation angles are used to compute the space-stabilized direction cosines, i_s, j_s, and k_s, as follows:

$$
\begin{bmatrix} i_s \\ j_s \\ k_s \end{bmatrix} = \begin{bmatrix} \cos\theta & -\sin\theta & 0 \\ \sin\theta & \cos\theta & 0 \\ 0 & 0 & 1 \end{bmatrix} \begin{bmatrix} \cos\epsilon & 0 & \sin\epsilon \\ 0 & 1 & 1 \\ -\sin\epsilon & 0 & \cos\epsilon \end{bmatrix} \begin{bmatrix} 1 \\ 0 \\ 0 \end{bmatrix}
$$

$$(4.1)$$

where i_s, j_s, and k_s constitute a right-handed coordinate system with k_s the local vertical, positive downward; i_s is the projection of the aircraft longitudinal reference line in the horizontal plane, positive forward; and j_s is positive in the direction of the right wing. Azimuth is measured positive to the right, and elevation is positive upward.

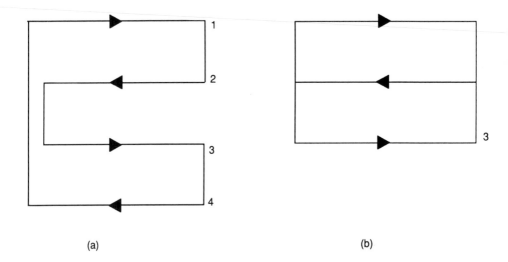

Figure 4.2 Typical elevation scan patterns: (a) 4-bar and (b) 3-bar.

Equation (4.1) can be stated in more compact notation as

$$[s] = [\theta][-\epsilon]1_r \qquad (4.2)$$

where $[s]$ represents the space-stabilized direction cosines, $[\theta]$ and $[-\epsilon]$ are coordinate rotation matrices, and the unit vector is 1_r.

The commanded antenna position space-stabilized direction cosines are then converted to direction cosines in the aircraft body coordinate system. Using the notation of (4.2), we can express the coordinate transformation as

$$[a] = [A_r][A_p][s] \qquad (4.3)$$

where $[a]$ represents the direction cosines in aircraft coordinates, A_r is the roll angle, and A_p is the aircraft pitch angle. A_r and A_p are usually supplied by the aircraft inertial reference system, which is not a part of the radar.

4.2 DISPLAYS

The two most common display formats are shown in simplified form in Figure 4.4. Many other pieces of information often presented on aircraft for which the radar observer is also the pilot (e.g., aircraft speed, artificial horizon) have been omitted for clarity. The B-scan, represented by a vertical range sweep, has been used historically in airborne fire control radars for air-to-air search modes because of the simple mechanization. The y-axis is range from the aircraft; the x-axis is the azimuth scan

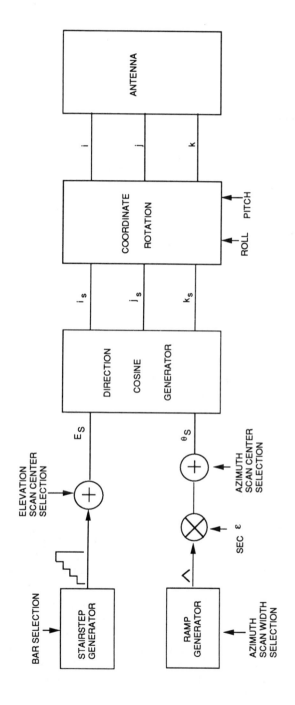

Figure 4.3 Scan generation and stabilization.

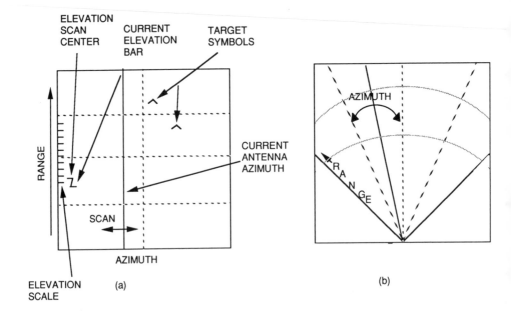

Figure 4.4 Common display presentations: (a) B-scan and (b) sector PPI.

angle. In early fire control radars, the amplitude of the radar return was intensity-modulated directly on the range sweep. A bright vertical line was quite evident and provided an indication of the instantaneous azimuth position of the radar antenna. In modern systems, the radar data are highly processed before display, and target detections are often represented by target symbols and not just simply "blips." For example, the returns from a batch of pulses are doppler-analyzed, and then the range-doppler map from several batches may be noncoherently integrated before the application of a constant false alarm rate detection process to declare the presence of a target. Often a vertical line is introduced into the display to provide the traditional, familiar display of current antenna position.

Horizontal range scale, vertical azimuth scale, and elevation angle markings may be etched into the cathode-ray tube (CRT) overlay or may be incorporated into the electronic symbology. Figure 4.4 shows a common method used in fire control radars to present the elevation angle of the antenna. One horizontal element of the Z-shaped symbol indicates the elevation scan center position, and the other denotes the position of the current antenna bar. The elevation angle is read using the elevation scale markings.

The display device is conceptually similar to the computer-driven monitor that has become very familiar since the introduction of the personal computer. Usually a raster-scanned CRT is driven by a scan converter that contains the radar data and other display symbology. When the B-scan display is used to display a ground map, the map is distorted, especially at near ranges. When ground map is the primary mode of the radar, the sector plan position indicator (PPI) display shown in Figure 4.4(b) is used.

4.3 RANGE PROFILE

An idealized low-PRF range profile is shown in Figure 4.5. There is no return for ranges less than the aircraft altitude. The return from the region below the aircraft is termed the *altitude return*. The antenna main beam is usually positioned within a few degrees of horizontal; therefore, the antenna gain in the vertical direction is quite low, often 30 to 50 dB below the gain of the main antenna beam. The large magnitude is the result of the near-normal incidence angle and the close range. The ground that is illuminated by the main beam of the antenna experiences the gain of the antenna pattern on both the transmit path and the receive path. The side-lobe clutter region extends from the altitude line through the main-lobe clutter region. Figure 4.5 pictures primarily the effect of elevation side lobes, but azimuth side lobes will also be present at the same ranges of main-lobe clutter. The decay in side-lobe clutter amplitude is the result of increasing range and decreasing the incidence angle.

The return from the aircraft at position A in Figure 4.5 has to compete only with side-lobe clutter for detection, since the main beam does not strike the ground at the range of A. The return from A may be detectable based on amplitude processing only, that is, noncoherent or nondoppler processing. Aircraft B must compete with main-lobe clutter and, in general, will not be detected without doppler processing. Target C represents a stationary water tower located 3 km ahead of the platform which must compete with side-lobe clutter. Target D represents a slow-moving ground vehicle which must compete with main-lobe clutter.

4.4 CLUTTER SPECTRA

Figure 4.6 depicts the true (i.e., unambiguous) clutter spectrum for three range gates of the range profile of Figure 4.5. The values were calculated using (2.1), (2.6), and (2.7) using $2/\lambda = 64$ Hz/m/s (19.2 Hz/ft/s), $V = 300$ m/s (984 ft/s), $h = 1,000$m (3,280 ft), and an antenna beamwidth of 3 degrees. For a range gate width of 150m, the area below the aircraft contained in a cone ±14.8 degrees from the vertical is in a single range gate. The doppler extent is ±9,481 Hz. The water tower is located 3 km directly ahead of the platform and is competing with side-lobe clutter from a depression angle of 18.434 degrees. The doppler extent of side-lobe clutter in the range gate containing the water tower is ±18,213 Hz. The return from target A is competing with side-lobe clutter from a depression angle of 45.07 degrees, and target A is closing at 376 m/s (1,234 ft/s). The main beam of the antenna is scanned in azimuth 45 degrees and depressed in elevation 2.87 degrees

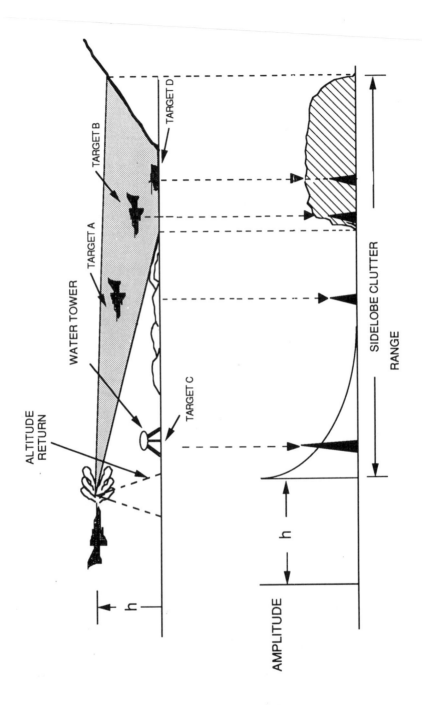

Figure 4.5 Low-PRF range profile.

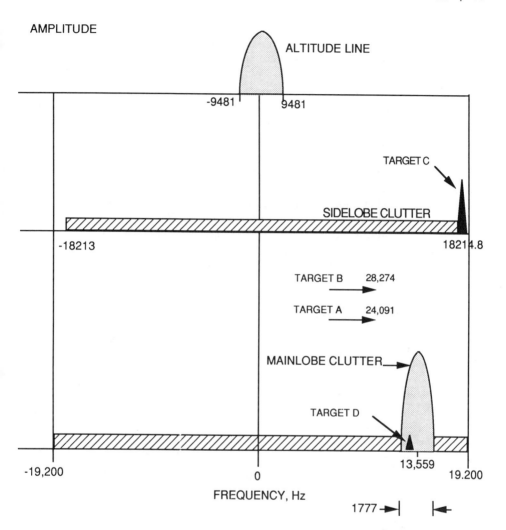

Figure 4.6 True clutter spectra.

such that the total angle from the velocity vector is 45.072 degrees. The range gates containing targets B and D have main-lobe clutter centered at 13,559 Hz with an extent of 1,777 Hz. Target B is closing at 441.78 m/s. The doppler of target B is 28,274 Hz, which is off the scale of Figure 4.6. The side-lobe clutter extent in the range gate containing target B is 19,094 Hz. Target D is assumed to be moving at 10 m/s (33 ft/s) relative to the ground. It is obscured because its return is only 451 Hz from the center of the main-lobe clutter. The side-lobe clutter extent in the range gate containing target D is 19,200 Hz.

Of course the spectra of Figure 4.6 are not actually observed with a low-PRF radar. A typical low PRF is 3,000 Hz, which provides a maximum unambiguous range of 50 km (26.7 nmi). Therefore, the maximum unambiguous doppler frequency is 3,000 Hz. All the spectral components of Figure 4.6 are folded within the 3,000-Hz bandwidth. The observed spectra is represented by Figure 4.7. The altitude line clutter fills the inter-PRF spectral region, and target detection is not possible. The side-lobe clutter level competing with target A is higher due to the spectral folding, but detection may still be possible. Targets B and D are not detectable because of main-lobe clutter. Note that there is a 1,223-Hz clutter-free region (888 to 2,111 Hz) where targets could be detected. The probability that the target appears in the clutter-free region is 1,223/3,000 = 0.40. Figure 4.8 illustrates the effect of clutter on the entire range-doppler spectrum. This plot was generated using a pulse-doppler radar simulation and the scenario described in Figure 4.5. The signal in range gate 7, which covers the entire frequency spectrum, is the altitude line. Main-lobe clutter has been shifted to 0 Hz to enable clutter cancellation. As discussed previously, main-lobe clutter covers a significant portion (approximately 60%) of the entire range-doppler spectrum.

4.5 LOW-PRF DOPPLER MODE

As discussed in Chapter 2, increasing the PRF (i.e., using a medium-PRF mode) is often the only acceptable method of increasing the extent of the clutter-free region. Multimode radars that have a medium-PRF mode may not provide an air-to-air doppler low-PRF mode. However, a low-PRF doppler mode will be described because many airborne radars have one.

Figure 4.9 shows the signal processing flow for the low-PRF doppler mode. The details of the processing elements are described in Chapter 7. A few of the processing considerations that are unique to the low-PRF mode will be described here. The gain of the receiver can be adjusted as a function of time (range) to prevent the strong side-lobe clutter from near ranges from saturating the receiver or the analog-to-digital (A/D) converter. The expected strength of the signal as a function of range can be adequately estimated, and a predetermined inverse gain function, called a *sensitivity time control,* can be used to maintain a more constant receiver output. This is because the range is unambiguous, and the side-lobe clutter strength usually decreases monotonically with range. If the range is ambiguous, then it is generally necessary to provide higher values of instantaneous dynamic range.

The reference signal translates the received spectrum so that main-lobe clutter is centered at zero doppler. The primary processing objective is to eliminate the main-lobe clutter from those range cells in which it exists. Side-lobe clutter is not the primary problem, since the target only competes with side-lobe clutter from the same range. Side-lobe clutter can exist throughout the entire spectral region, and rejection can only be accomplished by control of the antenna side lobes. The clutter canceler can be the only clutter rejection element of the processor, or it

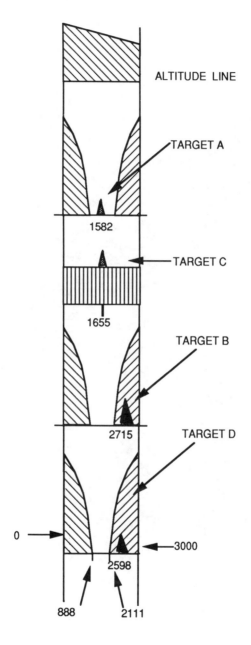

Figure 4.7 Observed low-PRF spectra.

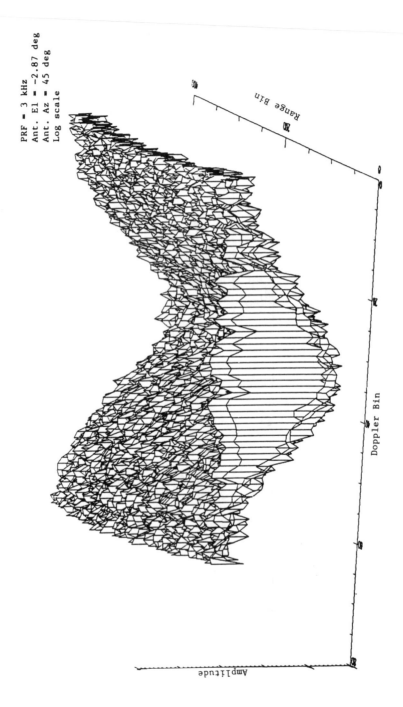

Figure 4.8 Simulated low-PRF range-doppler plot.

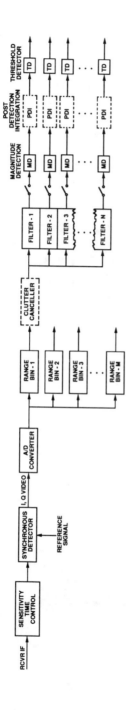

Figure 4.9 Signal processor for a low-PRF pulse-doppler radar.

may be used as a prefilter for a range-gated doppler analysis. Figure 4.10 illustrates the effect of a double delay line clutter canceler on the example presented earlier in the chapter. The altitude line is clearly shown in range gate 7. In addition, target C, the water tower, and target A are easily detected. However, targets B and D, because of their location in frequency, are removed, together with main-lobe clutter, by the clutter canceler. In order to increase the detectability of all targets and distinguish certain types of targets of interest, many low-PRF radars will vary their transmitted PRF.

Figure 4.11 represents the same scenario as that depicted in Figure 4.10; however, the PRF has been changed from 3 to 12 kHz, and the number of doppler filters has been increased from 64 to 256. The apparent doppler of each of the targets has changed, and target B is now outside the clutter notch. By transmitting multiple PRFs, the probability that a target is detected on any single PRF is increased.

PRF variation can also be used to help distinguish between ground moving targets and airborne targets. The true doppler frequency f_{dt} is given by

$$f_{dt} = n \text{ PRF} + f_{da}$$

where:

n = an integer;
f_{da} = the apparent doppler frequency.

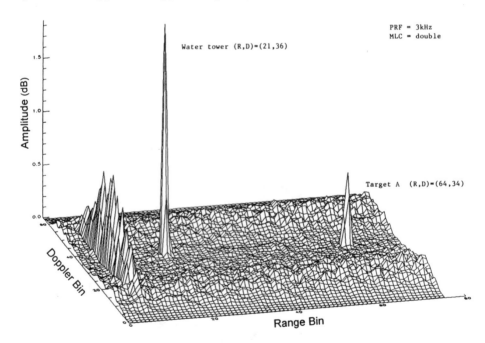

Figure 4.10 Low-PRF range-doppler plot with double delay line clutter canceler.

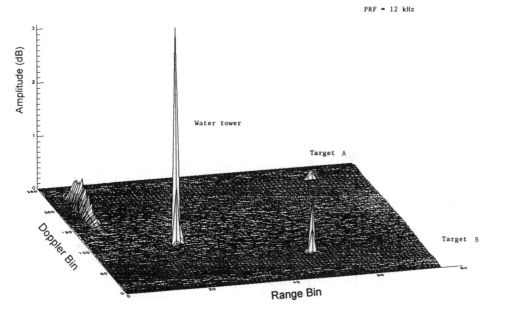

Figure 4.11 Simulated range-doppler plot at PRF = 12 kHz.

Usually $n = 0$ for ground moving targets and $n \geq 1$ for airborne targets. If the PRF is held constant for a batch of pulses and then changed for the next batch, the apparent doppler shift of the airborne target will change, but the doppler shift of the ground moving target will not.

PRF variations are sometimes used as an ECCM feature and to discriminate against multiple-time-around echoes. Multiple-time-around echoes are returns from ranges greater than the unambiguous range and can fall in the early range gates of the subsequent range sweep. Since the low-PRF waveform is supposed to be unambiguous in range, these returns can be falsely interpreted to be from near ranges. The PRF variations cause the multiple-time-around returns to appear in different range gates, reducing the likelihood that the returns will be noncoherently integrated, detected, and declared as a target.

4.6 NONCOHERENT MODES

The typical noncoherent modes of a multimode airborne radar will be described briefly for completeness. The modes are:

1. Ground map;
2. Terrain avoidance;

3. Air-to-air ranging;
4. Air-to-ground ranging;
5. Terrain following.

4.6.1 Ground Map

The antenna pattern of airborne pulse-doppler radars is usually a narrow pencil beam with a beam width between 2 degrees and 4 degrees. The pencil beam is not ideal for ground mapping because only a narrow patch in range is illuminated by the narrow elevation beamwidth, except during low-altitude flight when small elevation depression angles are used. Some systems provide a method of "spoiling" the elevation beam when mapping to provide a greater elevation beamwidth. Beam-spoiling methods include altering the reflector shape for reflector-type antennas or feeding only part of the array for planar array antenna systems.

The antenna elevation scan is usually a 1-bar raster, although other selections such as a 2-bar raster are often provided. The preferred display is the sector scan PPI of Figure 4.4(b), because it provides an undistorted map. The dynamic range of the radar returns from the ground far exceeds the contrast ratio available from the display device. Some dynamic range compression (e.g., logarithmic) is provided in the radar or display subsystem.

4.6.2 Terrain Avoidance

From the radar viewpoint, terrain avoidance is very similar to ground mapping. Terrain avoidance refers to the capability to fly close to the ground, usually to avoid enemy defenses, by flying around hills, as shown in Figure 4.12. In manual terrain avoidance, the pilot flies around terrain obstacles using a map display and other cues that describe the performance capabilities of the aircraft. The radar scan pattern may be controlled automatically by the system computer to ensure that the area being mapped and the data update rate is sufficient to provide adequate warning time. One method is to establish a clearance plane below the aircraft and highlight on the display any terrain projecting above the clearance plane, as shown in Figure 4.12(a).

4.6.3 Air-to-Air Ranging

Air-to-air ranging is used during short-range combat, often called a *dogfight*, when using guns or short-range missiles. Dogfights are primarily visual encounters, but the range information is valuable to calculate an accurate fire control solution, such as the proper lead angle. In the simplest implementation, a small fixed horn antenna pointed along the aircraft's longitudinal axis is used instead of the normal

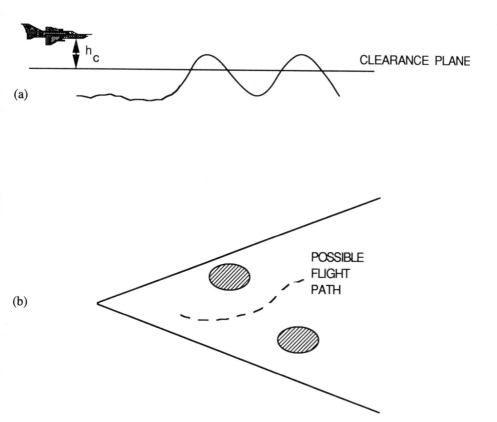

Figure 4.12 Terrain avoidance: (a) vertical profile and (b) plan view.

high-gain array. The horn "floods" the area, and angle tracking is not provided. The target is automatically acquired and tracked by the range tracker. In some systems, the normal antenna searches a narrow field of view corresponding to the pilot's gunsight or "heads-up" display. The target is automatically acquired and tracked in range and angle, thereby providing more information for fire control. Tracking is described in Chapter 13.

4.6.4 Air-to-Ground Ranging

The air-to-ground ranging function measures the range to the ground along the boresight of the antenna denoted by point A in Figure 4.13. Range is used in bomb release calculations and is essential to the terrain-following mode. Range tracking is accomplished by making use of the elevation monopulse characteristics of the antenna, as shown in Figure 4.14. Points on the ground at elevation angles below

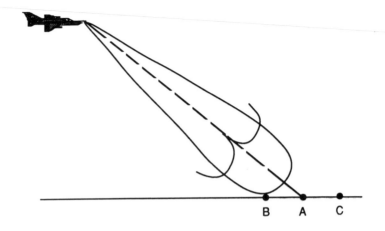

Figure 4.13 Air-to-ground ranging geometry.

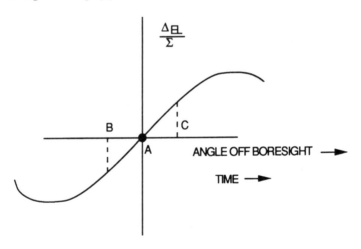

Figure 4.14 Monopulse response with respect to angle off boresight.

the boresight, such as point B, produce an elevation difference channel signal Δ_{el} that is out of phase with the sum signal Σ. Point A, located on boresight at the null of the difference pattern, is the desired range. Greater depression angles than boresight correspond to shorter ranges; lesser depression angles, such as point C, correspond to longer ranges. The horizontal scale of Figure 4.14 could be calibrated in (nonlinear) units of range or time delay. The range tracker tracks the range corresponding to the null of the difference pattern.

Accurate ranging requires that the elevation monopulse axis of the antenna remain vertical. This can be accomplished by providing a roll gimbal on the antenna.

Many air-to-air systems do not have a roll gimbal. A method of generating a roll-stabilized monopulse difference signal by electrically combining signals from the antenna elevation and azimuth channels is described in [1].

4.6.5 Terrain Following

In the terrain-following mode, the aircraft is flown at a constant heading while attempting to maintain a constant altitude above the terrain, as shown in Figure 4.15. The radar performs air-to-ground ranging as the antenna is scanned in elevation to profile the earth. Terrain following and terrain avoidance can be integrated to provide a mode in which the pilot has the option to go around or over a terrain feature.

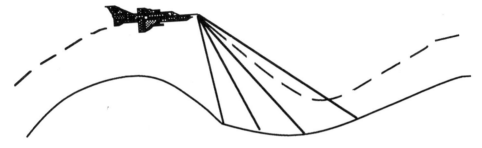

Figure 4.15 Terrain-following geometry.

Reference

[1] Zwagerman, P. K., "Air-to-Ground Ranging Using Electronic Roll Stabilization of Monopulse Data," *Proc. IEEE National Radar Conference*, 1988.

CHAPTER 5
High-PRF Mode
Linda L. Harkness

The high-PRF waveform ensures that the target doppler frequencies are unambiguously detected, but causes the target return to be highly ambiguous in range. The advantages and limitations of the high-PRF modes are summarized in Table 5.1. This chapter will address:

1. The salient characteristics of the waveform;

Table 5.1
Characteristics of the High-PRF Mode

Benefits	*Limitations*
For a fixed peak power, higher average power and thus longer detection range is achieved in high-PRF operation as a result of the higher duty factor	Highly ambiguous range
Unambiguous doppler, no blind zones except zero doppler and near main-lobe clutter	Eclipsing of target return interferes with initial detection and tracking
Good look-down nose-aspect detection (targets in clutter-free region)	Ranging methods more complicated, often less accurate
Illumination for semiactive missiles	Side-lobe clutter reduces tail-aspect detection sensitivity

2. The *velocity search* (VS) mode;

3. The *range-while-search* (RWS) mode;

4. Range-gated high PRF.

High-PRF tracking will be discussed in Chapter 13.

5.1 WAVEFORM CHARACTERISTICS

The pulse envelope of a typical high-PRF waveform is shown in Figure 5.1. The PRF, where PRF = $1/T$, is chosen to provide unambiguous doppler measurement and is on the order of 300 kHz for an X-band fire control radar. The true range profile is shown in Figure 5.2 and is, of course, the same as presented for the low-PRF mode in Chapter 4. The unambiguous range R_u is given by

$$R_u = cT/2 = c/(2 \text{ PRF})$$

where c is the speed of light.

$$R_u = 500\text{m for PRF} = 300 \text{ kHz}$$

The returns from all ranges are collapsed into the 500m range profile observed by the radar, which if drawn to the scale of Figure 5.2 would appear as a spike occupying the region from 0 to 0.27 nmi. All targets would be obscured by main-lobe and side-lobe clutter and would not be detected on the observed range profile.

The duty factor, $d_u = \tau/T$, usually ranges from 0.33 to 0.5. The average power, P_{ave}, and the power in the central line of the transmitter spectrum, P_{cl}, are related to the peak pulse transmitter power, P_p, by

$$P_{ave} = P_p d_u$$

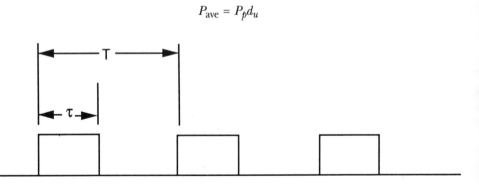

Figure 5.1 High-PRF waveform pulse envelope.

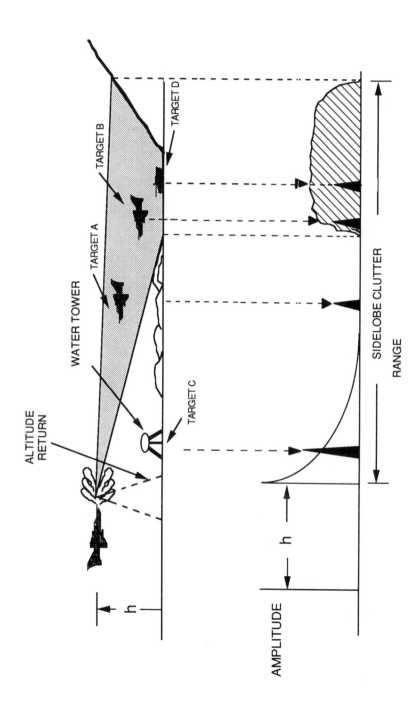

Figure 5.2 True range profile.

$$P_{cl} = P_p d_u^2$$

For one of the signal-processing methods that will be described later in this chapter, the effective signal power for target detection is proportional to P_{cl}. Therefore, the signal-to-noise ratio is enhanced by employing a duty factor approaching 0.5. Airborne radars transmit and receive using the same antenna, and therefore cannot transmit and receive simultaneously. The maximum receiver duty factor, d_r, is given by

$$d_r = 1 - d_u$$

Actual receiver duty factors must be somewhat less than the maximum to allow for finite rise and decay times of the transmit pulse and receiver gating function, which may be on the order of 50 to 100 ns. Many of the types of receiver protection devices (e.g., gas discharge transmit/receive (T/R) tubes) used on low-PRF radars are not suitable for high-PRF because their recovery times may be as long as 0.5 to 1.0 μs, which is an unacceptable fraction of the available receive time.

The true doppler spectra corresponding to several different ranges are shown in Figure 5.3, which are, of course, the same as previously shown for low-PRF. The velocities assumed for this example are velocity of the radar = 300 m/s, target A = 233 m/s, target B = 325 m/s, and target C = 10 m/s. The individual doppler spectra cannot be observed at each range because of the range ambiguities. However, the true spectra overlay without doppler ambiguity yields the spectrum shown in Figure 5.4, which contains the components from all ranges. Figure 5.4 is the spectrum actually observed by the high-PRF radar and is the one usually shown [1,2]. The returns from targets A and B are in the clutter-free region because each is approaching the radar (nose aspect), causing the total closing rate to be greater than the radar velocity. Target D is traveling at a slow velocity and is immersed in main-lobe clutter, while target C, the water tower, is located at the very edge of side-lobe clutter. Figure 5.5 illustrates a simulated high-PRF spectrum from a scenario containing the radar parameters from the example in Figure 5.4 and a single target traveling at 1,000 m/s. The frequency spectrum has been shifted such that main-lobe clutter is just to the left of 0 Hz, emphasizing the large "clutter-free" region.

5.1.1 Establishing the Minimum PRF

The minimum PRF is chosen to ensure that the highest velocity target of interest is in the clutter-free region. The criterion for the minimum PRF is depicted graphically in Figure 5.6. The most conservative approach is to choose the PRF sufficiently high that the target return is not in the back-lobe clutter region corresponding to the spectral line at f_0 + PRF. The PRF is given by

$$\text{PRF} > \frac{2(V_A + V_T)}{\lambda} + \frac{2V_A}{\lambda} = \frac{2(2V_A + V_T)}{\lambda} \tag{5.1}$$

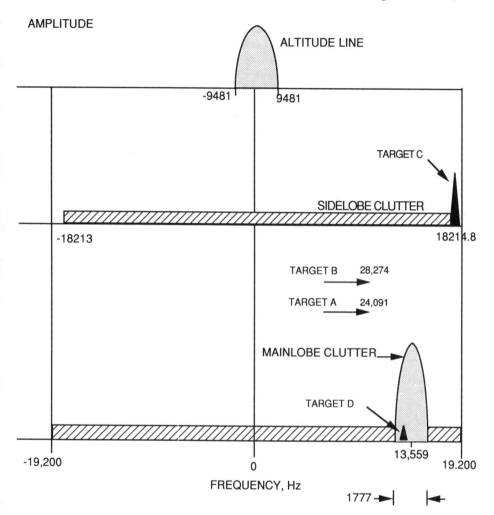

Figure 5.3 True doppler spectra at various ranges.

In some applications a minimum PRF may be chosen that is less than the value given by (5.1). For example, back-lobe clutter may be judged to be negligible because the antenna back lobes are highly attenuated by the aircraft structure.

5.1.2 Reducing the Effects of Eclipsing

Target returns that arrive while the receiver is turned off during transmit are described as eclipsed. The target returns move through these eclipse zones as the target range changes. The eclipse period, T_E, is given by

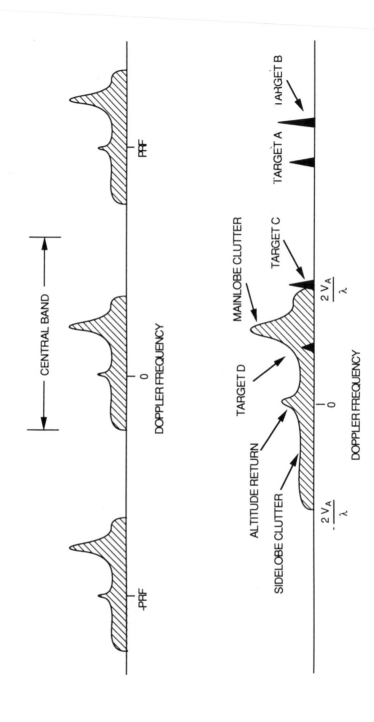

Figure 5.4 Observed high-PRF doppler spectrum.

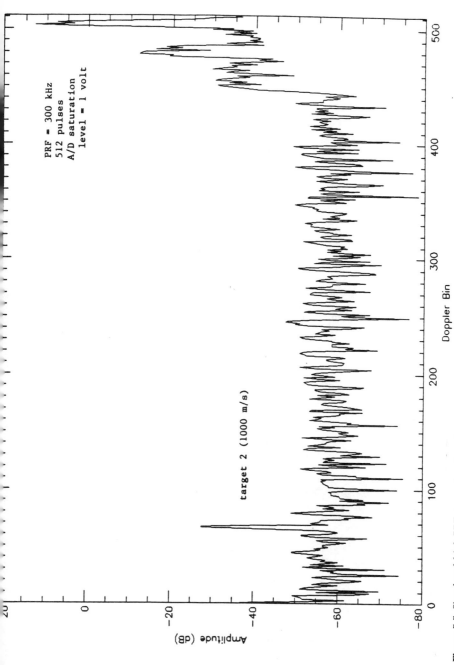

PRF = 300 kHz
512 pulses
A/D saturation
level = 1 volt

target 2 (1000 m/s)

Figure 5.5 Simulated high-PRF range-doppler plot.

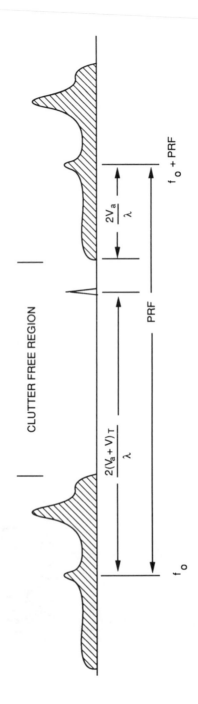

Figure 5.6 Establishing the minimum PRF.

$$T_E = R_u / V_R = c / (2PRF \ V_R)$$

where:

R_u = unambiguous range;
V_R = radial velocity (total closing rate).

Using a typical PRF = 250 kHz, T_E will range from about 0.6 sec for a nose encounter, total closing rate of 305 m/s (1,000 ft/s) to 6 sec for a tail approach of 93 m/s (305 ft/s).

Since receiver duty factors are typically 0.4, substantial losses in signal strength may occur during approximately one-third of the eclipse cycle, as shown in Figure 5.7. Signal loss for 1 to 2 sec is not acceptable during tracking, and several PRFs are provided to reduce the effects of eclipse. Two methods of employing multiple PRFs can be used:

1. The PRF may be changed in response to a decrease in signal strength.
2. The PRF may be cycled at a rapid rate (e.g., after each coherent dwell) through a set of PRFs chosen to ensure that the target is uneclipsed on at least one PRF.

The set of PRFs can be used in a search mode (using method 2) to reduce the probability that eclipsing will prevent target detection for long periods.

One method of choosing a second PRF (the minimum PRF perhaps being the first one, PRF_1) is as follows. Choose a range R_E at which it is acceptable for

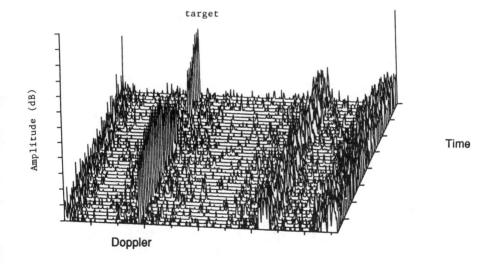

Figure 5.7 High-PRF spectra illustrating target eclipsing over time.

the target to be eclipsed at both PRFs. Then use (5.2) to establish an effective eclipsing repetition frequency f_E for the pair of PRFs.

$$f_E = c/2R_E \qquad (5.2)$$

Then the second PRF, PRF$_2$, is given by

$$PRF_2 = PRF_1 + f_E \qquad (5.3)$$

Figure 5.8 shows the eclipsing performance of a pair of PRFs. Note that for a single pair of PRFs there are substantial areas where the target can be eclipsed on both PRFs. Therefore, three PRFs (resulting in three PRF pairs) or four PRFs (resulting in six PRF pairs) are frequently provided to reduce the mutually eclipsed regions. The effect on an additional PRF pair is illustrated by the dashed line in Figure 5.8.

Additional criteria that are sometimes used to choose the ensemble of PRFs include hardware considerations and use of the PRFs for range ambiguity resolution. Range ambiguity resolution for medium-PRF modes is discussed extensively in Chapters 11 and 12. The application of multiple-PRF ranging to high-PRF mode is similar in principle. Mooney and Skillman (in [1, Section 19, p. 13]) describe a detailed approach suitable for high PRF. The hardware considerations affecting PRF selection include ease of generation (e.g., digital countdown from a single reference oscillator) and control of the harmonics of the PRF to prevent generation of spurious signals at harmful frequencies. For example, the following choice of PRFs can be generated by counting down from a 30-MHz oscillator:

$$PRF_1 = 30{,}000/120 = 250.000 \text{ kHz}$$
$$PRF_2 = 30{,}000/121 = 247.934 \text{ kHz}$$
$$PRF_3 = 30{,}000/122 = 245.902 \text{ kHz}$$

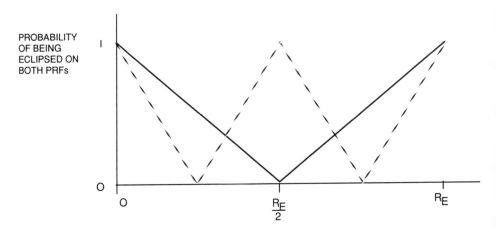

Figure 5.8 Eclipsing performance of a pair of PRFs.

If the radar IF of 30 MHz corresponds to zero doppler, then the 120th, 121st, and 122nd harmonics of PRF_1, PRF_2, and PRF_3, respectively, fall at zero doppler and are rejected with the altitude line. The value of f_E corresponding to the PRF pair (PRF_1, PRF_2) is 2,066 Hz (250,000 − 247,934). The eclipsing performance can be explored using (5.2).

5.2 VELOCITY SEARCH MODE

The VS mode is the radar mode that usually provides the longest detection range. No attempt is made to resolve the range ambiguities, but several PRFs may be used to reduce the eclipse periods, as previously discussed. A typical VS mode display is shown in Figure 5.9. Other data often presented with the radar data are omitted

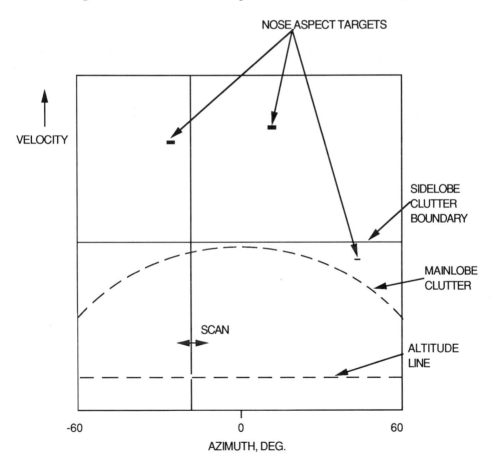

Figure 5.9 Velocity search display.

for clarity. The presentation of velocity versus azimuth is similar to the range versus azimuth B-scan display of low-PRF modes (Figure 4.4). The VS display is sometimes called a *B-prime display* because of this similarity. The dashed lines denoting main-lobe clutter and altitude line do not generally appear on the display, because they are rejected by the doppler processing. Recall that the doppler frequency of the main-lobe clutter varies with the antenna pointing angle. Occasional false alarms caused by the main-lobe clutter notch filter as it moves in frequency may be noted along the arc. Any target detected above the main-lobe clutter arc has a closing rate that is greater than the component of aircraft velocity in that direction; therefore, we are approaching the target in its nose hemisphere. Heretofore, we might have inferred that all nose-on target encounters result in the target doppler frequency being in the clutter-free region, but that is not strictly true for large azimuth angles.

The envelope of the signal received from a point target versus time is depicted in Figure 5.10. The envelope has the shape of the two-way antenna pattern. The total on-target time is usually defined as the time between the −3-dB points on the one-way antenna pattern (−6-dB two-way). The total on-target time is divided into a number of coherent dwell periods, usually between three and four periods. A

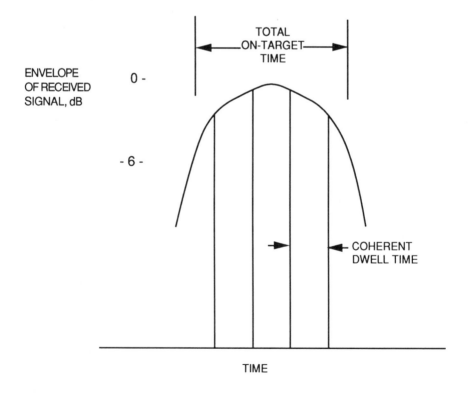

Figure 5.10 Signal strength versus on-target time.

number of pulses are stored (e.g., 512 or 1,024) and analyzed using an FFT. The filter outputs for all of the coherent dwells during the total on-target time are further processed by a postdetection integrator, which provides the increased detection sensitivity relative to the range-while-search mode.

Figure 5.10 illustrates only one relative position between the start of the coherent dwell time and the beginning of the total on-target time. Coherent dwell intervals that occur when the target is at the beam edge suffer significant loss relative to the main beam. Since the target location is unknown, the relative position is a uniformly distributed random variable. It is customary to calculate an average azimuth beam shape loss, which is then used in performance predictions. Similarly, an average elevation beam shape loss can be calculated assuming the target elevation is uniformly distributed between elevation bars.

5.3 RANGE-WHILE-SEARCH MODE

The range ambiguities are resolved in the RWS mode, resulting in computer-generated range versus azimuth display, either a PPI or a B-scan similar in appearance to Figure 4.4. Linear frequency modulation (FM) ranging is widely used to resolve the range ambiguities; one such method is illustrated in Figure 5.11. Three consecutive coherent dwell periods are needed. The transmit signal is at constant frequency during dwell A. The received frequency differs from the transmitted frequency by the doppler shift f_D. The transmitter frequency is changed at a constant rate during dwell B. Because the transmitter changes frequency during the round-trip propagation time T_D, the difference between the transmitted and the received frequency is shifted by an additional amount that is proportional to the delay. Linear FM ranging schemes do not attempt to measure T_D directly because the target is not detectable until after doppler processing, but instead measure the three frequencies associated with the three dwell periods. The difference frequency during dwell B, f_1, is given by

$$f_1 = f_D - (df_0/dt)_1 T_D = f_D - (df_0/dt)_1 (2R/c) \tag{5.4}$$

where:

$(df_0/dt)_1 =$ transmitter sweep rate;
$R \qquad =$ target range.

Equation (5.4) can be solved to yield

$$R = \left(\frac{c}{2}\right) \frac{f_D - f_1}{\left(\dfrac{df_0}{dt}\right)_1} \tag{5.5}$$

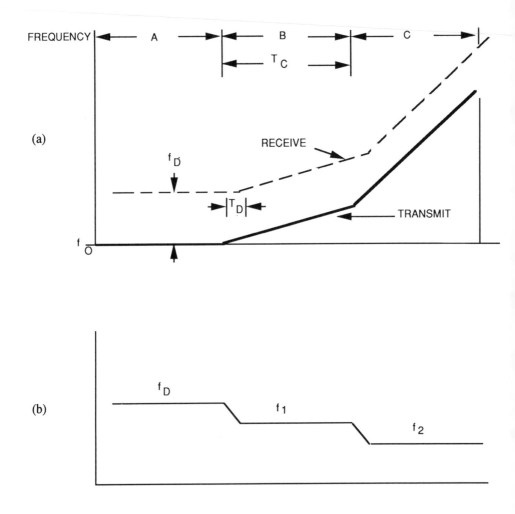

Figure 5.11 Linear FM ranging with two positive slopes: (a) transmitted and received frequencies versus time and (b) difference frequency.

The target range can be solved directly using (5.5) because f_1 and f_D are measured, and $(df_0/dt)_1$ is a known design parameter. However, if two or more target returns are present during each dwell, then it is possible to generate extraneous solutions, referred to as *ghosts*. For this reason, a third dwell with a different sweep rate is used to ensure an accurate solution for the two-target case. One method is to compute the range independently using dwell pairs (A,B) and (A,C) and require that the target be detected at the same range in both pairs. Another method uses the solution of the two equations of the form of (5.4) corresponding pair (B,C) to yield the expected doppler, f_{DC}:

$$f_{DC} = \frac{f_1\left(\dfrac{df_0}{dt}\right)_2 - f_2\left(\dfrac{df_0}{dt}\right)_1}{\left(\dfrac{df_0}{dt}\right)_2 - \left(\dfrac{df_0}{dt}\right)_1} \tag{5.6}$$

Every possible combination of the target detections f_1 and f_2 observed during dwells B and C, respectively, is used to calculate a value of f_{DC}. The sets of f_1 and f_2 whose computed values of f_{DC} are equal to a measured doppler value observed during dwell A, f_D, are accepted as valid target observation pairs. The corresponding ranges are then computed from

$$R = \left(\frac{c}{2}\right)\frac{f_1 - f_2}{\left(\dfrac{df_0}{dt}\right)_1 - \left(\dfrac{df_0}{dt}\right)_2} \tag{5.7}$$

5.3.1 Range Resolution

Equation (5.5) can provide an insight into the range resolution available with linear FM ranging. Consider two objects at ranges R_1 and R_2. From (5.5),

$$R_1 - R_2 = \Delta R = \left(\frac{c}{2}\right)\frac{f_{11} - f_{12}}{\left(\dfrac{df_0}{dt}\right)} \tag{5.8}$$

The smallest frequency difference $(f_{11} - f_{12})$ that can be measured is the doppler filter bandwidth, Δf, given by

$$\Delta f = 1/T_c \text{ for } T_c \gg T_D \tag{5.9}$$

where T_c is the coherent dwell time.

Also, the total frequency excursion of the transmitter, B, is given by

$$B = T_c\left(\frac{df_0}{dt}\right)_1 \tag{5.10}$$

Combining (5.8), (5.9), and (5.10) yields

$$\Delta R = \frac{c}{2T_c\left(\dfrac{df_0}{dt}\right)_1} = \frac{c}{2B} \tag{5.11}$$

Equation (5.11) is the same fundamental relationship that governs the resolution of pulsed and pulse compression systems. Note also that resolution is a function of the FM sweep linearity.

5.3.2 Range Quantization

Equation (5.11) expresses not only the resolution but the quantization of the range measurements. For typical values of $(df_0/dt) = 10$ MHz/s and $T_D = 8$ ms, then

$$\Delta R = 1{,}875 \text{ m } (1 \text{ nmi})$$

$$B = 80 \text{ kHz}$$

Although 1.8 km or even greater quantization may be satisfactory for initial detection, finer quantization is often required. Track-while-scan systems employ Kalman filters to provide a smoothed estimate of range. Some radars provide two or more selections of range resolution (sweep rate). Because the FM sweep rates greatly affect the clutter spectrum, the radar is usually operated with the minimum FM consistent with the mission resolution and quantization requirements.

5.3.3 Effect of Linear FM Ranging on Clutter

Equation (5.4) can be used to determine the effect on the clutter. Each frequency component of the side-lobe clutter spectrum is the result of ground patches located at a different angle away from the velocity vector. Different angles usually imply different ranges; thus, the amount of frequency shift due to the ranging modulation will also be different. The frequency shift of key clutter region elements for one specific example is illustrated in Table 5.2 and depicted graphically in Figure 5.12.

5.4 RANGE-GATED HIGH PRF

All the previous discussion of high-PRF operation has assumed a single range gate. Certainly, single-range-gate systems are the most common in operation today. However, with the availability of small, powerful, low-cost digital signal processing, system concepts that were inappropriate for airborne radars can now be considered. The possible benefits of a range-gated high-PRF mode are listed in Table 5.3. By providing more than one range gate, the amount of side-lobe clutter with which a

Table 5.2
Example of Linear FM Ranging Effects on Clutter

Clutter Component	Range km (nmi)	Doppler Frequency, Hz No FM	10 MHz/s
Altitude line	9.3 (5)	0	−608
Main-lobe clutter	93 (50)	19,900	13,824
Side-lobe clutter:			
Upper edge	370 (200)	20,000	−4,304
Lower edge	370 (200)	−20,000	−44,304

target must compete is reduced. The clutter or noise is spread across multiple range gates, thus increasing the signal-to-noise ratio and the maximum detection range of the system.

The increased range resolution might permit closely spaced aircraft flying in formation (i.e., having the same doppler frequency) to be recognized as multiple targets instead of appearing as one. Linear FM or multiple-PRF ranging may be used to resolve the range ambiguities. The smaller range gate size also means that the amount of clutter power in each range gate is proportionally reduced.

The transmitter duty factor must be reduced to provide a greater number of matched-time-duration receiver range gates. Assuming that the PRF has been chosen near the minimum necessary to achieve the required unambiguous velocity coverage, then additional range gates are provided by reducing the pulse width. The reduced pulse width (and duty factor) results in less average power and less power in the central spectral line if the radar is constrained to the same peak transmitter power in the range-gated high-PRF mode as is used in the single range-gated, high-duty-factor, high-PRF mode. However, a dual-mode transmitter can be designed that provides both a low peak power, high-duty-factor mode and a higher peak power, range-gated high-PRF mode with similar average powers.

A typical implementation of a range-gated high-PRF mode will include between one and six PRFs ranging in frequency from 100 to 200 kHz. The multiple PRFs may be used for fine range resolving, or FM ranging may be implemented on each PRF for range confirmation. The number of range gates is usually between 10 and 40 with a range resolution of between 150 and 250 ft using low-ratio pulse compression. The signal-processing functions of a range-gated high-PRF mode can be mechanized in either analog or digital circuitry. Each of the parallel range gate channels must replicate the doppler processing that is provided in the single-channel system, as illustrated in Figure 5.13. In conventional high-PRF analog processors, the bandpass filter passes only the central band doppler frequencies. In contrast, the samples from range-gated high-PRF digital processors represent the power in all doppler bands passed by the receiver's IF amplifier. However, since

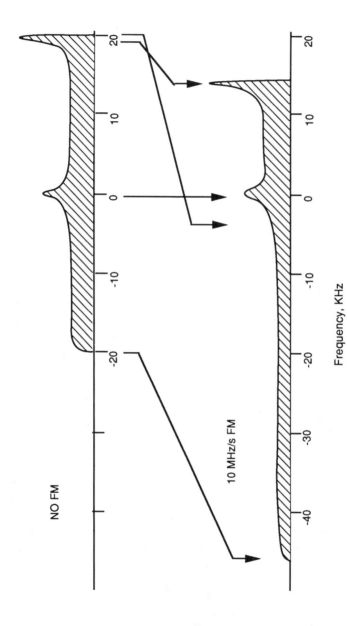

Figure 5.12 Effects of linear FM on clutter.

Table 5.3
Characteristics of Range-Gated High-PRF Operation

Benefits	*Limitations*
Reduced side-lobe clutter	Increased complexity
Increased range resolution	Higher peak power transmitter or reduced average power

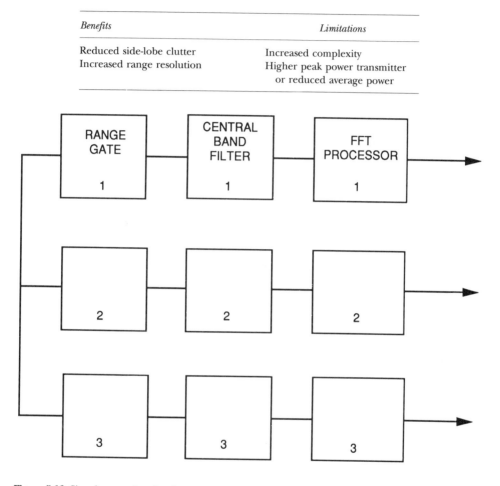

Figure 5.13 Signal processing for three range gates.

the power includes both signal and clutter, the signal-to-noise ratio is essentially the same as when central band processing is employed.

References

[1] Skolnik, M. I., *Radar Handbook*, New York: McGraw-Hill, 1970.
[2] Stimson, G. W., *Introduction to Airborne Radar*, Hughes Aircraft Company, El Segundo, CA, 1983.

CHAPTER 6

Medium-PRF Mode

Linda L. Harkness

The medium-PRF mode has become one of the most important modes of modern airborne pulse-doppler radar, primarily because of the capability that it provides to detect slow-closing rate targets from a high-speed platform. The characteristics of the medium-PRF mode are presented in Table 6.1. Because of the importance of the medium-PRF mode and because it is not widely documented, Chapters 10 and 11 treat the detailed design aspects of selecting an ensemble of medium PRFs and analyzing the resulting system performance. This chapter serves as an overview of the medium-PRF mode concepts.

The equations for unambiguous velocity V_u and unambiguous range coverage R_u presented originally as (1.10) and (1.11), respectively, are both functions of the PRF and therefore cannot be independently chosen. It can be easily shown that R_u and V_u are related by

Table 6.1
Medium-PRF Mode Characteristics

Benefits	*Limitations*
Better low-altitude tail chase mode than high-PRF	No velocities clear of side-lobe clutter
Accurate ranging	Many PRFs and pulse widths; complex processing needed to cope with range and doppler ambiguities
Good nose aspect performance	Rejection of large targets in side lobes

$$V_u R_u = c\lambda/4 \qquad (6.1)$$

For X-band radars, $V_u R_u \sim 2.5 \times 10^6$ m^2/s or 2.5×10^7 ft^2/s. Figure 6.1 is a plot of (6.1). The choice of PRF for a given mode determines the operating point on the hyperbola and thus the ambiguity characteristics. The boundaries of the shaded area are the maximum velocity and range of interest. If the entire area were below the hyperbola, then it would be possible to choose a single PRF that provides both unambiguous velocity and range. However, the shaded area shown is more representative of air-to-air radars. Two points are highlighted on the curve. One shows the highest value of PRF that permits low-PRF operation; the second depicts the lowest PRF that permits high-PRF operation. In the domain between the points, ambiguities exist in both range and velocity. We have defined this domain as medium-PRF operation.

6.1 WAVEFORM CHARACTERISTICS

6.1.1 Range Profile

The true range profile, shown in Figure 6.2, is independent of the PRF, of course. The same example used in Chapters 4 and 5 to illustrate the high- and low-PRF modes, respectively, will be used to describe the effect of the medium-PRF waveform on the apparent range and doppler profiles. The unambiguous range R_u is also shown. Target A is visible in this illustration on the basis of amplitude only; that is, there is no doppler processing, because its return competes with a very low value of side-lobe clutter. The apparent range profile is shown in Figure 6.3. The returns from all the unambiguous range intervals are added together. Main-lobe clutter extends throughout virtually all of the range gates. As a result, the return from target A is obscured by the main-lobe clutter and altitude line and is no longer detectable (without doppler processing).

6.1.2 Doppler Profile

The true doppler profiles at three ranges are shown in Figure 6.4. The true doppler is not a function of the PRF. The apparent doppler profiles of the range gates containing targets A, B, and C are shown in Figure 6.5. The PRF is 10 kHz. The target A range gate does not contain the peak of the zero doppler altitude line, but there is some power in the spectral region near zero. The 1,777-Hz spectral width of the main-lobe clutter is the same as the low-PRF mode example of Figure 4.6. The range gate containing target A in the true range profile did not contain main-lobe clutter. The ambiguous main-lobe clutter has been aliased in both the range and doppler profiles and now appears in the range gate containing target A, which is closing at 376 m/s (24,091 Hz true doppler; 4,091 Hz apparent

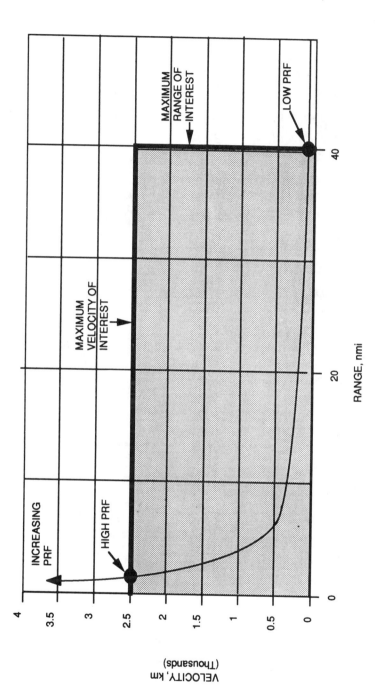

Figure 6.1 Unambiguous range versus unambiguous velocity.

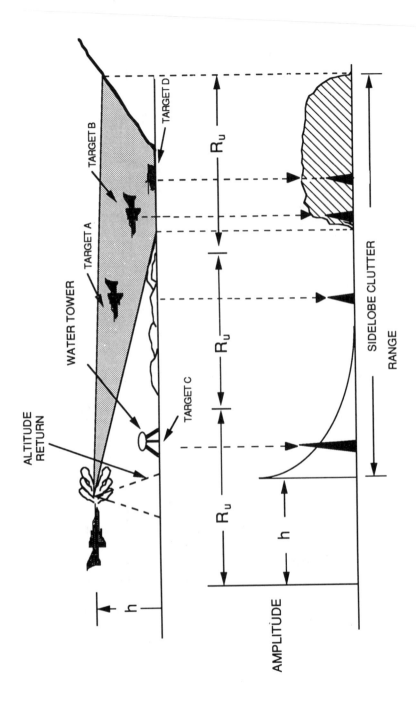

Figure 6.2 True range profile.

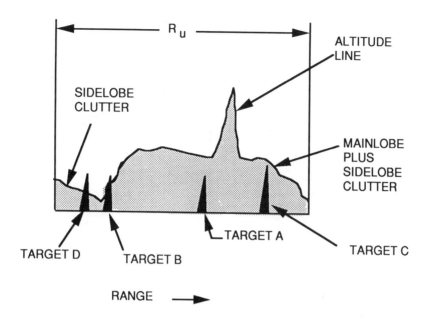

Figure 6.3 Apparent range profile.

doppler). Target B, which is closing at 441.78 m/s (28,274 Hz true doppler; 8,274 Hz apparent doppler), is also unobscured. The spectral region that is clear of main-lobe clutter is 8,223 Hz. The probability of the target appearing in the clutter-free region in this example is 8,223/10,000 = 0.822. Figure 6.6 illustrates the full range-doppler plot for the scenario including targets A and C. There is a much more extensive doppler region that is free of main-lobe clutter (approximately 82%). However, only the water tower is visible at range bin 20; target A is obscured by main-lobe clutter. Figure 6.7 represents the same scenario with a double delay line clutter canceler and postdetection integration to further enhance the signal-to-noise ratio.

The spectral width and true center frequency of main-lobe clutter are a function of antenna pointing angle and aircraft speed. The apparent center frequency of main-lobe clutter, and thus the obscured doppler frequencies, are a function of the PRF. Several PRFs are used during a single on-target time to ensure that the target return will not be obscured by aliased main-lobe clutter. Analyzing the doppler-clear regions is facilitated using the computer-generated blind zone charts described in Chapter 12. Chapter 11 discusses the use of multiple PRFs to resolve the range ambiguities.

6.1.3 Range Blind Zones

The medium-PRF mode has an eclipsing phenomenon similar to the high-PRF mode but to a much lesser degree. Target returns cannot be received when transmitting. In

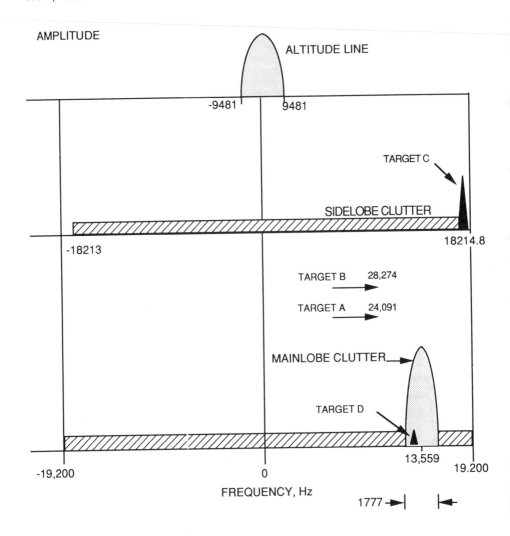

Figure 6.4 True doppler profile.

practice, the receiver is off for 1 or 2 extra range gate positions after the transmitter pulse. The maximum fraction of the interpulse interval available for target reception $d_r = (1 - d_t)$, where d_t is the transmitter duty factor. For typical values of PRF = 10 kHz and transmitter pulse width $\tau = 1$ μs, $d_r = 0.99$. Clearly, range blind zones are not a major consideration. However, if the effective pulse width of 1 μs is achieved by compressing a transmitter pulse $\tau = 13$ μs, then $d_r = 0.87$. The extent of the blind range with a pulse compression system is not negligible. Multiple PRFs can be used during a single on-target time in a manner somewhat similar to that used to reduce obfuscation by main-lobe clutter so that targets are uneclipsed in

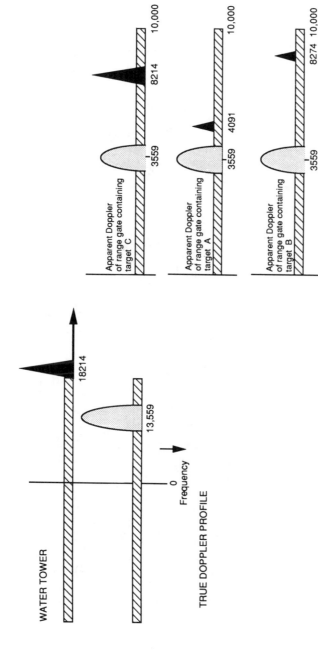

Figure 6.5 Apparent doppler profile of range gates containing targets A and B.

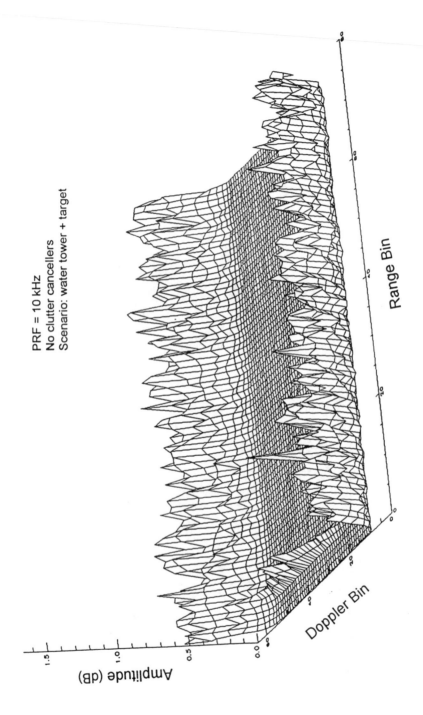

PRF = 10 kHz
No clutter cancellers
Scenario: water tower + target

Figure 6.6 Simulated medium-PRF range-doppler plot.

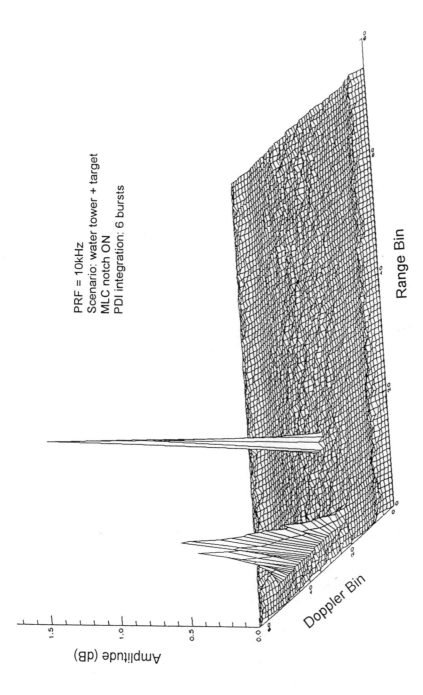

PRF = 10kHz
Scenario: water tower + target
MLC notch ON
PDI integration: 6 bursts

Amplitude (dB)

Range Bin

Doppler Bin

Figure 6.7 Medium-PRF range-doppler plot with main-lobe clutter canceler and postdetection integration.

range. Therefore, three criteria, not including hardware considerations, that must be met by the PRF selection are:

1. Reduced doppler blind zones;
2. Reduced range blind zones;
3. Compatible with range ambiguity resolving.

6.2 SEARCH MODE

The antenna azimuth and elevation scan pattern selections employed are the same as those described for the low-PRF mode and depicted in Figures 4.1 and 4.2. The target is generally ambiguous in both range and doppler on any single PRF. However, several PRFs, usually seven to nine, are transmitted during a single sweep of the antenna beam over the target. Automatic detection is performed after each of the PRF dwells using some type of constant false alarm rate (CFAR) circuit. Theoretically, the unambiguous range of a single target can be determined using the detections on two PRFs. However, detection on three PRFs is required to reduce the probability of spurious solutions caused by the presence of two targets. Only targets for which the unambiguous range has been determined (i.e., targets that have been detected on three or more PRFs) are declared detections and displayed. A typical display is shown in Figure 6.8. The vertical axis is unambiguous range, and therefore the display has an appearance similar to that of the low-PRF doppler mode display.

6.2.1 PRF Programming

The radar cycles through its ensemble of PRFs during a single on-target time. The PRF programming will be illustrated using representative values. Assume that seven PRFs, as shown in Table 6.2, are necessary to meet the range and doppler blind zone reduction requirements. Furthermore, assume that 16 valid pulses must be collected at each PRF for doppler analysis and that the range of interest is 80 nmi. The round-trip propagation time corresponding to 80 nmi is 1 ms. Each time that the PRF changes, the apparent ranges of the ambiguous reflectors, targets and clutter, change. Typically, some number of fill pulses is used to allow the system to stabilize, reset the automatic gain control (AGC), and so on. In this example, the dead time used between PRF changes is 1.5 ms and is intended to allow for both round-trip propagation time and settling. Table 6.2 shows that the total time required for the PRF program is 24 ms. If the antenna beamwidth is 2.4 degrees and the scan rate is 60 deg/s, then the available on-target time = 2.4/60 = 40 ms. Therefore, it is possible to have two repetitions of the PRF program if performed as shown in Figure 6.9, which minimizes the dead time.

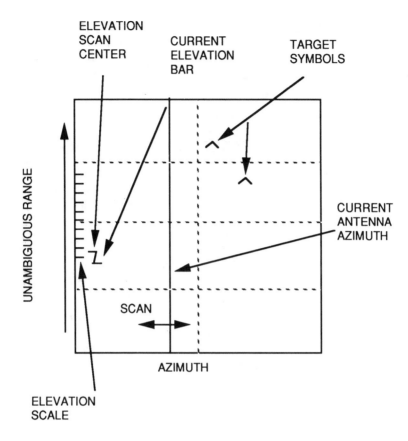

Figure 6.8 Medium-PRF B-scan display.

Table 6.2
PRF Program Timing

PRF, KHz	*Time Required to Transmit 16 Pulses, ms*	*Dead Time, ms*
8	2.00	1.5
9	1.78	1.5
10	1.60	1.5
11	1.45	1.5
12	1.33	1.5
13	1.23	1.5
14	1.14	1.5
15	1.07	1.5
Total	11.68	12.0

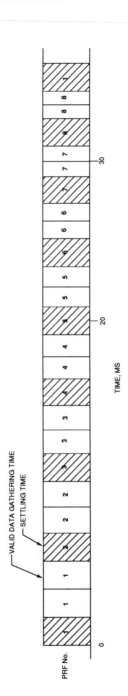

Figure 6.9 Medium-PRF program.

6.2.2 Pulse Width Diversity

The ensemble of PRFs shown in Table 6.2 are typical and span nearly an octave. If a constant transmitter pulse width and peak power are used, then the average power will vary proportionally. If the total number of pulses processed during a coherent processing interval is constant, which is usually the case for FFT processors, the dwell time varies, but the probability of detection on each dwell is not affected. If it is desired to operate the transmitter near its maximum average at all PRFs, then a number of pulse widths, perhaps a different one for each PRF, that will maintain the transmitter duty factor nearly constant can be provided. Of course, the receiver video bandwidth should be changed also to match the pulse width. If constant pulse width, peak power, and dwell time are used, as might be the case for non-FFT processors, then the probability of detection will decrease as the PRF is decreased. The effect on detection of the duty factor variation is discussed in detail in Chapter 11.

Pulse width diversity also requires corresponding changes in processing. For example, the spacing between range gates is normally set approximately equal to the pulse width. Using a single range gate spacing with a 2:1 variation in pulse width is undesirable. Range gates set too far apart result in unacceptable range gate straddling losses and perhaps even missing samples.

Oversampling increases the multiple detections caused by a single target. The extraneous detections must eventually be eliminated in the processing. Also, the range-ambiguity resolving algorithms compensate for the differences in range gate spacing.

6.2.3 Processing

Figure 6.10 shows the signal processing flow for the medium-PRF doppler mode. The details of the processing elements are described in Chapter 7. A few of the processing considerations that are unique to the medium-PRF mode will be described here. The range is unambiguous, and the side-lobe clutter strength generally decreases monotonically with range. Therefore, the gain of the receiver can be adjusted as a function of time (range), referred to as *sensitivity time control* (STC), to prevent the strong side-lobe clutter of near ranges from saturating the A/D converter. In the medium-PRF mode, the return in a range gate is the vector sum of the returns from several ranges, and it is necessary to provide a higher value of instantaneous dynamic range. When the PRF is changed, the vector sum in each range gate changes. Therefore, STC and slow-acting AGC are not appropriate. Figure 6.10 illustrates a digital AGC mechanization. The first few pulses received after the change of PRF are used to establish individual gains for each of the range gates so that nominally the full A/D converter dynamic range can be used without saturation.

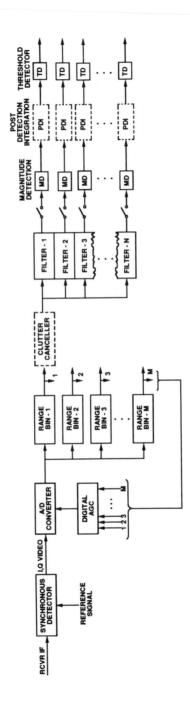

Figure 6.10 Medium-PRF mode processing.

6.2.4 Preventing False Detections of Side-Lobe Clutter

The peak antenna side lobes of a high-performance airborne radar are typically 30 dB below the main-beam gain in the angular region 5 to 10 beamwidths from the peak and taper to 40 or 45 dB in the far side-lobe region. Thus, the antenna can provide 60- to 90-dB attenuation of the side-lobe clutter relative to main-lobe clutter. Despite this careful attention to antenna side lobes to control the average level of side-lobe clutter, the return from large reflecting objects on the ground can enter through the side lobes and cause false detections. A simple numerical example illustrates the plausibility of side-lobe detection. Assume that the radar is capable of detecting a target with a radar cross section (RCS) of 1 m^2 at 40 nmi. Table 6.3 shows that a large reflector, such as a metal building, returns a stronger signal to a low-flying aircraft than the target and thus can be detected. The target is not in the main beam, and its doppler frequency will not be the same as that of main-lobe clutter. Therefore, the building will not be rejected by the main-lobe clutter filter. The water tower from the example shown in Figure 6.2 is an example of this phenomenon.

The guard channel illustrated in Figure 6.11 provides one method of rejecting side-lobe detections. The relative gains of the main antenna and the guard horn are shown in Figure 6.12. The guard horn has less gain than the antenna main beam but more than the antenna side lobes. Therefore, a signal entering through the side lobes will produce a greater amplitude response in any given range-doppler cell of the guard channel than in the corresponding cell in the main channel. The decision logic rejects main-channel detections, which have a main-to-guard amplitude ratio that is too small. A receiver and signal-processing channel must be provided that is similar to the main channel. At first, this replication might appear to be a high price to pay for false alarm control; however, radars that use the guard channel approach during search use the same channel as a monopulse angle tracking channel during track modes. Thus, the true additional cost to a monopulse radar is minimal. Figure 6.13 illustrates the range-doppler plot of the guard channel for the scenario described in Figure 6.2, containing targets A and C. The water

Table 6.3
Relative Signal Strength of Building Having 1,000 m^2 RCS

Factor	Building	Target	Relative Strength, dB
Range, nmi	2	40	52
RCS, m^2	1,000	1	30
Side-lobe suppression, dB	−70	0	−70
Total, dB			12

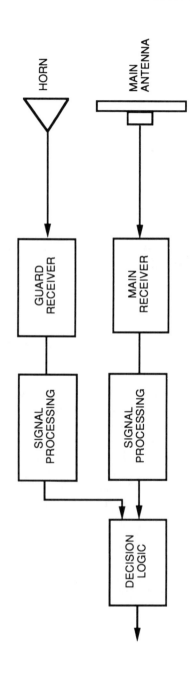

Figure 6.11 Guard channel block diagram.

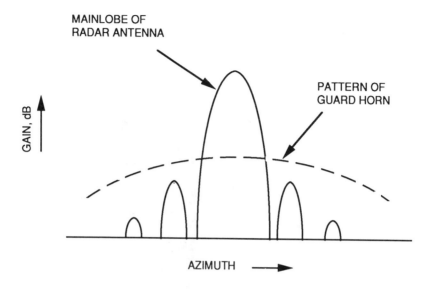

Figure 6.12 Patterns of the main antenna and guard horn.

tower has a stronger response in the guard channel than it does in the sum channel shown in Figure 4.11. As a result, the ratio of the main-to-guard amplitude will be much smaller than that for a main-lobe target and the side-lobe discrete (i.e., the water tower) will be rejected.

A second approach can be used for single-channel (i.e., nonmonopulse) radars. The range ambiguity resolution function will accurately determine the range of the side-lobe target. Detectable side-lobe targets are usually associated with very short ranges. The amplitudes of all the candidate detections are compared to a range-variable threshold. Short-range targets entering through the main beam are expected to be very large and detected. The side-lobe target, although having a large RCS, suffers from the relative attenuation of the two-way antenna side lobes and may be rejected by the range-variable threshold.

6.3 INTERLEAVED MODES

Many airborne radars implement both high-PRF and medium-PRF modes. The high-PRF mode can typically provide 50% greater detection range against nose aspect targets than a medium-PRF mode transmitting the same average power. For tail aspect targets, the situation is reversed for aircraft altitudes less than 10,000 ft: the medium-PRF mode is superior [1]. High-PRF and medium-PRF waveforms can be interleaved to ensure the highest probability of detection independent of target heading. A second benefit is that accurate ranging and higher range resolution are achieved in the medium-PRF mode. One method of interleaving is to scan one

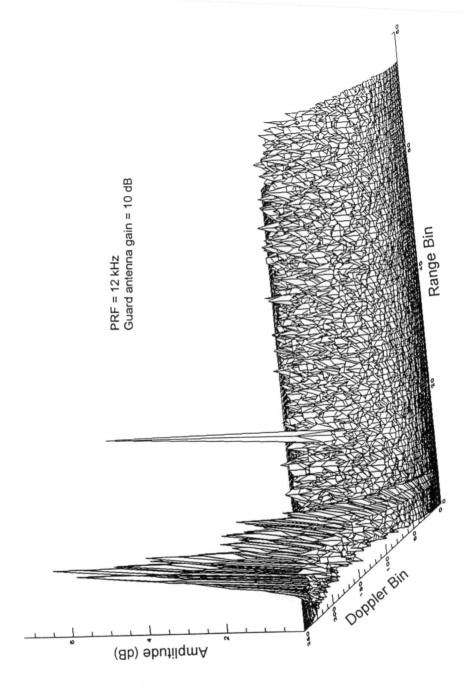

PRF = 12 kHz
Guard antenna gain = 10 dB

Figure 6.13 Guard channel range-doppler plot.

elevation bar at a high-PRF mode and then retrace it on the same bar or an intermediate elevation at a medium-PRF mode. Interleaving frame-to-frame is another alternative.

Additional information on the medium-PRF mode can be acquired from additional readings listed in the references [2–4].

References

[1] Aranoff, E., and N. M. Greenblatt, "Medium PRF Radar Design and Performance," *20th Tri-Service Radar Symposium*, Environmental Research Institute of Michigan, Ann Arbor, 1974, pp. 53–57.

[2] Hovanessian, S. A., "Medium PRF Performance Analysis," *IEEE Trans. Aerospace and Electronic Systems*, Vol. AES-18, No. 3, May 1982, p. 286.

[3] Long, W. H., and K. A. Harriger, "Medium PRF for the AN/APG-66 Radar," *Proc. IEEE*, Vol. 73, No. 2, February 1985, p. 301.

[4] Ringel, B., D. H. Mooney, and W. H. Long, "F-16 Pulse Doppler Radar (AN/APG-66) Performance," *IEEE Trans. Aerospace and Electronic Systems*, Vol. AES-19, January 1983, p. 147.

CHAPTER 7

Phased-Array Pulse-Doppler Radar

Melvin L. Belcher, Jr.

7.1 INTRODUCTION

This chapter addresses the application of phased-array technology to airborne intercept radar. The principles of operation of a phased-array antenna are first summarized with emphasis on its relation to previously introduced pulse-doppler radar parameters. A brief overview of system engineering and operational/application issues is then provided to afford insight into user requirements. Phased-array antennas and technologies are then described in more detail along with airborne interceptor (AI) radar implementation considerations. The closing section of this chapter addresses the interaction of pulse-doppler signal synthesis and processing with phased-array antennas.

7.2 ELECTRONIC SCANNING

7.2.1 Principles of Implementation

The primary benefit of using a phased-array antenna is electronic beam agility. Rather than depending on gimbaled planar arrays driven by electric or hydraulic motors to implement search and track operations, the beam of a phased array can be repositioned within a period as short as a few microseconds. Mechanically scanned array antennas possess a fixed beamforming network. The antenna is focused at infinity along its boresight, that is, the direction orthogonal to the plane of the antenna. Only signals arriving along this boresight direction will be coherently summed through the beamforming network.

In contrast, a phased-array antenna possesses element-level phase control that can be adjusted as depicted in Figure 7.1. Given that a received (or transmitted) sinusoidal signal can be perceived as a plane wave, the pointing direction of the antenna is determined by the element-to-element phase shift [1]. The far-field pattern of an electronically scanning linear array of N elements as a function of angle $E(\theta)$, can be expressed as

$$E(\theta) = E_e(\theta) \sum_{n=1}^{N} A_n \exp\left[-j\frac{2\pi}{\lambda} nd(\sin\theta - \sin\theta_0)\right] \qquad (7.1)$$

where:

A_n = magnitude of element weight;
n = element index;
d = element spacing;
θ = angle;
θ_0 = commanded beam pointing direction.

The element pattern, $E_e(\theta)$, is the pattern of a single radiator in the array environment. The element pattern determines the polarization of the array and determines the gain roll-off as a function of scan angle. The summation term is the array factor and includes aperture tapering as well as the element-to-element phase shift that determines the scan angle. An element-to-element phase shift of 0 degrees corresponds to θ_0 set equal to zero so that the beam is scanned along the boresight orthogonal to the plane of the antenna. In the absence of a taper ($A_n = 1$), the

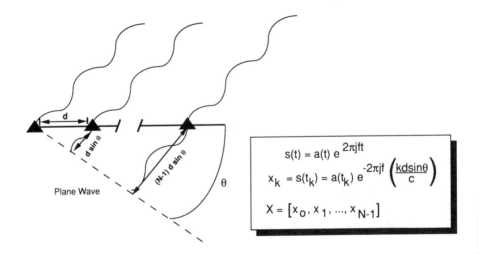

Figure 7.1 Phased-array operation.

antenna gain is equal to the product of the single-element gain and the number of elements N.

Both passive and active phased arrays are employed for airborne radar applications. Passive arrays employ a conventional central transmitter and receiver so that individual elements are driven only by passive devices such as a phase shifter. An active array is implemented so that each element is driven by an independent transmit amplifier and receive amplifier, as well as passive devices such as the phase shifter and possibly an attenuator. An active array can provide significantly greater flexibility and pattern control than a passive array, though at substantially higher per-element costs.

Both passive and active phased arrays require element-level phase shifters. At X-band, ferrite phase shifters are typically used for passive arrays to minimize loss. Solid-state diode phase shifters afford faster switching speed and more compact mechanization, but have greater loss. Active arrays employ diode phase shifters since these can be placed before the transmit amplifier and after the first receive amplifier so that sensitivity is not significantly impacted. Typically, a distributed beam-steering computer is implemented to ensure that electronic beam scanning capability is not compromised by delays in computation and communication of the individual phase shifter commands to the individual elements.

7.2.2 Effects on Radar Sensitivity

The two-way gain of a phased array decreases with increasing electronic scan angle. Geometry dictates that the projected antenna aperture extent decreases with increasing scan angle, with a corresponding decrease in transmit gain and receive antenna aperture. In addition, reflective mismatch increases with increasing scan angle, resulting in additional signal-to-noise ratio (SNR) loss. The composite effect on two-way SNR is generally estimated by taking the cosine-cubed of the scan angle. The system loss factor described in Chapter 14 is then augmented for phased-array systems by the term L_{scan}, such that

$$L_{scan} = -10 \cdot \log(\cos^3 \theta_0) \tag{7.2}$$

Peak and average scan loss is depicted in Figure 7.2. Peak loss corresponds to (7.2), while average loss is computed on the premise that the scan angle is uniformly distributed between $0°$ and the abscissa value. Generally, average SNR losses are used in estimating system performance. However, the designer should ensure that a fixed-boresight phased-array radar, such as that used in AI applications, can meet minimum sensitivity requirements at the maximum scan angle demanded by operational requirements.

Beamwidth also increases with increasing scan angle. The theoretical resolution of an antenna is given by λ/D, where D is the effective aperture extent along

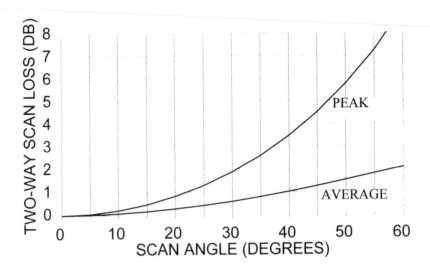

Figure 7.2 Phased-array scan loss.

a given dimension. The effective extent of the antenna in a given scan plane decreases with increasing scan angle due to the foreshortening effect that also degrades gain. Hence, the one-way azimuth and elevation beamwidths of a phased-array antenna can be estimated as

$$\beta_{az} = \frac{\lambda}{D_{az} \cdot \cos(\theta_{az})} \tag{7.3a}$$

$$\beta_{el} = \frac{\lambda}{D_{el} \cdot \cos(\theta_{el})} \tag{7.3b}$$

where the subscripts indicate azimuth and elevation, respectively. As a consequence of this main-lobe broadening with scan angle, the spectral extent of main-lobe clutter will vary with increasing scan angle independent of the clutter doppler geometry. Assuming a phased-array antenna with a boresight mechanically aligned with the center line of the aircraft and the associated velocity vector, (2.7) can be modified to estimate main-lobe clutter doppler bandwidth as

$$\delta f_{mlc} |_{pa} = \frac{2va}{\lambda} \tan(\phi_{ant}) \delta\phi_0 \tag{7.4}$$

where $\delta\phi_0$ is the phased-array beamwidth at boresight

The calculation of the main-lobe clutter magnitude outlined in Chapter 15 must be modified to reflect both the cosine-cubed scan loss and the beamwidth broadening that increases the main-beam-illuminated surface area. The net results of these effects is that received main-lobe clutter power varies with the cosine-squared of the electronic scan angle. This relationship implies that the ratio of signal power to main-lobe clutter power varies with the cosine of the electronic scan angle.

7.2.3 Antenna Side-Lobe Suppression

As noted above, the beam pointing direction caused by electronic scanning is determined by the element-to-element phase shift in that plane. However, the phase shifters that impose element-level phase shift are limited to a finite number of bits. There is a corresponding random quantization error that can be estimated in radians as

$$\sigma_{\theta q} = \frac{\pi}{\sqrt{12}\ 2^{n_b}} \tag{7.5}$$

where n_b is the number of bits in the phase shifter. Phase quantization error is a dominant source of array factor error in some implementations. As addressed subsequently, quantization error does not directly affect pulse-doppler operation if the phase center of the array is not shifted during a coherent processing interval (CPI), as is normally the case for medium-PRF and high-PRF processing.

It is often important to estimate the root-mean-square (rms) antenna side-lobe performance. Medium-PRF performance is largely determined by the capability of the antenna to reject side-lobe clutter. Side-lobe clutter rejection corresponds to the two-way T/R side-lobe level. Susceptibility to electronic countermeasures (ECM) and unintentional electromagnetic interference are largely determined by the degree of side-lobe suppression attained by the radar on receive.

The far-field pattern of an aperture antenna is given by the Fourier transform of the antenna aperture illumination function. This principle manifests itself in phased-array scanning, since the element-to-element phase shift corresponds to a shift in the far-field position of the main lobe, just as one would anticipate from the Fourier transform pair relationship. As in any aperture antenna, the nominal side-lobe level is determined by the illumination function taper. Suppression of antenna side lobes via tapering is mathematically equivalent to imposing a weighting function across a CPI in order to suppress the doppler filter frequency side-lobe response. For example, a Hamming taper may be employed to provide the illumination function of the aperture.

Any given antenna has an error side-lobe floor analogous to the spurious doppler response imposed by amplitude/phase reference noise in the receiver, as described in Chapter 15. In the case of the phased-array antenna, random element-to-element amplitude/phase tracking error imposes a corresponding spurious side-lobe response that cannot be significantly decreased by tapering. Mechanical displacement error of the radiating element and element/channel failure also contribute significantly to the spurious side-lobe floor of the antenna.

The antenna pattern can be interpreted as the Fourier transform of the sum of the desired antenna illumination function and an illumination error function that represents deviation from that function. The worst case consists of periodic errors across the antenna face that impose a spatially coherent illumination error function that manifests as discrete side lobes of significant magnitude. In modern antennas designed to achieve suppressed side lobes, the residual errors after calibration are generally randomized to minimize formation of such discrete spurious side lobes.

The rms side-lobe level can be estimated presupposing that the illumination error function can be characterized by normally distributed amplitude, phase, and mechanical displacement errors resulting in randomly distributed side lobes. Wang has provided useful expressions to estimate the resulting error side-lobe level for array antennas with such randomly distributed errors as well as randomly distributed element failures [2]. Adapting his expressions yields the rms side-lobe level magnitude in units of dBi, SLL, as

$$
\text{SLL} \cong -10 \cdot \log \left\{ 0.5 + \frac{P_{go}}{\pi \left[(1 + a^2) e^{\delta^2 + \left(\frac{2\pi\sigma_d}{\lambda} \right)^2} - P_{go} \right]} \right\} \tag{7.6}
$$

where:

a = element-level amplitude error;
δ = element-level phase error;
σ_d = three-dimensional rms displacement error of element;
P_{go} = probability of given element being functional.

In general, transmit and receive side-lobe levels should be computed separately due to associated differences in amplitude/phase tracking error and failure distributions. This expression only applies to the side-lobe error component of the antenna pattern. Close to the main lobe, the Fourier transform of the illumination factor taper will determine side-lobe magnitude, since this component will be significantly larger than the side-lobe error component. Pulse-doppler clutter rejection capability may be limited by these near-in side lobes that are largely determined by the antenna taper. The diffraction side-lobe component imposed by the taper decreases with increasing angle off the beam pointing direction in contrast to the side-lobe error component that is distributed throughout the antenna field of view (FOV).

The phase quantization error defined in (7.5) forms a lower bound on element-level error. For example, consider a phased array employing five-bit phase shifters that has been precisely calibrated so that the random phase shifter quantization in phase is the dominant error in the illumination function. Using (7.5), five bits would impose an rms quantization error of about 28 mrads. With all modules in operation, this would provide an rms error side-lobe level of about −26 dBi. However, the side-lobe floor would increase to about −15 dBi with 1% element failures and to −8 dBi with 5% failure.

The peak spurious side-lobe level of the antenna may be determined by the radome and interaction between the radar antenna and local scattering centers. Radome mismatch problems that develop over time due to moisture penetration and material degradation can impose major degradation effects such as the back-lobe reflection noted in Chapter 15. Similarly, objects embedded in the radome or mounted in front of the antenna also degrade the antenna pattern, producing spurious side-lobe components.

In contrast to a continually scanning pulse-doppler radar, phased-array pulse-doppler radars transmit and receive discrete coherent bursts at the antenna pointing angle of a specified CPI in·duration. A given transmission or reception to or from a given beam position is termed a *dwell*. A pulse-doppler dwell consists of a coherent burst of duration equal to the radar CPI. High-PRF and medium-PRF coherent bursts are typically transmitted and received within a single dwell. Alternating transmission and reception is performed over the dwell, as is done for conventional pulse-doppler radars. In low-PRF operation, a dwell generally consists of a single pulse with separate transmit and receive actions potentially required because the radar may interleave other T/R actions between low-PRF pulses.

7.3 SOFTWARE FUNCTIONAL IMPLEMENTATION

7.3.1 Radar Resource Management

In order to fully exploit the capabilities afforded by beam agility, the radar control process must be highly automated. Given a typical operational scenario in which a phased-array AI radar is interleaving tens or hundreds of separate target tracks while supporting multiple search and acquisition operations, even a dedicated operator could not maintain pace with the required decision rate to choose beam pointing direction and waveform parameters.

Resource management varies markedly depending on the individual radar system designer as well as the digital software/hardware architecture of the radar system. The underlying control loop is depicted in Figure 7.3. This radar control software is typically implemented on a general-purpose processor with a real-time operating system. The operator typically designates targets of interest or establishes the overall functional operating mode of the radar which may include emission control parameters as well as specific mission requirements. The resource management software schedules specific dwells, including beam pointing direction and

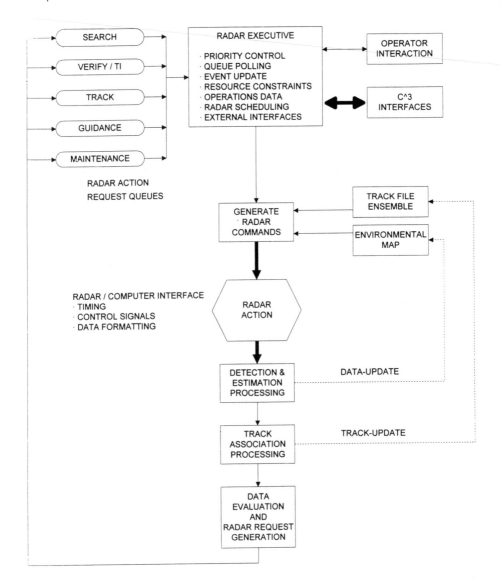

Figure 7.3 Phased-array radar resource management.

waveform selection so as to satisfy these user requirements. The data processor supporting the resource management software generally runs asynchronously from the radar. Hence, the subsystem timing and control signals necessary to implement T/R radar actions are generally generated by some manner of dedicated interface that also buffers received radar data prior to inputting it to the data processor.

The central feature of phased-array radar resource management is the existence of an ensemble of queues of competing requests for radar actions. Each commanded search process is continuously requesting illumination of the associated search raster. Search dwells are generally lower in priority than verify and track dwells. Each target under track is assigned a track file incorporating some measure of track update priority as well as the estimated target position and velocity estimated by the track filtering algorithms. Compared to surface-based pulse radars, pulse-doppler phased-array radars possess a relatively low beam rate due to the necessity of supporting a CPI of adequate duration in each beam position.

Search or track performance can be increased at the expense of degrading the other function's performance by shifting time and average power resources between the two. The capability of a phased-array radar to interleave search, track, and other functions is limited by average transmit power, antenna aperture, and other hardware parameters. A given search volume Ω_s is defined in steradians as

$$\Omega_s = \Delta_{az}(\sin(\theta_{el\ max}) - \sin(\theta_{el\ min})) \tag{7.7}$$

where:

Δ_{az} = azimuth extent of search in radians;
$\theta_{el\ max}$ = maximum elevation scan angle of search extent;
$\theta_{el\ min}$ = minimum elevation scan angle of search extent.

Given a solid-angle beamwidth of Ω_b and a search dwell duration of τ_s, it follows that the time required to accomplish a given search operation γ_s is given approximately by

$$\gamma_s = \frac{\Omega_s \tau_s}{\Omega_b O_s} \tag{7.8}$$

where O_s is the portion of occupancy devoted to search. Occupancy is the portion of radar timeline devoted to transmit or receive actions, and approaches unity for high-PRF and medium-PRF operations. T/R switching time and beam-steering time detract from the achievable occupancy, since these operations effectively impose "dead time," in which neither the transmit nor receive functions are active but are typically of negligible value. The search occupancy is that portion of the radar timeline devoted to search T/R operations. For example, if three out of four CPIs were devoted to search dwells, then O_s would be equal to 0.75, presupposing negligible dead time. The solid beamwidth of the antenna can be approximated as the product of the azimuth and elevation beamwidths in radians. The search dwell duration is determined by detection sensitivity requirements, including clutter rejection. For fixed operating parameters, a given search volume can be probed at intervals of γ_s so that this period defines the search frame time, which is the revisit rate for probing a given volume of space-seeking targets.

A corresponding estimation of the time required to support track operations can be computed as follows. Given a track dwell duration of τ_t and N_t targets under independent track, the corresponding track frame time is given by

$$\gamma_t = \frac{N_t \tau_t}{O_t} \qquad (7.9)$$

where O_t is the portion of occupancy devoted to track. The track dwell duration may be different than that used by search because of variations in the CPI imposed by waveform tailoring or the degree of noncoherent integration. The measurement update rate on a given target under track will be the inverse of γ_t, assuming uniform track rates.

In addition to search and track, a multifunction AI radar will likely support other functions such as navigation, air-to-surface weapon delivery, and target identification. It is unlikely that the resource management process can support unity occupancy if it must coordinate interleaving of high-PRF, medium-PRF, and low-PRF dwells. It follows that

$$O_s + O_t = \frac{\Omega_s \tau_s}{\Omega_b \gamma_s} + \frac{N_t \tau_t}{\gamma_t} \le 1 \qquad (7.10)$$

where the unity bound may not be achieved in practice. The occupancy allocated to search and track must also support verification and track initiation dwells. As the number of target tracks is increased, the search frame time must be increased or the search volume must be decreased.

In the usual case that search and track are the dominant radar functions, the limits of resource management are imposed by the associated timeline demands. The upper bound on search frame time is imposed by the need to maintain situational awareness such that a given search volume must be swept in less time than a threatening target can transit it or penetrate within some critical distance of the radar platform. Fire control requirements for track retention or target illumination constrain the maximum track frame time between target measurement updates.

7.3.2 Search

Phased-array beam agility allows the radar designer to devise optimized search scans. This capability is in contrast to conventional AI radar operation, where mechanical inertia constrains the shape, density, and extent of search and acquisition scanning. In contrast, a phased array can support multiple interleaved search operations throughout the antenna FOV. Search scans can be optimized for target acquisition cued from other sensors so as to minimize radar emissions as well as maximize the radar timeline and energy available to perform other functions.

The antenna can be commanded so as to compensate for rotational motion imposed by aircraft maneuvers. With an inertial measurement unit to monitor aircraft attitude, the beam can be automatically adjusted to compensate for aircraft banking. In AI applications, it may be desirable to designate search sectors in terms of some theater coordinate system based on a priori data detailing threat ingress/egress corridors or airborne early warning data from broad-area surveillance sensors. Ideally, this type of threat sector data is transmitted directly to the radar controller so that there is minimum demand for pilot intervention to establish radar parameters.

The detection efficiency of a phased-array radar is typically enhanced by the use of a two- or three-stage detection process. Two-stage detection/verification or alert-confirm processing is commonly employed. Essentially, the radar detection threshold is lowered relative to that necessary to maintain a specified false alarm rate so as to improve search sensitivity. Other than this threshold adjustment, search is carried out nominally with the antenna illuminating a programmed raster beam by beam. False detections are rejected with a second, dedicated dwell transmitted into raster positions where an initial detection report originated. Typically, the verification beam rate relative to the search beam rate runs between 0.1% and 5.0%, depending on search detection threshold tuning and local interference and clutter conditions.

In general, the verification beam should be transmitted close in time and frequency to the original dwell providing the search detection. This strategy is motivated by the desire to observe the target within the correlation interval of the RCS fluctuation [3]. PRF parameters may be varied in order to resolve range-doppler ambiguities so as to support track initiation. For example, medium-PRF verification dwells could be used in conjunction with high-PRF search dwells. Two-stage detection processing can increase search sensitivity by some 1 to 2 dB under realistic conditions.

Mechanically scanned AI radars can also employ search/verify processing. The fundamental role of the verify dwell is to reject false alarms. The phased-array radar simply injects an extra dwell into an ongoing search raster to accommodate verification. However, a mechanically scanned antenna must terminate search or TWS scanning in order to verify the search detection, which may lead to unacceptable threat coverage outages. The capability of phased-array radars to interleave multiple search/verification operations is of major significance in attempting to acquire small-RCS, low-flying aircraft under heavy clutter conditions.

In general, a phased-array radar only achieves marginal improvements in detection range against a target over that afforded by a mechanically scanned radar of equivalent power-aperture area. The combination of search-verify processing and scan tailoring can produce significant improvement under some circumstances. However, the major advantage for a phased-array radar in search is that it can attain track initiation range on par with its initial detection range [4]. A mechanically scanned radar may require three to five scans of 2- to 10-sec duration each to establish track on an approaching target. In contrast, a phased-array radar is typically configured to implement verification and track initiation dwells within tens of milliseconds of the initial detection. Hence, a phased-array radar can initiate firm track on targets at a greater range than an equivalent power-aperture mechanically scanned radars.

7.3.3 Track

Most modern AI radars support TWS operation, where a repetitive sequence of PRFs is transmitted as the antenna is scanned over one of several search patterns. The resulting detections are correlated into tracks over multiple scans. A phased-array radar can support track-during-scan (TDS) operation, where dedicated track beams are interjected into the normal TWS raster. The phased-array radar sequentially measures the position of the targets under track using dedicated track beams with pulse-doppler waveform parameters optimized using the estimated target RCS and range as well as target/clutter geometry. TDS can support much higher update rates than TWS (say, 5.0 Hz versus 0.5 Hz), as well as enable waveform optimization to ensure adequate $S/(N + C)$. TDS affords obvious advantages for supporting multiple-target fire control and tracking small, maneuvering targets.

Determining the traffic capacity of a phased-array radar can be problematical. Obviously, the data processor must be sized to possess adequate throughput and memory to support the number of track files required. Given that traffic capacity is increased by minimizing update rates, it is frequently desirable to implement relatively sophisticated Kalman track filters that incorporate aircraft flight path characteristics. However, the track capacity is fundamentally limited by the beam rate that can be supported by the radar, as indicated in the preceding subsection. As the traffic loading increases, the radar should automatically delay search operations in order to support multiple-target engagement. In the limit, the maximum number of independent tracks is given by the inverse of the product of the required single-target track rate and the average CPI. For example, given a 10-ms CPI and 5-Hz track rate, a maximum of 20 independent tracks could be supported with radar occupancy totally dedicated to track. In practice, track capacity is increased by only tracking targets under engagement at high update rates and illuminating multiple targets with a single dwell when they are spaced sufficiently close to permit it. It is imperative in AI applications that such formation tracking techniques be sufficiently robust to support immediate track initiation on targets that emerge from the parent cluster.

These track capacity augmentation techniques require adaptive resource management implementation, which also generally mandates some manner of estimating real-time track quality such as the inverse of the covariance magnitude produced by a Kalman filter. The track update interval can be estimated from the real-time estimated covariance so as to provide a specified tracking accuracy or probability of track retention. In addition, Kalman filtering can readily incorporate known target flight path characteristics, which can be exploited to reduce the required track update rate.

In general, tracking accuracy requirements for AI fire control are readily supported by phased-array antennas. Tracking accuracy must be sufficient to maintain track and support periodic target illumination if required for semiactive homing by air-to-air missiles. The accuracy of modern fighter cannons tends to be limited by factors other than radar tracking performance such as projectile velocity dispersion and recoil impact on aircraft orientation control. Typically, antenna side-lobe suppression requirements drive element-level amplitude/phase control accuracy and the associated alignment and calibration requirements.

Emissions control (EMCON) requirements to minimize radar transmissions may place a premium on maximizing the interval between track updates by the radar. In a sparse target environment, the driving requirement in determining the maximum radar update interval is to ensure that the target is within the beamwidth of a given track dwell. Under this scenario, the maximum track update interval will be jointly determined by radar measurement accuracy and the acceleration characteristics of the target, which may evolve over an engagement. In a dense target environment, the correct association of measurements with track files can become a significant challenge. Similarly, rejection of false measurements produced by ECM can be important in some scenarios. Blackman provides an excellent summary of these issues in the context of multiple-target tracking [5].

The capability of a phased-array AI radar to simultaneously track a target and interceptor can potentially be exploited in similar fashion with surface-based air defense systems. Electronic counter-countermeasures (ECCM) can potentially be enhanced by using precision track data to combat track deception ECM via a midcourse data link to an air-to-air missile. More generally, joint target-missile tracking can be used to support trajectory shaping to extend fly-out range and ensure an advantageous end-game intercept geometry.

Operational requirements dictate a large FOV for a fixed-boresight AI phased array. The radar must be capable of maintaining target illumination during own-ship maneuvers. Maximum electronic scan angles exceeding 60 degrees are desirable. Sensitivity and metric accuracy are generally allowed to degrade significantly toward the edge of the FOV, since operations in this regime are generally associated with near-in targets.

7.3.4 Function Interleaving and Sensor Fusion

The previous subsections have stated that dwell-to-dwell interleaving of different functions is a principal benefit of phased-array radar operation. In particular, the benefits of search/verify processing and TDS operation have been delineated. In addition, phased-array AI operation permits operations that require relatively low measurement rates but extended coherent processing duration to be multiplexed with high-rate, short-CPI functions. This capability is essential to maintaining situational awareness.

For example, spotlight synthetic-aperture radar (SAR) imaging and high-resolution noncooperative target recognition (NCTR) operations mandate extended observation of specific targets. In a dense threat environment, dedicating a conventional mechanically scanned radar to such functions can impose potentially hazardous interruptions to search and track operations. By operating the radar in a TDS fashion, these low-rate functions can be interleaved with target track and search operations to ensure that the aircrew is kept apprised of local threat positions.

Similar considerations apply to sensor fusion. EMCON to mitigate detection by enemy electronic support measures (ESM) and multispectral applications such as NCTR benefit significantly from sensor fusion. An advanced aircraft could be designed to merge raw measurements or processed track data files from a combination of AI radar, ESM sensors, identification-friend-or-foe (IFF) receiver, and infrared search and track (IRST) sensor data. The advantage of phased-array operation in a sensor fusion operation is that beam pointing and dwell duration can be jointly tailored to minimize radiation while supporting target tracking throughout the radar FOV. Radar updates on a given target can be minimized to the extent necessary to provide a required track accuracy or maintain track retention. Tracks can be silently (if perhaps coarsely) updated with measurements from the other onboard sensors. In general, the other platform sensors are relatively dependent on target characteristics compared to radar, and they measure range grossly at best. Phased-array radar operation enables sensor fusion even in spatially diverse threat environments, which would preclude multiple target tracking support by mechanically scanned radars.

7.4 PHASED-ARRAY HARDWARE IMPLEMENTATION

7.4.1 Airborne Phased-Array Implementation

Despite the elimination of mechanical scanning, AI phased-array radars present significant mechanical engineering challenges. As in any airborne pulse-doppler implementation, a phased-array radar must operate in a relatively hostile environment characterized by extreme vibration. Mechanical and acoustic isolation from the airframe in combination with robust component design is imperative for the suppression of spurious sidebands so as to permit detection of small, low-flying targets, as addressed in Chapter 15.

The need for compactness dictates corporate-feed beamforming for AI applications. The weighted outputs of each element are summed in a constrained feed typically constructed from waveguide or coaxial cable. Constrained feed implementation of the beamformer also facilitates conformal array implementation as required for some applications.

Angle measurement accuracy may be enhanced by implementing monopulse so that difference-azimuth and difference-elevation channels have to be implemented in parallel with the sum channel [6]. The difference-pattern channels are used to estimate the position of a target return to a precision typically on the order of 1/10 to 1/20 of a beamwidth. Additional receive channels may also be implemented to facilitate search and to support ECCM.

7.4.2 Examples of Airborne Phased-Array Implementation

The E-3 Sentry radar supports high-PRF and low-PRF operation using a one-dimensionally scanning phased array. AI operation mandates two-dimensional electronic scanning nominally in the azimuth and elevation planes. In contrast, electronic elevation scanning in conjunction with mechanical azimuth rotation is attractive for airborne early warning radars such as that of the E-3. Aircraft rotational motion can be compensated by appropriate adjustment in the beam-steering computer using onboard inertial navigation data. This implementation also lends itself to low-azimuth-side-lobe antenna design.

Westinghouse Electric Corporation developed the first deployed U.S. airborne two-dimensional phased-array radar, the APQ-164 used in the B-1B bomber. This radar employs a passive array to perform terrain following/avoidance, navigation, and weapon delivery functions. The APQ-164 achieved initial operational capability in 1986.

The Flashdance radar used on board the MiG-31 holds the distinction of being the first operational phased-array AI radar. The Arrowhead array, as it is known in Russia, is a passive array that has reportedly been operational since 1979 and was developed by the Research Institute of Equipment Design. This system possesses a relatively high effective radiated power for an AI system. It is reportedly capable of tracking ten targets simultaneously and guiding missiles to four targets simultaneously. The array supports integrated L-band IFF operation as well as X-band AI functions.

In contrast, U.S. AI phased-array efforts emphasize development of solid-state active arrays such as the APG-77, developed jointly for the F-22 aircraft by Westinghouse Electric Corporation and Texas Instruments, Inc. The F-22 radar system implementation strongly emphasizes automated resource management with highly integrated sensor fusion operation. The emphasis on sensor fusion is driven by EMCON restrictions to minimize detection range against the host aircraft by potentially hostile sensors and support of noncooperative target recognition. Support of beyond-visual-range engagements demands that firm track is maintained once an aircraft is identified via ESM or other means. Hence, the integrated resource management process must minimize radar emissions while ensuring high probability of track retention from the multiple sensor measurements. These requirements dictate large traffic capacity in the track filtering process.

Prior to the development of the APG-77 radar, three generations of developmental solid-state arrays were developed by the U.S. Air Force between 1964 and 1988. The Solid State Phased Array (SSPA) developed by TI was the final of these prototype arrays and the first to employ gallium arsenide devices as required for efficient operation at X-band. The array was designed so as to provide sensitivity comparable to that provided by modern conventionally scanned AI radars. The SSPA possesses 1,980 X-band modules distributed in an equilateral triangular grid with 1.71-cm element spacing. The array is circular with a diameter of 0.81m. The corporate feed beamformers are constructed from stripline.

7.4.3 Motivations for Solid-State Active Array

Phased arrays afford significant advantages for AI applications. As noted previously, phased-array radar supports simultaneous engagement of multiple targets. Rapid interleaved operation of air-to-air and air-to-ground radar modes over spatially diverse regions helps maintain the situational awareness of the aircrew while simultaneously supporting navigation and weapon delivery functions. In addition, a phased-array antenna can be implemented to provide a low-RCS and thus supports low-RCS aircraft design.

Analyses based on current technology by Texas Instruments personnel indicate that a passive phased array employing a traveling wave tube (TWT) transmitter uses about twice as much prime power and imposes a weight penalty some three times greater for the antenna-transmitter-power supply ensemble than a solid-state active array offering equivalent performance [7]. A major advantage of a solid-state active array over a passive array is the reduction of beamforming losses between the transmitter/receiver and free space. Improvement of power conversion efficiency of solid-state transmit components continues to be emphasized in technology development. In addition to improving system-level efficiency, improving power conversion efficiency decreases array cooling requirements. Reduction in waste heat handling requirements also permits increasing module transmit power.

Solid-state active arrays lend themselves to high-reliability designs. Historical experience with AI radars suggests that most major system failures are associated with either microwave power generation (the TWT or associated high-voltage power supply) or the antenna gimbaling servomechanisms. Solid-state active arrays eliminate both mechanisms. In addition, active arrays possess inherent redundancy, since a substantial number of modules must fail before substantially degrading performance. Tube-based microwave power modules have also been developed that are expected to afford significantly greater reliability than historically associated with conventional airborne transmitter designs.

In practice, depot work may be paced by the necessity of maintaining side-lobe suppression below a given level, which in turn limits the number of modules that can be allowed to fail. Hence, it is important for active arrays to employ some manner of bit in test (BIT) capability to support element-level failure detection. BIT is also desired to support element-level alignment and calibration of amplitude/phase response.

Active array development has been driven by a desire to achieve over an order of magnitude increase in mean time between critical failures (MTBCF). In addition to the inherent redundancy of an active array, it has been necessary to develop high-reliability subsystems, such as dc power supplies, and to implement some degree of redundancy in the signal/data-processing chain. The SSPA has reportedly demonstrated an MTBCF on the order of 2,500 hours.

Active arrays can potentially be designed to support operation over very wide operational bandwidths. Wideband operation enhances radar resilience to ECM and potentially offers improvement of target detection and clutter rejection performance. In addition, this capacity potentially permits frequency diverse functions such as IFF and ESM to be integrated into radar system. However, occupancy generally constrains implementation of other functions in an AI radar.

The perennial concern over solid-state array technology is cost. Solid-state active arrays are inherently complex compared to conventional AI radar designs. Major module builds with strict cost control constraints have led to estimated acquisition costs on the order of a few hundred dollars per module. Proponents of solid-state arrays prefer to focus on life cycle cost, since this is generally more in line with conventional AI radar estimates than fly-away cost would be.

7.4.4 Solid-State Array Technology

The SSPA active array is illustrated in Figure 7.4. As previously described, this design represents the first modern X-band active array. Note the impressive packaging density employed in the array.

A typical solid-state T/R module is depicted in Figure 7.5. The phase shifter and attenuator are depicted as shared, though separate transmit and receive chains have also been used. The number of bits in the phase shifter is typically between five and seven. The attenuator is primarily used on receive to impose a taper for side-lobe suppression, since the transmit amplifiers are typically operated in saturation for maximum power conversion efficiency. Depending on the sophistication of the design, the timing and control logic may employ a look-up table compiled during the array alignment process to correct fixed amplitude and phase errors as a function of frequency and other parameters. Compensation data are generally stored and implemented at the element level or in intermediate processors, each supporting multiple phased-array elements.

Each module contains an independent transmitter and receiver front-end. Current X-band high-power amplifiers (HPA) generally possess transmit power levels between 1W and 10W. Current low-noise amplifiers (LNA) generally possess noise figures on the order of 2 to 2.5 dB. Beamforming is conducted at the carrier frequency so that frequency conversion and associated filtering operations are performed in a central receiver as in a conventional AI radar design.

Module vulnerability to overload or physical damage from excessive RF input is often of concern in the design process. As evident in Figure 7.5, the designer must depend on the combined performance of the circulator and T/R protector to shield the LNA from harm. It is imperative that neither main-lobe clutter nor ECM signals are of sufficient magnitude to drive the LNA into saturation, which

Figure 7.4 The SSPA active array antenna. (Photo courtesy of Texas Instruments.)

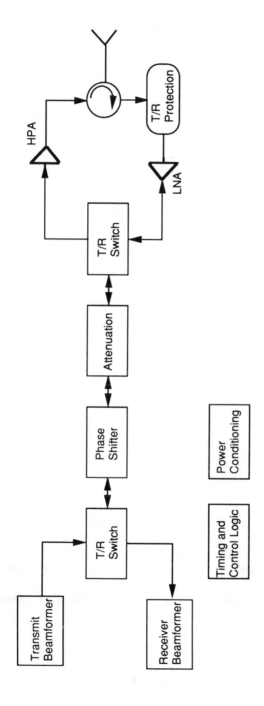

Figure 7.5 Typical solid-state T/R module.

would significantly degrade clutter rejection capability. On-module power conditioning is used to suppress power supply noise that would otherwise induce spurious sidebands, as addressed in Chapter 15.

7.5 EFFECTS ON PULSE-DOPPLER OPERATION

7.5.1 Electronic Beam-Scanning Effects

7.5.1.1 Clutter Spreading

As noted previously, high-PRF and medium-PRF pulse-doppler dwells are generally transmitted and received with the beam pointing direction held constant. Hence, any residual phase error in the beam-steering process is suppressed, since it will be fixed over the CPI. Moreover, the only clutter spreading will be induced by platform motion, since the antenna is stationary over the CPI as opposed to the beamshape modulation imposed by a continually scanning antenna over a CPI.

In contrast, the beam will likely be resteered between low-PRF pulses. The beam-steering system can be designed to compensate for transitional and rotational platform motions over the PRI in order to minimize clutter spreading. Array-based motion compensation is implemented via displaced phase center antenna (DPCA) techniques, where the phase center of the antenna is electronically shifted between pulses so as to effectively suppress clutter spreading induced by platform motion.

7.5.1.2 Clutter Fill

A negative consequence of electronic beam scanning is that clutter fill time is required for each new beam position. Clutter fill is the portion of the dwell necessary to allow maximum range clutter returns to arrive back at the radar and is excluded from the CPI. In order to ensure adequate main-lobe clutter rejection, no data are input to the doppler filter until this period has expired. Otherwise, the main-lobe clutter will receive only a partial time domain weighting so that the doppler sidebands will be significantly higher than the nominal value and potentially obscure target returns. A similar transient period must also be allowed for a main-lobe clutter rejection filter processor if implemented to reduce dynamic range requirements on the doppler filter processor.

For example, an aircraft at an altitude of 20,000 ft can potentially receive significant clutter returns from out to a range of 322 km, even in the absence of major mountain ranges on the horizon. This range would correspond to a clutter fill time of some 2.15 ms. The requisite clutter fill time is generally determined by the radar designer on the basis of radar sensitivity and anticipated clutter characteristics.

The truncation effect is illustrated in Figure 7.6, which denotes the Fourier transform of a 40-dB Taylor weighting for several data aperture fills. An aperture

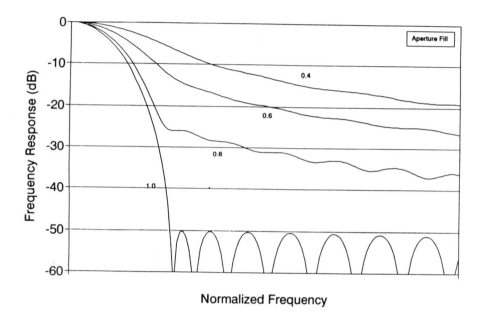

Figure 7.6 Truncation effects on –50-dB Taylor weighting performance.

fill of unity corresponds to each pulse in a dwell being weighted within the applied Taylor taper. The corresponding peak side-lobe magnitude for several Taylor weights as a function of aperture fill is depicted in Figure 7.7.

This composite clutter fill time must be treated as a SNR loss, since this energy does not contribute to increasing SNR. The resulting clutter fill loss in SNR can be represented as

$$L_{cf} = -10 \cdot \log\left(\frac{\text{CPI}}{\text{CPI} + T_{cf}}\right) \tag{7.11}$$

where T_{cf} is the clutter fill time. Clutter fill time is sometimes referred to as *burn time* to denote that the radar is expending power that does not contribute to SNR. For example, given a 5-ms dwell duration with the 2.15-ms fill time estimated above, this would result in a clutter fill loss of about 2.4 dB in SNR.

Any pulse-doppler radar must incorporate allowance for clutter fill time when switching PRFs. However, the relatively wide spacing of phased-array beam positions mandates that each beam position include clutter fill if significant long-range clutter returns are anticipated. In contrast, a mechanically scanning radar operating at a constant PRF, such that a given point in space will be illuminated by several CPIs as the beam sweeps, effectively maintains clutter fill.

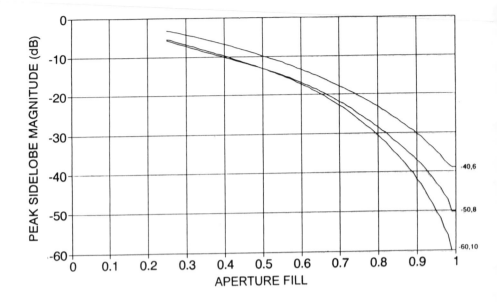

Figure 7.7 Peak side-lobe suppression degradation due to weighting truncation.

7.5.2 Spatial/Temporal Coherence

In Section 7.2.3, the effects of element-level amplitude/phase error on antenna side lobes were described. In a passive array, there is generally no impact on the temporal coherence of the signal from the effects of individual element amplitude/phase errors. These errors are fixed over a high-PRF/medium-PRF CPI. Hence, there is no undesirable amplitude or phase modulation that induces spurious sidebands, as described in Chapter 15. Dwell-to-dwell phase center stability is required to prevent degradation of low-PRF operations such as MTI or spotlight SAR, where the beam is resteered between pulses.

In general, spatial side-lobe suppression and spectral sideband suppression are decoupled for the case of passive arrays. Active arrays are more complex, since each element contains an independent transmitter (HPA) and receiver front end (LNA). Element-to-element variation in amplitude/phase tracking response induced by these active components produces spurious spatial side lobes as well as introducing spurious spectral sidebands. The amplitude/phase error of the active array induces spurious spectral sidebands that are spatially coherent so as to achieve the full spatial directivity of the antenna when the modulation error is highly correlated from module to module.

Power supply ripple and switching noise can induce spurious sidebands in solid-state modules and are common to all the modules, assuming a single power supply bus. If the T/R amplifiers all possess the same sensitivity to dc power fluctuations, then the antenna pattern will not be affected, even if significant spurious sidebands are induced. In contrast, if there is significant variation among amplifiers in their sensitivity to supply voltage variation, spurious antenna side lobes will be generated, but the spurious spectral sidebands measured within the antenna main beam will decrease in relative magnitude, since the uncorrected spurious response components will not receive the full spatial gain of the array.

Iglehart devised a series of useful spurious signal models representing active-array hardware effects on phase stability [8]. Gostyukhin has devised an expression characterizing the joint spatial-spectral response of an active array given that element-to-element amplitude/phase tracking can be parameterized in terms of mean and standard deviation of the element-level active elements in response to spurious driving modulation such as power supply ripple [9]. Adapting his notation leads to an expression of characterizing the joint spectral-spatial antenna pattern $E(f, \theta)$ as

$$|E(f, \theta)|^2 = |E_a(\theta)|^2 \left| \sum_{n=1}^{N} A_n e^{j\psi_n} \right|^2 |\delta(f)|$$

$$+ \left| \sum_{n=1}^{N} A_n e^{j\psi_n} \right|^2 (\alpha^2 + \beta^2) S(f)$$

$$+ \sum_{n=1}^{N} |A_n|^2 (\alpha^2 \sigma_a^2 + \beta^2 \sigma_\beta^2) S(f) \qquad (7.12)$$

where:

ψ_n = $\dfrac{2\pi}{\lambda} nd(\sin \theta - \sin \theta_0)$;

α, β = complex sensitivity coefficient of active component;

σ_a, σ_β = standard deviation of sensitivity coefficient;

$S(f)$ = power spectrum of driving modulation error (such as power supply ripple).

The first two terms of this equation are common to passive and active phased-array radars. The third term is unique to active arrays and can be interpreted as contributing to the antenna error side lobes as well as the spurious sidebands. Note that in the absence of driving modulation error, the first term corresponds to (17.1), the desired antenna pattern, multiplied by an impulse function in the frequency domain. As developed in Chapter 15, the ideal composite frequency reference of a coherent radar system is an impulse function in the frequency domain. Note that the first two terms receive the full coherent gain of the antenna as indicated by the summation of the current elements. In contrast, the third term is multiplied by a summation of the current element power implying noncoherent spatial integration.

An interesting consequence of this interaction between spatial and spectral signal components is that the spurious sideband magnitude relative to the desired signal can be greater in the antenna side lobes than in the main beam. The spurious signal component is distributed throughout the antenna FOV, in contrast to the desired signal response component which is focused into the main lobe. Given random element-to-element amplitude/phase error, the power of the spurious sideband component in the main lobe is suppressed by a factor of N, where N is the number of elements. Hence, a 2,000-element active array with uniform module modulation error statistics would possess composite spurious sideband magnitude suppression in the main beam of some 33 dB relative to the mean spurious sideband magnitude associated with a single module.

In practice, active component tolerances are specified so as to support low phase-noise active-array designs. In addition to the factor-of-N suppression factor described above, the noise figure of a solid-state HPA is generally lower than that of a conventional radar traveling-wave tube, which results in lower spurious sidebands being generated. Active arrays can support extremely demanding clutter rejection requirements.

References

[1] Mailloux, R. J., *Phased Array Antenna Handbook*, Norwood, MA: Artech House, 1994.

[2] Wang, H. S. C., "Performance of Phased-Array Antennas With Mechanical Errors," *IEEE Trans.*, Vol. AES-28, April 1992.

[3] Dana, R. A., and D. Moraitis, "Probability of Detecting a Swerling 1 Target on Two Correlated Observations," *IEEE Trans.*, Vol. AES-17, No. 5, September 1981.

[4] Billam, E. R., "Phased Array Radar and the Detection of 'Low Observables,'" IEEE International Radar Conference, 1990.

[5] Blackman, S. S., "*Multitarget Tracking With an Agile Beam Radar,*" in Y. Bar-Shalom, ed., *Multitarget-Multisensor Tracking: Applications and Advances*, Vol. 2, Norwood, MA: Artech, 1992.

[6] Nessmith, J. T., and W. T. Patton, "Tracking Antennas," in *Antenna Engineering Handbook*, R. C. Johnson and H. Jasik, eds., New York: McGraw-Hill, 2nd edition, 1984.

[7] McQuiddy, D. N., et al., "Transmit/Receive Module Technology for *X*-Band Active Array Radar," *Proc. IEEE*, Vol. 79, March 1991.

[8] Iglehart, S. C., "Noise and Spectral Properties of Active Phased Arrays," *IEEE Trans.*, Vol. AES-11, No. 6, November 1975.

[9] Gostyukhin, V. L., "Spectral Characteristics of Active Phased Antenna Arrays," originally published in *Izvestiya VUZ. Radioelektronika*, Vol. 30, No. 2, 1987 (English translation: 1987, New York: Allerton Press).

CHAPTER 8

Doppler Processing

Mark A. Richards

Doppler processing is the term applied to filtering or spectral analysis of the signal received from a fixed range over a period of time corresponding to several pulses. In general, the spectrum of the signal from a single range bin consists of noise, clutter, and one or more target signals. Figure 8.1 shows a notional generic doppler spectrum. As discussed in Chapter 2, this spectrum is periodic with a period equal to the pulse repetition frequency (PRF), so only the *principal period* from $-PRF/2$ to $+PRF/2$ is shown. Receiver noise is spread uniformly throughout the spectrum. Main-lobe and side-lobe clutter occupy a portion of the spectrum. Stationary and moving targets can occur anywhere in the spectrum, as appropriate to their radial velocity relative to the radar.

In many situations, the relative amplitudes of the clutter, target, and noise signals are as shown: the target returns are above the noise floor ($S/N \gg 1$), but

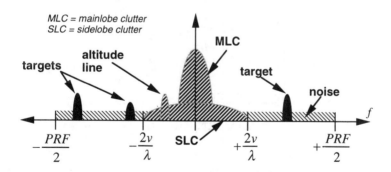

Figure 8.1 The principal period of a notional generic doppler spectrum containing noise, clutter, and target components. Preprocessing has been used to center the main-lobe clutter at zero doppler.

147

below the clutter ($S/C \ll 1$). In this case, targets cannot be detected reliably based on amplitude alone. Doppler processing is then used to separate the target and clutter signals in the frequency domain. The clutter can be explicitly filtered out, leaving the target return(s) as the strongest signal present; or the spectrum can be computed explicitly so that targets outside of the clutter region can be located by finding frequency components that significantly exceed the noise floor.

In this chapter we describe the two major classes of doppler processing, MTI and pulse-doppler processing. We consider only coherent doppler processing using digital implementations, since this is the approach taken in most modern radars. Alternative systems using noncoherent doppler processing and implementations based on analog technologies are described in [1–5].

8.1 MOVING-TARGET INDICATION

Figure 8.2 illustrates the two-dimensional data matrix that can be formed from the coherently demodulated baseband returns from a series of N pulses. The samples in each column are successive samples of the returns from a single pulse (i.e., successive range bins). Each element of a column is one complex number, representing the real and imaginary (I and Q) components of one range bin. Consequently,

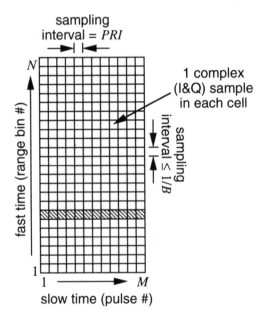

Figure 8.2 Notional two-dimensional data matrix. Each cell is one complex number. Doppler processing is applied to a series of samples in the slow time dimension for each fixed range, as indicated by the shaded row.

each row represents a series of measurements from the same range bin over successive pulses. The range dimension (vertical in Figure 8.2) is often called *fast time* because the sampling rate in this dimension is at least equal to the transmitted pulse bandwidth, and therefore is on the order of hundreds of kilohertz to tens or even hundreds of megahertz. The pulse number (horizontal in Figure 8.2) dimension is referred to as *slow time* because successive samples are separated by the PRI of the radar. Thus, the sampling rate in this dimension is the PRF and is therefore on the order of ones to tens, and sometimes hundreds of kilohertz. As indicated by the shading, doppler filtering operates on rows of this matrix.[1]

Since the matrix is filled by columns, but doppler processing is performed by rows, the physical memory used to accumulate a group of pulses must accommodate both column- and row-oriented access. Memories that permit fast access to the data matrix in either dimension are called *corner turn* memories.

MTI processing applies a linear filter to the slow time data sequence in order to suppress the clutter component. Figure 8.3 illustrates the concept. In this figure, we have assumed that knowledge of the target motion and scenario geometry has been used to center the clutter spectrum at zero doppler frequency. Clearly, some form of highpass filter is needed to attenuate the clutter without filtering out moving targets in the clear portions of the doppler spectrum.

The output of the highpass MTI filter will be a new slow time signal, which is then passed to a detector. If the amplitude of the filtered signal exceeds the detector threshold, a target will be declared. Note that in MTI doppler processing, the presence of a moving target is the only information obtained. The filtering process of Figure 8.3 does not provide any estimate of the doppler frequency at which the target energy causing the detection occurred, or even of its sign; thus, it "indicates" the presence of a moving target, but does not determine whether the target is approaching or receding, or at what radial velocity. Furthermore, it provides no indication of the number of moving targets present. On the other hand, MTI processing is very simple and computationally undemanding. Despite its simplicity, a well-designed MTI can improve the S/C ratio by tens of decibels.

8.1.1 Pulse Cancelers

Suppose a fixed radar illuminates a moving target surrounded by perfectly stationary clutter. In the absence of noise or other interference, the clutter component of the echo signal from each pulse would be identical, while the phase of the moving-target component would vary due to the changing range. Subtracting the echoes from successive pairs of pulses would cancel the clutter components completely. The target signal would not cancel in general due to the phase changes.

---

[1]Not all digital processors necessarily form a data matrix similar to that in Figure 8.2 explicitly. MTI processors in particular can be implemented more simply. However, the data matrix is used explicitly in many other processors, and is useful for illustration of doppler filtering concepts.

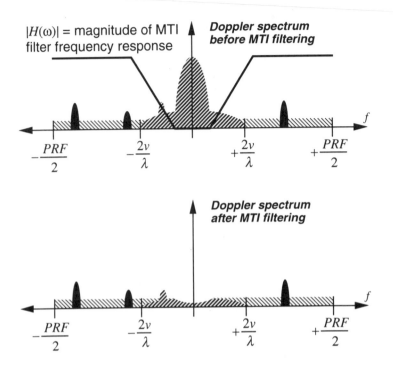

Figure 8.3 The concept of MTI filtering.

This observation motivates the two-pulse MTI canceler, also referred to as the *single* or *first-order canceler*. Figure 8.4(a) illustrates a flow graph of a two-pulse canceler. The input data are a sequence of baseband complex (I and Q) data samples from the same range bin over successive pulses, forming a discrete-time sequence $x[n]$ with an effective sampling interval T equal to the PRI. The discrete time transfer function of this linear finite impulse response (FIR, also called *tapped delay line* or *nonrecursive*) filter is simply $H(z) = 1 - z^{-1}$. The frequency response is obtained by setting $z = e^{j2\pi fT}$ [6]. Figure 8.5(a) plots the magnitude of the frequency response. Note that the filter does indeed have a null at zero frequency to suppress the clutter energy. Spectral components representing moving targets may either be partially attenuated or even amplified, depending on their precise location on the doppler frequency axis. Also note that, like all discrete-time filters, the frequency response is periodic with a period of 2π in the normalized frequency variable ω, corresponding to a period in actual frequency of PRF Hz. We will return to the implications of this periodicity in Section 8.1.3.

The two-pulse canceler is a very simple filter; its implementation requires no multiplications and only one subtraction per output sample. As Figure 8.4(a)

$$H(z) = 1 - z^{-1}$$

(a)

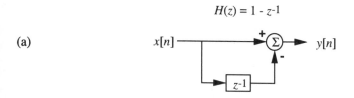

$$H(z) = 1 - 2z^{-1} + z^{-2}$$

(b)

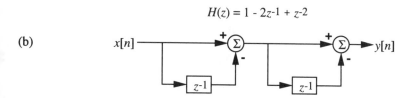

Figure 8.4 Flow graphs and transfer functions of basic MTI cancelers: (a) two-pulse canceler and (b) three-pulse canceler.

showed, however, it is a poor approximation to an ideal highpass filter for clutter suppression. The next traditional step up in MTI filtering is the three-pulse (second order or double) canceler, obtained by cascading two two-pulse cancelers. The flow graph and frequency response are shown in Figures 8.4(b) and 8.5(b). The three-pulse canceler clearly improves the null depth at zero doppler, but it does not improve the consistency of response to moving targets at various doppler shifts away from zero.

The idea of cascading two-pulse canceler sections to obtain higher order filters can be extended to the N-pulse canceler, obtained by cascading $N - 1$ two-pulse canceler sections. The filter coefficients that result are given by the binomial series:

$$h[n] = \binom{N}{n} = \frac{N!}{n!(N-n)!}, \ n = 0, \ldots, N-1 \tag{8.1}$$

Other types of digital highpass filters could also be designed for MTI filtering. For example, an FIR highpass filter could be designed using standard digital filter design techniques such as the window method or the Parks-McClellan algorithm [6]. Alternatively, infinite impulse response (IIR) highpass filters could be designed. Many operational radar systems, however, use simple two- or three-pulse cancelers for the primary MTI filtering.

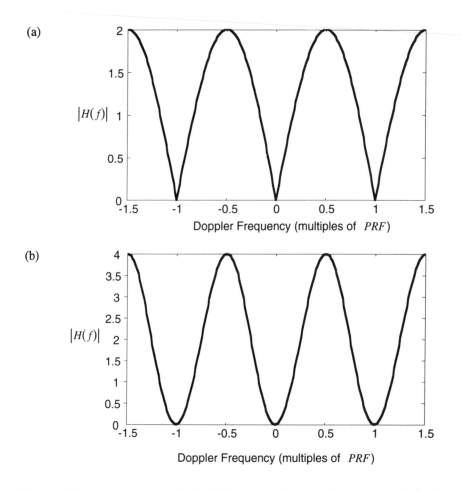

Figure 8.5 Frequency response of basic MTI cancelers: (a) two-pulse canceler and (b) three-pulse canceler.

8.1.2 Matched Filters for Clutter Suppression

It is well known that the linear filter that optimizes detection performance in the presence of white noise is the FIR matched filter, and that the coefficients of the filter are given by the matrix equation

$$\mathbf{h} \equiv \begin{bmatrix} h[1] \\ \vdots \\ h[N-1] \end{bmatrix} = \mathbf{R}_I^{-1}\mathbf{s}^* \tag{8.2}$$

where **h** is the column vector of filter coefficients, $\mathbf{R_I}$ is the covariance matrix of the interference, **s*** is the column vector representing the desired signal to which the filter is matched, and the asterisk denotes the complex conjugate [4]. Equation (8.2) can be applied to the MTI problem to design more optimal MTI filters than the N-pulse canceler. For a simple example, consider the first-order ($N = 2$) matched filter. The interference consists of noise of power σ_n^2, which is uncorrelated from one pulse to the next, and clutter of power σ_c^2, which has a correlation of ρ from one pulse to the next, where $|\rho| < 1$. For this case,

$$\mathbf{R_I} = \begin{bmatrix} \sigma_c^2 + \sigma_n^2 & \rho\sigma_c^2 \\ \rho^*\sigma_c^2 & \sigma_c^2 + \sigma_n^2 \end{bmatrix} \tag{8.3}$$

so that

$$\mathbf{R_I^{-1}} = k \begin{bmatrix} \sigma_c^2 + \sigma_n^2 & -\rho\sigma_c^2 \\ -\rho^*\sigma_c^2 & \sigma_c^2 + \sigma_n^2 \end{bmatrix} \tag{8.4}$$

where k absorbs constants resulting from the matrix inversion. To compute **h**, we now need a model for the assumed target signal **s**. For a moving target, the desired signal is just a sinusoid at the appropriate doppler frequency ω_d:

$$\mathbf{s} = A[e^{j2\pi f_d t_0} \quad e^{j2\pi f_d(t_0+T)}]' = \hat{A}[1 \quad e^{j2\pi f_d T}]' \tag{8.5}$$

where the prime symbol represents the transpose operation. In MTI, however, the target velocity and therefore doppler shift is unknown; the target could be anywhere in the doppler spectrum. We thus use the expected value of **s** over one period ($2\pi/T \,\mathrm{rad/sec} = PRF\,\mathrm{Hz}$) of the doppler spectrum. The signal model then becomes simply

$$\mathbf{s} = \hat{A}[1 \quad 0]' \tag{8.6}$$

Finally, combining (8.4) and (8.6) in (8.2) gives the coefficients of the optimum two-pulse filter:

$$\mathbf{h} = \hat{k}[\sigma_c^2 + \sigma_n^2 \quad -\rho\sigma_c^2]' \tag{8.7}$$

To interpret this result, consider the case in which the clutter is the dominant interference and is highly correlated from one pulse to the next. Then σ_n^2 is negligible compared to σ_c^2, and ρ is close to 1. The matched filter coefficients are then 1 and approximately −1, that is, nearly the same as the two-pulse canceler.

We see that despite its simplicity, the two-pulse canceler is nearly a matched filter for MTI processing when the clutter-to-noise ratio is high.

Equations (8.2) to (8.6) can be extended in a straightforward way to higher order MTI filters. As the order increases, the corresponding N-pulse canceler becomes a poorer approximation to the matched filter [4].

8.1.3 Blind Speeds and Staggered PRFs

The frequency response of all discrete-time filters is periodic, repeating with a period of 2π rad in the normalized frequency, corresponding to a period of $PRF = 1/T$ Hz of doppler shift. Figure 8.5 illustrated this for the two- and three-pulse cancelers. Since MTI filters are designed to have a null at zero frequency, they will also have nulls at doppler frequencies that are multiples of the PRF. Consequently, a target moving with a radial velocity that results in a doppler shift equal to a multiple of the PRF will be suppressed by the MTI filter. Velocities that result in these unfortunate doppler shifts are called *blind speeds* because the target return will be suppressed; the system is "blind" to such targets.

Blind speeds could be avoided by choosing the PRF high enough so that the first blind speed exceeds any actual velocity likely to be observed for targets of interest. Unfortunately, higher PRFs correspond to shorter unambiguous ranges. As discussed in Parts I and II of this book, it is frequently not feasible to operate at a PRF that allows unambiguous coverage of both the range and doppler intervals of interest. The use of *staggered PRFs* is an alternative approach that raises the first blind speed significantly without significantly affecting unambiguous range.

PRF staggering can be performed on either a pulse-to-pulse or block-to-block basis. The latter case is common in airborne pulse-doppler radars. In block-to-block PRF stagger, a dwell of N pulses is transmitted at a fixed PRF. A second dwell is then transmitted at a different fixed PRF. In some systems, as many as eight PRFs may be used. Because the blind speeds are different for each PRF used, a target that falls in a blind speed of one PRF will be visible in the others. The first velocity that is completely blind to all of the PRFs is the least common multiple (LCM) of the individual blind speeds, which will be much higher than any one of them. Target detections are accepted and passed to subsequent processing (e.g., tracking) only if they occur in some minimum fraction of the PRFs used, for example one out of two or three, or three out of eight PRFs. In medium PRF systems, particularly those using a small number of PRFs, each of the "major" PRFs used for extending the unambiguous doppler region may be accompanied by one or two additional "minor" PRFs to resolve range ambiguities as well. Selection of these PRFs is discussed in Chapter 12.

The advantages of a block stagger system are that multiple-time-around clutter can be canceled using coherent MTI and the radar system stability, particularly in the transmitter, is not as critical as with a pulse-to-pulse stagger system [4]. The disadvantage is that the overall velocity response may not be very good, and the transmission of multiple dwells consumes large amounts of the radar timeline.

Pulse-to-pulse PRF stagger varies the PRI from one pulse to the next within a single dwell. This has the advantage of achieving increased doppler coverage with a single dwell. One disadvantage is that the data are now a nonuniformly sampled sequence, making it more difficult to apply coherent doppler filtering to the data and complicating analysis. Another is that ambiguous main-lobe clutter causes large pulse-to-pulse amplitude changes as the PRF varies. Consequently, pulse-to-pulse PRF stagger is generally used only in low PRF modes.

Figure 8.6 illustrates a pulse-to-pulse stagger PRI sequence. If we choose the PRIs as multiples of a small base PRI, $T_n = k_n T_b$, then the first blind doppler frequency will occur at [4]

$$f_{b1} = \frac{1}{T_b} \tag{8.8}$$

The *stagger ratio* between any two PRIs is just the ratio $k_m : k_n$. Furthermore, if we define the average PRI as

$$T = \frac{k_1 + k_2 + \ldots + k_N}{N} T_b \tag{8.9}$$

then the first blind speed can be expressed as a multiple of the blind speed corresponding to this average PRI, which is $f_b = 1/T$:

$$f_{b1} = \frac{k_1 + k_2 + \ldots + k_N}{N} f_b \tag{8.10}$$

For example, a two-PRI system with a stagger ratio of 5:4 would have a first blind speed 4.5 times that of a system using a fixed PRI equal to the average of the two individual PRIs.

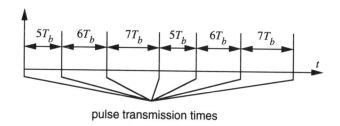

pulse transmission times

Figure 8.6 Sequence of PRIs in a pulse-to-pulse three-period stagger system with a stagger ratio of 5:6:8.

Because of the nonuniform time sampling, the frequency response of a pulse-to-pulse staggered system is typically computed by determining the response to a pure sinusoid of arbitrary initial phase for the MTI filter structure of interest. Repeating for each possible sinusoid frequency, we can determine the frequency response point by point [4,7]. For a two-pulse canceler, the result is

$$|H(f)|^2 = 1 - \frac{1}{N}\sum_{n=1}^{N} \cos(2\pi f T_n) \tag{8.11}$$

In (8.11) we have used the actual frequency in hertz rather than normalized frequency because the nonuniform sampling rate invalidates the usual definition of normalized frequency. Figure 8.7 illustrates the frequency response for a two-PRI system using a two-pulse canceler as the MTI filter and a stagger ratio of 31:33. Note that the first blind speed occurs at 32 times the blind speed that would have been obtained using a fixed average PRI.

8.1.4 MTI Figures of Merit

The goal of MTI filtering is to suppress clutter. In doing so, it also attenuates or amplifies the target return, depending on the particular target doppler shift. The change in signal and clutter power then affects the probabilities of detection and false alarm achievable in the system in a manner dependent on the particular design of the detection system.

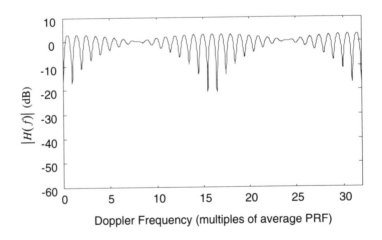

Figure 8.7 Frequency response of a pulse-to-pulse stagger system using a two-pulse canceler and a stagger ratio of 31:33.

There are three principal MTI filtering figures of merit in use. *Clutter attenuation* measures only the reduction in clutter power at the output of the MTI filter as compared to the input, but is simplest to compute.

Improvement factor quantifies the increase in signal-to-clutter ratio due to MTI filtering; as such, it accounts for the effect of the filter on the target as well as on the clutter. *Subclutter visibility* is a more complex measure which also takes into account the detection and false alarm probabilities and the detector characteristic. Because of its complexity, it is less often used. In this chapter, we will concentrate on clutter attenuation *CA* and improvement factor *I*.

Clutter attenuation is simply the ratio of the clutter power at the input of the MTI filter to the clutter power at the output:

$$CA = \frac{\sigma_{ci}^2}{\sigma_{co}^2} = \frac{\int_{-PRF/2}^{PRF/2} S_c(f)\,df}{\int_{-PRF/2}^{PRF/2} S_c(f)|H(f)|^2 df} \tag{8.12}$$

where σ_{ci}^2 and σ_{co}^2 are the clutter power at the filter input and output, respectively; $S_c(f)$ is the sampled clutter power spectrum; and $H(f)$ is the discrete-time MTI filter frequency response. Since the MTI filter presumably reduces the clutter power, the clutter attenuation will be greater than 1. In fact, clutter attenuation can be several tens of decibels. However, it also depends on the clutter itself; it is not entirely under the designer's control, and a change in clutter power spectrum due to changing terrain or weather conditions will alter the achieved clutter cancellation.

Improvement factor *I* is defined formally as the signal-to-clutter ratio at the filter input divided by the signal-to-clutter ratio at the filter output, averaged over all target radial velocities of interest [8]. Considering for the moment only a specific target doppler shift, we can write the improvement factor in the form [7]

$$I = \frac{(S/C)_{\text{out}}}{(S/C)_{\text{in}}} = \left(\frac{S_{\text{out}}}{S_{\text{in}}}\right)\left(\frac{C_{\text{in}}}{C_{\text{out}}}\right) = G \cdot CA \tag{8.13}$$

where *G* is the *signal gain*. Figure 8.5 makes clear that the effect of the MTI filter on the target signal is a strong function of the target doppler shift. Thus, *G* is a function of target velocity, while clutter attenuation CA is not.

To reduce *I* to a single number instead of a function of target doppler, the definition calls for averaging uniformly over all target dopplers "of interest." If a target is known to be at a specific velocity, then the improvement factor can be obtained by simply evaluating (8.13) at the known target doppler. It is more common to assume the target velocity is unknown *a priori* and use the average target gain over all possible doppler shifts, which is just

$$G = \int_{-PRF/2}^{PRF/2} |H(f)|^2 \, df \qquad (8.14)$$

Table 8.1 gives the improvement factor for two- and three-pulse cancelers for the case of a Gaussian clutter power spectrum. If the clutter spectrum is narrow compared to the PRF, then the improvement factor can be 20 dB or more even for the simple two-pulse canceler.

Expressions for improvement factor equivalent to (8.12) through (8.14) can be developed in terms of the autocorrelation function of the clutter and the MTI filter impulse response. For low-order filters such as two- or three-pulse cancelers and clutter power spectra with either measured or analytically derivable autocorrelation functions, the resulting equations can be easier to evaluate than the frequency domain versions. Derivations and examples are given in [3,7].

8.1.5 Limitations to MTI Performance

The basic idea of MTI processing is that repeated measurements (pulses) of a stationary target yield the same echo amplitude and phase; thus, successive pulses, when subtracted from one another, should cancel. Any effect internal or external to the radar that causes the received echo from a stationary target to vary will cause imperfect cancellation, limiting the improvement factor. Perhaps the simplest example is transmitter amplitude instability. If two transmitted pulses differ in amplitude by 10% (equivalent to 0.83 dB), then the signal resulting from subtracting the two echoes from a perfectly stationary target will have an amplitude that is 10% that of the individual echoes. Consequently, clutter attenuation can be no better than $20 \log_{10}(1/0.1) = 20$ dB. For a two-pulse canceler with a peak signal gain G of 2, the maximum achievable improvement factor is 23 dB. Other sources of limitation due to radar system instabilities include instability in transmitter or oscillator frequencies, transmitter phase drift, coherent oscillator locking errors, PRI jitter, pulse width jitter, and quantization noise. Simple formulas to bound the

Table 8.1
Improvement Factor for Gaussian Clutter Power Spectrum

Standard Deviation of Clutter Power Spectrum	Improvement Factor (dB)	
	Two-Pulse Canceler	Three-Pulse Canceler
PRF	4	6.3
PRF/10	23	43
PRF/100	43	83

achievable clutter attenuation due to each of these error sources are given in [1,4]. External to the radar, the chief limiting factor is simply the width of the clutter spectrum itself. Wider spectra put more clutter energy outside of the MTI filter null so that less of the clutter energy is filtered out. This effect was illustrated in Table 8.1. The effective clutter spectrum width can be increased by radar system instabilities or by measurement geometry and dynamics. For instance, a scanning antenna adds some amplitude modulation due to antenna pattern weighting to the clutter return, increasing the spectral width somewhat. As described in Chapter 2, platform motion can drastically increase the clutter spectral spread.

8.2 PULSE-DOPPLER PROCESSING

Pulse-doppler processing is the second major class of doppler processing approaches. Recall that in MTI processing, the fast time/slow time data matrix is highpass filtered in the slow time dimension, yielding a new fast time/slow time data sequence in which the clutter components have been attenuated. Pulse-doppler processing differs in that filtering in the slow time domain is replaced by explicit spectral analysis of the slow time data for each range bin. Thus, the result of pulse-doppler processing is a data matrix in which the dimensions are fast time and doppler frequency. The spectral analysis is most commonly performed by computing the discrete Fourier transform (DFT) of each slow time row of the data, but other techniques can also be used [9]. The doppler spectrum is then analyzed at each frequency sample to decide whether a moving target is present or not.

The advantages of pulse-doppler processing are that it provides at least a coarse estimate of the radial velocity component of a moving target, including whether the target is approaching or receding, and that it provides a way to detect multiple targets, provided they are separated enough in doppler to be resolved. The chief disadvantages are greater computational complexity of pulse-doppler processing as compared to MTI filtering and longer required dwell times due to the use of more pulses for the doppler measurements.

8.2.1 The Discrete Time Fourier Transform of a Moving Target

To understand the behavior of pulse-doppler processing, it is useful to consider the Fourier spectrum of an ideal moving point target. Consider a radar illuminating a moving target over a dwell of N pulses, and suppose a moving target is present in a particular range bin. If the target's velocity is such that the doppler shift is f_d Hz, then the slow time received signal after quadrature demodulation is

$$x[n] = Ae^{j2\pi f_d nT}, \quad n = 0, \ldots, N-1 \tag{8.15}$$

where T is the radar's PRI, which is the effective sampling interval in slow time.

The discrete time equivalent to the conventional Fourier transform of analog signals is the *discrete time Fourier transform*, or DTFT, defined in normalized frequency coordinates by [6]

$$X(\omega) = \sum_{n=-\infty}^{\infty} x[n] e^{-j\omega n} \tag{8.16}$$

Unlike the DFT, which is the subject of the next subsection, the DTFT has a continuous frequency variable. For the finite-length ideal complex sinusoid of (8.15), the DTFT becomes a geometric series for which we can find a closed form solution. Using the relation $\omega = 2\pi f_d T$ to put the frequency variable in units of hertz, we have

$$X(f) = A \frac{\sin[\pi(f - f_d)NT]}{\sin[\pi(f - f_d)T]} e^{-j\pi(N-1)(f-f_d)T} \tag{8.17}$$

The magnitude of this function is illustrated in Figure 8.8 for the case where $f_d = PRF/4$ and $N = 20$ pulses. Because its form is so nearly identical to the sinc function $\sin x/x$, the expression of (8.17) is often called a *digital sinc* or *aliased sinc* (asinc) function. As one would expect, the main lobe of the response is centered at $f = f_d$ Hz. For numbers of pulses N greater than 4, the two-sided width of the main lobe at the -3-dB points is $0.89/NT$ Hz. At the 4-dB points, it is $1/NT$ Hz, and the null-to-null main-lobe width is $2/NT$ Hz. The first side lobe is 13.2 dB below the response peak. These main-lobe width measures determine the doppler resolution of the radar system. Note that, whichever measure is used, they are all inversely proportional to NT, which is the total elapsed time of the set of pulses

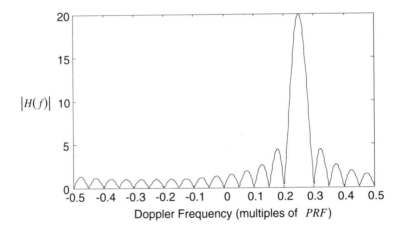

Figure 8.8 Magnitude of the DTFT of a pure complex sinusoid representing an ideal moving-target slow time data sequence with $f_d = PRF/4$ and $N = 20$ pulses.

used to make the spectral measurement. Thus, doppler resolution is determined by the observation time of the measurement. Longer observation allows finer doppler resolution.

Because of the high side lobes, it is common to use a data window to weight the slow time data samples $x[n]$ prior to computing the DTFT. To analyze this case, replace $x[n]$ with $w[n]x[n]$ in (8.16) and again use (8.15) for the particular form of $x[n]$. After again converting from normalized frequency to hertz and recognizing that $x[n]$ is finite length, (8.16) becomes

$$X(f) = \sum_{n=0}^{N-1} w[n]e^{-j2\pi(f-f_d)nT} = W(f-f_d) \tag{8.18}$$

This is simply the Fourier transform of the window function itself, shifted to be centered on the target doppler frequency f_d rather than at zero. We saw this in Figure 8.8; the asinc function is just the Fourier transform of the rectangular window function (equivalent to no window). Harris [10] gives an extensive description of common window functions and their characteristics. In general, nonrectangular windows cause an increase in main-lobe width, a decrease in peak amplitude, and a decrease in signal-to-noise ratio in exchange for large reductions in peak side-lobe level. Table 8.2 summarizes these four key properties of several common windows.[2] Details of the computation of these figures of merit, as well as a much more extensive table, are in [10].

Table 8.2
Key Properties of the Fourier Transforms of Some Common Data Windows

Window	Main-Lobe Width (Relative to Rectangular Window)	Peak Gain (dB Relative to Rectangular Window)	Peak Side Lobe (dB)	Signal-to-Noise Ratio Loss (dB)
Rectangular	1.0	0.0	−13	0
Hann	1.62	−6.0	−32	1.76
Hamming	1.46	−5.4	−43	1.34
Kaiser, $\alpha = 2.0$	1.61	−6.2	−46	1.76
Kaiser, $\alpha = 2.5$	1.76	−8.1	−57	2.17
Dolph-Chebyshev (50-dB equiripple)	1.49	−5.5	−50	1.43
Dolph-Chebyshev (70-dB equiripple)	1.74	−6.9	−70	2.10

---

[2] These data are from Harris [10], who uses a slightly different definition of many of the windows than is used in most data analysis and simulation packages, for reasons having to do with DFT symmetry properties. Effectively, Harris's version of an N-point-shaped window (e.g., Hamming) constitutes the first N points of the $N + 1$-point symmetrical version more commonly used. The difference is of minor consequence, particularly as N gets large. See [10] for details.

The losses in peak amplitude may appear excessive, but it is important to remember that the window also weights the interference components in the signal, and for detection purposes it is the reduction in signal-to-noise ratio that is more important. This can be shown to be [7]

$$\Delta SNR = \frac{\left(\sum\limits_{n=0}^{N-1} w[n]\right)^2}{N\sum\limits_{n=0}^{N-1} w^2[n]} \tag{8.19}$$

As an example, for the Hamming window, the change in signal-to-noise ratio is only −1.36 dB.

8.2.2 Sampling the DTFT: The Discrete Fourier Transform

In practice, we do not compute the DTFT because its frequency variable is continuous. Instead, we compute the DFT, which is given by

$$X[k] = \sum\limits_{n=0}^{N-1} x[n]e^{-j2\pi nk/N} \tag{8.20}$$

Comparing (8.16) and (8.20), we see that for a finite-length data sequence, $X[k]$ is just $X(\omega)$ evaluated at $\omega = 2\pi k/N$ rad. Thus, the DFT computes N samples of the DTFT evenly spaced across one period of the DTFT. Converted into doppler frequency units, these samples occur at frequencies of $k/NT = k(PRF/N)$ Hz.

The DFT is invariably computed using a fast Fourier transform (FFT) algorithm. Most common is the radix 2 or radix 4 Cooley-Tukey algorithm [6], which in addition to being fast has a number of good structural properties for implementation in either hardware or software. However, there is no one FFT algorithm; a wide class of such algorithms exists. They differ in the data set length they are suited for; whether they optimize the number of additions, multiplications, or their sum; the regularity of their computational flow; how efficiently they use memory; and their quantization noise and signal scaling properties. Burrus and Parks [11] give a thorough derivation and summary of most of the algorithms in current use. The DTFT of an ideal moving target always has the same functional form, that is, it looks the same. It is merely translated to the appropriate center frequency, and if a data window is used, its main lobe is widened and attenuated and the side lobes are reduced. The DFT, however, computes samples of the DTFT only at fixed doppler frequencies. The appearance of a plot of the doppler spectrum computed using a DFT can be a strong function of the relation between the actual signal frequency and the DFT frequency samples. This is illustrated in Figure 8.9. In both

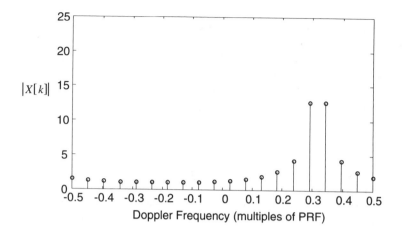

Figure 8.9 The DFT of an ideal moving-target signal: (a) magnitude of the 20-point DFT of a 20-point sinusoidal sequence with frequency 0.25*PRF* (Figure 8.8 shows the DTFT of the same sequence) and (b) magnitude of the 20-point DFT when the signal frequency is changed to 0.275*PRF*.

parts of the figure, the signal is 20 samples of a pure complex sinusoid; consequently, in both cases the DTFT of the signal is the asinc function of (8.17). In Figure 8.9(a), the center frequency is f_d = 0.25*PRF* Hz, and we have used an N = 20-point DFT, so that the DFT samples occur at multiples of 0.05*PRF* Hz. Because the asinc function is centered on one of the DFT sample frequencies *and* the DFT size equals the data set length, one of the DFT samples occurs at the peak of the asinc, while all of the others happen to fall on zeros of the asinc function. Consequently, the DFT is a single impulse function. In Figure 8.9(b), all conditions are the same except that f_d = 0.275*PRF*, which falls halfway between two DFT sample frequencies. The DTFT main lobe is now represented by two DFT samples, but these each fall to the side of the peak of the DTFT, causing an apparent loss of gain and a broadening of the response, since we now have two samples instead of one defining the peak. Furthermore, the other DFT samples now fall near the peaks of the side lobes rather than on the zeros between them. The plot of this DFT looks very different from that of Figure 8.9(a), yet in both cases the underlying DTFT is identical except for a translation on the frequency axis.

One solution to this problem is to compute a denser set of samples of the DTFT, so that DFT does a better job of tracing out the DTFT. Since the DFT computes samples at intervals of *PRF/N* Hz, where N is the DFT size, this requires increasing the DFT size so it is greater than the data set size. To use a size M DFT on a length N data sequence where $M > N$, the data sequence must be defined for $n = N, \ldots, M - 1$. This is done by simply appending $M - N - 1$ zeroes to the end of $x[n]$ to extend it to an M-point sequence, a process known as *zero padding*. Note that zero padding does not improve the resolution of the doppler measurement. The underlying DTFT is not changed because we still have only N actual measurements of the target signal. Zero padding simply makes it possible to evaluate this DTFT at a denser set of frequencies, providing a more complete representation of the underlying DTFT. Figure 8.10 illustrates the effect on the DFT of zero padding to a length of 64 samples for the same ideal target signal used in Figure 8.9(a). The asinc structure is beginning to become evident. As we zero pad and use larger DFTs, the DFT output approaches the DTFT shown in Figure 8.8.

In some situations, the number of data samples available can be greater than the desired DFT size, (i.e., $M < N$). This occurs when there is a need to reduce the DFT size for computational reasons, or when the radar timeline permits the collection of extra pulses and it is desirable to use them to improve the signal-to-noise ratio of the doppler spectrum measurement. A procedure known as *data turning* allows the use of an M-point DFT while taking advantage of all of the data.

In this process, the N-point data sequence is divided into contiguous M-point subsequences; if N is not an integer multiple of M, the last subsequence is padded with zeros to complete it. The subsequences are then added together sample by sample to form a single M-point sequence, and a conventional M-point DFT is computed. It can be shown that the DFT computed in this way produces exactly the samples of the DTFT of the full N-point sequence, despite the overlapping and adding of the subsequences. Figure 8.11 illustrates the actual *data turning* process of breaking down a sequence into subsequences, overlapping, and adding them.

Some caution is needed in applying a data window when the data are modified by zero padding or turning. In either case a length N window should be applied to the data before they are either zero-padded or turned. Applying an M-point window to the full length of a zero-padded sequence has the effect of multiplying

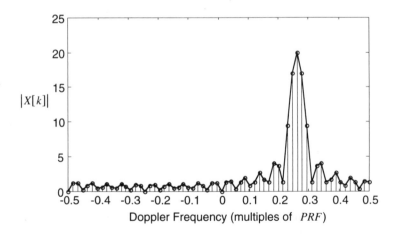

Figure 8.10 Effect of zero padding to 64 points on the DFT of the 20-point pure sinusoid of Figure 8.9(a).

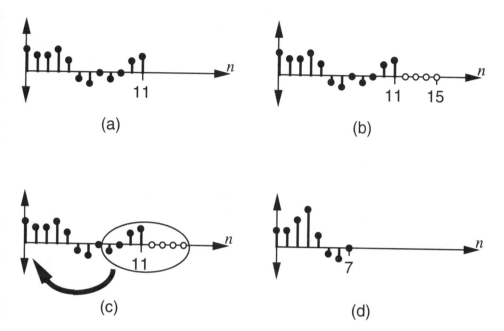

Figure 8.11 Data turning. A 12-point sequence is reduced to an 8-point sequence: (a) original 12-point time sequence; (b) sequence after padding to 16 samples; (c) the second group of 8 samples is overlapped and added to the first group; and (d) final 8-point sequence.

the data by a truncated, asymmetric window (the portion of the actual window that overlaps the N nonzero data points), resulting in greatly increased side lobes. Applying a shortened M-point window to a turned data sequence does not increase side lobes, since the full, symmetric window comes into play, but it does reduce spectral resolution and results in DFT samples that are not exactly equal to samples of the DTFT of the windowed N-point original data sequence. However, this effect is much milder than the effect of windowing errors in the zero padding case.

It was pointed out above that the peak value of the DFT obtained for a moving-target signal is greatest when the doppler frequency coincides exactly with one of the DFT sample frequencies, and decreases when the target signal is between DFT frequencies. This reduction in amplitude is called a doppler *straddle loss*. The amount of loss depends on the particular window used, but is always greatest for signals exactly halfway between DFT sample frequencies. The straddle loss can be computed by evaluating (8.20) with $x[n] = w[n]$ and $k = 1/2$; this is the gain at the halfway point between DFT bins. The computation is repeated with $k = 0$ to get the peak gain, and the ratio is evaluated. For a rectangular window, the maximum straddle loss is 3.91 dB, while for a Hamming window it is 1.74 dB. Thus, while any nonrectangular window causes a reduction in peak gain, typical windows have the desirable property of having less variability in gain as the doppler shift of the target varies. This effect is illustrated in Figure 8.12, which shows the maximum DFT output amplitude as a function of the target doppler shift for rectangular and Hamming windows for the case of $N = 16$. Note the general reduction in peak amplitude for the Hamming-windowed data compared to the unwindowed data. On the other hand, the variation in amplitude is significantly less for the windowed data; the amplitude response is more consistent.[3]

8.2.3 Matched Filter and Filter Bank Interpretations of Pulse-Doppler Processing With the DFT

Equation (8.2) defined the coefficients of the matched doppler filter. In MTI filtering, it is assumed that the target doppler shift is unknown. The resulting signal model of (8.6) leads to the pulse canceler as a near-optimum MTI filter. In contrast, DFT-based pulse-doppler processing attempts to separate target signals based on their particular doppler shift. Assume that the signal is a pure complex sinusoid (ideal moving target) at a doppler shift of f_d Hz. Following the approach of (8.5) and (8.6), the model of the signal vector is then

$$\mathbf{s} = \hat{A}[1 \quad e^{j2\pi f_d T} \quad \ldots \quad e^{j2\pi f_d (N-1) T}]' \tag{8.21}$$

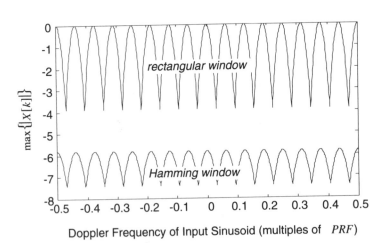

Figure 8.12 Variation of DFT output amplitude as a function of complex sinusoid input frequency for two different data analysis windows.

---

[3]Comparing this figure to Table 8.2 illustrates some effects of the difference between Harris's [10] and more conventional definitions of the Hamming window. The symmetric version used for the figure has a peak gain loss relative to the rectangular window of 5.8 dB instead of 5.4 dB, and the maximum straddle loss for the Hamming window is 1.61 dB instead of 1.74 dB.

If the interference consists only of white noise (no correlated clutter), then $\mathbf{R_I}$ reduces to $\sigma_n^2 \mathbf{I}$, where $\mathbf{I}$ is the identity matrix. It then follows that for an arbitrary data vector $\mathbf{x}$, the output $\mathbf{h'x}$ of the matched filter becomes

$$\mathbf{h'x} = \tilde{A} \sum_{n=0}^{N-1} x[n] e^{-j2\pi f_d nT} \tag{8.22}$$

When $f_d = k/NT = k\text{PRF}/N$ for some integer k, (8.22) becomes simply the N-point DFT of the data sequence $x[n]$. Consequently, the DFT is a matched filter to ideal moving-target signals provided that the doppler shift equals one of the DFT sample frequencies and the interference is white. Furthermore, since the DFT computes N different outputs from each input vector, it effectively implements a bank of N matched filters at once, each tuned to a different doppler frequency.

In fact, the relation between the DFT and a bank of filters can be made more explicit. Consider a signal $x[n]$ obtained with a long series of pulses and an N-point window function $w[n]$. The window function can be slid along the data sequence to select a portion of the data for spectral analysis, as shown in Figure 8.13. The DTFT of the resulting sequence $w[m - M] \, x[m]$ is

$$\begin{aligned}
X_M(f) &= \sum_{m=-\infty}^{\infty} w[m - M] x[m] e^{-j2\pi f mT} \\
&= e^{-j2\pi f MT} \sum_{m=-\infty}^{\infty} w[m - M] e^{-j2\pi f(m-M)T} x[m] \\
&= e^{-j2\pi f MT} \{\{ w[m] e^{-j2\pi f mT} * x[m] \}\}_{m=M}
\end{aligned} \tag{8.23}$$

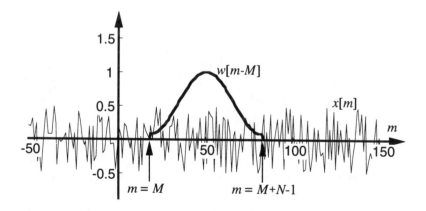

Figure 8.13 Relationship between data sequence $x[n]$ and sliding analysis window $w[n]$. The window length is N.

Equation (8.23) shows that the DTFT at a particular frequency is equivalent to the convolution of the input sequence and a modulated window function, evaluated at time M. Furthermore, if $W(f_d)$ is the DTFT of $w[n]$, then the Fourier transform of $w[m]e^{-j2\pi f_d m T}$ is $W(f - f_d)$, which is simply the Fourier transform of the window shifted so that it is centered at doppler frequency f_d Hz. This means that measuring the DTFT at a frequency f_d is equivalent to passing the signal through a bandpass filter centered at f_d and having a passband shape equal to the Fourier transform of the window function. Since the DFT evaluates the DTFT at N distinct frequencies at once, it follows that pulse-doppler spectral analysis using the DFT is equivalent to passing the data through a bank of bandpass filters.

Of course, it is possible to build a literal bank of bandpass filters, each one perhaps individually designed. In airborne pulse-doppler radars, however, the DFT is almost invariably used for doppler spectrum analysis. This places several restrictions on the effective filter bank design. There will be N filters in the bank, where N is the DFT size; the filter center frequencies will be equally spaced, equal to the DFT sample frequencies; and all the passband filter frequency response shapes will be identical, differing only in center frequency. The advantages to this approach are simplicity and speed with reasonable flexibility. The DFT provides a simple and computationally efficient implementation of the filter bank. The number of filters can be changed by simply changing the DFT size, and the filter shape can be changed by simply choosing a different window.

8.2.4 Fine Doppler Estimation

Peaks in the DFT output that are sufficiently above the noise level to cross an appropriate detection threshold are interpreted as responses to moving targets, that is, as samples of the peak of an asinc component of the form of (8.17). As we have seen, there is no guarantee that a DFT sample will fall exactly on the asinc function peak. Consequently, the amplitude of the DFT sample giving rise to a detection and its frequency are only approximations of the actual amplitude and frequency of the asinc peak. In particular, the estimated doppler frequency of the peak can be off by as much as one-half doppler bin, equal to $PRF/2N$ Hz.

If the DFT size N is significantly larger than the number of pulses (data sequence length) M, then several DFT samples will be taken on the asinc main lobe, and the largest may well be a good estimate of the amplitude and frequency of the asinc peak. Frequently, however, $N = M$ and sometimes, with the use of data turning, we even have $N < M$. In these cases, the doppler bins are large and a half-bin error may be intolerable. One way to improve the estimate of the true doppler frequency f_d is to interpolate the DFT in the vicinity of the detected peak. Theoretically, the correct interpolation involves using all of the DFT data samples and an asinc interpolation kernel, but this approach is computationally expensive. A simpler but very serviceable concept is illustrated in Figure 8.14. For each detected peak

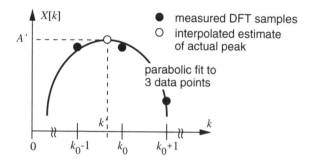

Figure 8.14 Refining the estimated target amplitude and doppler shift by interpolation around the DFT peak. k' is the estimated "bin" number of the peak, and A' is the estimated amplitude.

in the DFT output, a least-squares fit of a parabola is made to that peak and the two adjacent data samples. Once the parabola coefficients are known, the amplitude and frequency of its peak are easily found by differentiating the formula for the parabola and setting the result to zero.

8.3 ADDITIONAL DOPPLER PROCESSING ISSUES

8.3.1 Combined MTI and Pulse-Doppler Processing

It is not unusual to have both MTI filtering for gross clutter removal and pulse-doppler filter banks for detailed analysis of the clutter-canceled spectrum. Since both operations are linear, the order in which they are applied would appear to make no difference to the final doppler spectrum used for detection. However, differences in signal dynamic range can make their order significant in considering hardware effects, particularly when finite-word-length hardware is used.

Clutter is usually the strongest component of the signal; it can be several tens of decibels above the target signals of interest. If the signal is applied to a pulse-doppler filter bank prior to MTI filtering, the side lobes of the response from the clutter around zero doppler may swamp potential target responses at near-in velocities, masking these targets from possible detection. If the processor dynamic range is limited as well, then the strong clutter signal may drive the target signal amplitude below the minimum detectable signal of the processor, effectively filtering out the target.

For these reasons, the MTI filter is generally placed first if both processes are used. The MTI filter will attenuate the clutter component selectively so that the target signals become the dominant components. Subsequent finite-word-length processing will adapt the dynamic range to the targets rather than the now-absent clutter. In floating-point processors, dynamic range is less of an issue.

8.3.2 Transient Effects

All of the discussion in this chapter has assumed a steady-state scenario in the sense that the clutter spectrum is stationary and that filter transient effects have been ignored. As was seen in earlier chapters, in the range-ambiguous medium- and high-PRF modes, each received signal sample (range gate) contains contributions from multiple ranges because of the multiple contributing pulses. Whenever the radar PRF changes, several pulses, known as *clutter fill pulses*, must be transmitted before a steady-state situation is achieved. For example, suppose that in steady state each range gate contains significant contributions from four pulses (four range ambiguities). Then the fourth pulse is the first one for which steady-state operation is possible. The first three pulses are clutter fill pulses, and may not be used in pulse-doppler processing. Additional pulses may be used to set the automatic gain control of the receiver and are also not used for doppler processing.

Steady-state operation of the digital filters used for MTI processing occurs when the output value depends only on actual data input values, rather than any initial (typically zero-valued) samples used to begin the processing. For FIR filters of order N (length $N + 1$), the first N outputs are transients and are discarded in some systems. For simple single or double cancelers, this is only one or two samples.

8.4 OVERVIEW OF DISPLACED PHASE CENTER ANTENNA PROCESSING

MTI filtering and pulse-doppler processing provide an effective way to detect moving targets whose doppler shift is in the clear region of the spectrum on at least one PRF. Airborne targets can generally be detected in this manner. However, slow-moving ground targets having actual doppler shifts only slightly higher than the ground clutter will appear in the skirts of clutter spectrum at all PRFs and are therefore very difficult to detect. Recall that platform motion spreads the ground clutter spectrum as described in (2.7), repeated here for convenience:

$$\delta f_{mlc} = \frac{2v_a}{\lambda} \sin \phi_{ant} \delta\phi \qquad (8.24)$$

where δf_{mlc} is the main-lobe clutter spectral width, ϕ_{ant} is the total angle between the velocity vector and the antenna line-of-sight vector, and $\delta\phi$ is the antenna beamwidth. This spread of main-lobe clutter exacerbates the problem, raising the minimum velocity at which slow-moving ground targets can be detected.

Displaced phase center antenna (DPCA) processing is a technique for countering the platform-induced clutter spectral spreading. By minimizing the clutter spectral width, DPCA improves the probability of detection for slow-moving targets. The basic concept is to make the antenna appear stationary even though the platform is moving forward by electronically moving the receive aperture backwards during operation.

Figure 8.15 illustrates the concept using an electronic antenna that has two subapertures. The entire antenna is used on transmission for maximum gain, so the phase center for transmission is the point T in the middle of the antenna. Each half of the antenna has its own receiver, so there are in effect two receive apertures, having respective phase centers R1 and R2 that are Δx units from the transmit phase center.

If the transmit phase center is located at coordinate x_0 on the first pulse transmitted, then the forward receive phase center is at $x_0 + \Delta x$ and the aft receive phase center is at $x_0 - \Delta x$. The effective phase centers for the complete transmit-receive paths for the common full-array transmit apertures and the two receive apertures are $x_0 + \Delta x/2$ and $x_0 - \Delta x/2$. Now consider the motion of the platform over N pulses. If the PRI is T and the velocity is v_a, then the effective transmit-receive phase centers move forward by $v_a NT$ meters. We can show that if the T-R1 phase center is at position $x_0 + \Delta x/2$ on the first pulse, then the T-R2 transmit phase center will be in the same position N_s pulses later, where

$$N_s = \frac{\Delta x}{2vT} \tag{8.25}$$

N_s is the *time slip* in pulses. To get a feeling for the values involved, consider the case where $\Delta x = 3$m, $v_a = 200$ m/s, and $T = 2$ ms. Then $N_s = 3.75$ pulses.

The significance of the time slip given by (8.25) is that the data stream received on the aft receive aperture is geometrically equivalent to the data stream received on the forward receive aperture N_s pulses earlier.

Consequently, two-pulse cancellation can be implemented by taking each sample from the R1 data stream and subtracting the corresponding sample from the R2 data stream taken 3.75 pulses later. Even though these data samples were

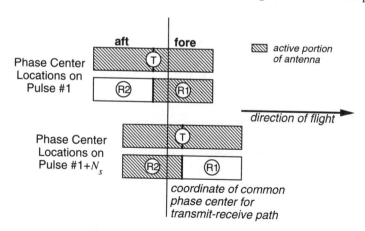

Figure 8.15 Relationship of transmit and receive aperture phase centers in DPCA processing.

collected on different receive apertures and more than one pulse apart in time, their effective transmit-receive phase centers are the same, so they appear equivalent to successive pulses from a *stationary* antenna. The effective stationarity of the antenna then implies that the clutter spectral width is not spread by the platform motion, therefore improving the detection of slow-moving ground targets.

There are many details to DPCA processing that we have not touched on. Most obviously, we do not have access to data taken 3.75 pulses later; the number of pulses must be an integer. A common approach to basic DPCA processing is to time slip one data stream by an integer number of samples to roughly align the two, and then use an adaptive two-pulse canceler to minimize the clutter power, effectively compensating for the remaining fractional-PRI time slip. In addition, DPCA processing may be applied independently after doppler processing in each doppler bin to improve performance at the expense of greater computation.

References

[1] Skolnik, M. I., *Introduction to Radar Systems*, New York: McGraw-Hill, 1980.

[2] Skolnik, M. I., ed., *Radar Handbook*, New York: McGraw-Hill, 1970.

[3] Nathanson, F. E., *Radar Design Principles*, 2nd edition, New York: McGraw-Hill, 1991.

[4] Schleher, D. C., *MTI and Pulsed Doppler Radar*, Boston: Artech House, 1991.

[5] Eaves, J. L., and E. K. Reedy, eds., *Principles of Modern Radar*, New York: Van Nostrand Reinhold, 1988.

[6] Oppenheim, A. V., and R. W. Schafer, *Discrete-Time Signal Processing*, Englewood Cliffs, NJ: Prentice-Hall, 1989.

[7] Levanon, N., *Radar Principles*, New York: John Wiley & Sons, 1988.

[8] *IEEE Standard Radar Definitions*, IEEE Standard 686-1982, Institute of Electrical and Electronics Engineers, New York.

[9] Kay, S. M., *Modern Spectral Estimation*, Englewood Cliffs, NJ: Prentice-Hall, 1988.

[10] Harris, F. J., "On the Use of Windows for Harmonic Analysis With the Discrete Fourier Transform," *Proc. IEEE*, Vol. 68, No. 1, January 1978, pp. 51–83.

[11] Burrus, C. S., and T. W. Parks, *DFT/FFT and Convolution Algorithms*, New York: John Wiley & Sons, 1985.

CHAPTER 9

Pulse Compression in Pulse-Doppler Radar Systems

Marvin N. Cohen

9.1 INTRODUCTION

Pulse compression encompasses various signal modulation and processing techniques used in pulse-doppler and other radar systems that allow the transmission of relatively long-duration waveforms while retaining the advantages inherent in high-range-resolution waveforms. The range resolution achievable with a given radar system is

$$\delta_r = c/(2B) \tag{9.1}$$

where c is the speed of light (3×10^8 m/s) and B is the bandwidth of the transmitted waveform. For a simple (noncoded on transmit and, therefore, not compressed on receive) pulse-radar system, $B = 1/T$, where T is the transmitted pulse duration. Thus,

$$\delta_r = cT/2 \tag{9.2}$$

for a simple pulse system.

This can be exhibited fairly simply in the time (range) domain by observing Figure 9.1. The figure is meant to represent a time sequence display of the relative positioning of an impinging radar pulse (of duration T) coming from the left, two reflectors separated by one-half the pulse length ($cT/2$), and the echo return formed as the pulse passes by and impinges on the two targets. The targets are

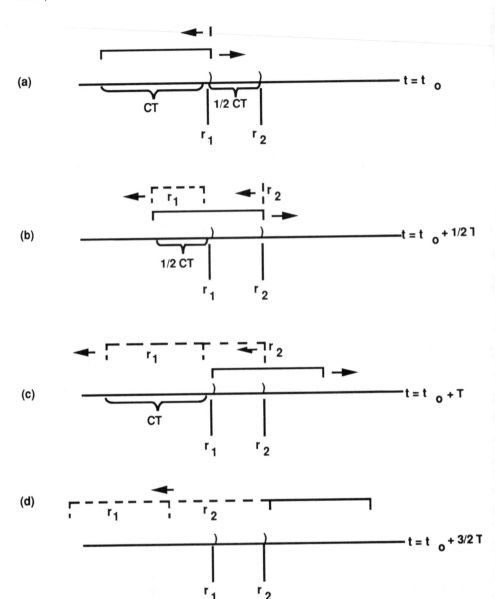

Figure 9.1 Time sequence display of (a) the transmit pulse beginning to impinge on the forward target at $t = t_0$, (b) the transmit pulse impinging on the rearward target at $t = t_0 + T/2$, (c) the transmit pulse halfway past the rearward target at $t = t_0 + T$, and (d) the transmit pulse completely past both targets.

presumed to be separated by precisely one-half a pulse length. Our aim is to demonstrate that the resulting echo pulses will be completely resolvable, but that any shorter reflector spacing would not allow full resolution of the return echoes. Figure 9.1(a) simply establishes the scenario at the time $t = t_0$, when the transmit pulse just begins to impinge on the forward target. Figure 9.1(b) depicts the situation at $T/2$ sec later, when the transmit pulse just begins impinging on the rearward target and the reflection from the forward target has been traveling back toward the radar for $T/2$ sec (and thus extends over $cT/2$ units in range). Figure 9.1(c) depicts the situation yet another $T/2$ (that is, $t = t_0 + T$) sec later, when the transmit pulse is completing its traverse past the forward target and is halfway past the rearward target. At this moment, the echo from the rearward target is just reaching the forward target and the echo formation from the forward target is about to cease. Finally, Figure 9.1(d) completes the picture, where both echoes are fully formed, not overlapping in space (time), and thus fully resolvable yet contiguous (which shows that no closer spacing of targets could yield fully separated returns).

In a pulse compression system, the transmitted waveform is modulated in phase or frequency so that $B \gg 1/T$. Let $\tau = 1/B$. Then, from (9.1),

$$\delta_r = c\tau/2 \tag{9.3}$$

where τ represents the effective pulse duration of the system after pulse compression. Thus, a pulse compression radar can use a transmit pulse of duration T and yet achieve range resolution equivalent to that of a simple pulse system with a transmit pulse of duration τ, where $T \gg \tau$, that is, incorporation of pulse compression in a radar system allows the system to use a transmit pulse of relatively long duration and low peak power to attain the range resolution and detection performance of a short-pulse, high-peak-power system. This is accomplished via modulating (or coding) the transmit waveform and compressing the resulting received waveforms.

The ratio of the transmitted pulse length T to the system's effective (compressed) pulse length is termed the *pulse compression ratio* and is given by

$$CR = T/\tau \tag{9.4}$$

Average power is given by $P \times T \times PRF$, where P is peak power, PRF is the pulse repetition frequency, and T is pulse duration. It follows that CR is the ratio of the average power transmitted by the pulse compression system to the average power transmitted by a simple pulse system, assuming that they both attain the same peak power and achieve the same range resolution. Since $\tau = 1/B$,

$$CR = T \times B \tag{9.5}$$

That is, the compression ratio also equals the time-bandwidth product of the system. Pulse compression systems can, for many purposes, be characterized by their time-bandwidth products [1–3].

Figure 9.2 illustrates a general, conceptual implementation of pulse compression processing in a radar system. The dispersive delay line and the pulse compression filter are of primary interest in this discussion. The RF source generates a short pulse of duration τ, which then passes through a dispersive delay line. The output of the dispersive delay line is a pulse of duration T ($T \gg \tau$), whose bandwidth B is $1/\tau$—the bandwidth of the input pulse. This signal is then amplified and transmitted through the radar antenna. When received, the signal is suitably processed and then passed through the pulse compression filter. Assuming that this filter is matched to the transmitted waveform, the result is a compressed pulse of effective duration $\tau = 1/B$, which provides the system with the ability to resolve targets separated by as little as $c \times \tau/2$ in range. The compressed signal is then further processed, amplified, and displayed.

The compressed pulse is the result of a signal of duration T passing through its matched filter. The time extent of the response out of the matched filter is thus on the order of $2T$, not the compressed pulse length τ, as indicated in Figure 9.2. The responses outside of $|t| < \tau$ are termed *range side lobes*. Since range side lobes from a given range bin may appear as signals in adjacent range bins, they must be controlled. The first two of the following three measures are often used to quantify the level of these side lobes. The third serves to quantify the loss in SNR performance due to the use of a mismatched filter in the receiver. All three are usually given in decibels.

1. Peak side-lobe level (PSL): PSL = 10 log (maximum side-lobe power/peak response power).
2. Integrated side-lobe level (ISL): ISL = 10 log (total power in the side lobes/peak response power).
3. Loss in processing gain (LPG): LPG = 20 log (CR/peak response voltage).

The PSL is closely associated with the probability that a false alarm in a particular range bin is due to the presence of a target in a neighboring range bin. A low PSL is especially important in scenarios where a high density of targets of different cross sections are expected or when attempting to generate range profiles of specific targets. The ISL, a measure of the total energy distributed among the side lobes, is also important in dense target scenarios, as well as when distributed clutter is present. A receiver filter matched to the transmit waveform provides maximum SNR in the presence of white noise, but one may employ mismatched pulse compression filters in the receiver to reduce ISL and PSL. The loss in SNR due to mismatched filtering as opposed to matched filtering is the LPG of the system. Various mismatched filtering techniques are discussed in Section 9.5.

Radar applications that can be aided by high range resolution include, for example, detection, tracking, classification, terrain mapping, accurate ranging, and distributed clutter suppression. High range resolution can be achieved either by transmitting a high peak power and short duration pulse, or by transmitting a lower

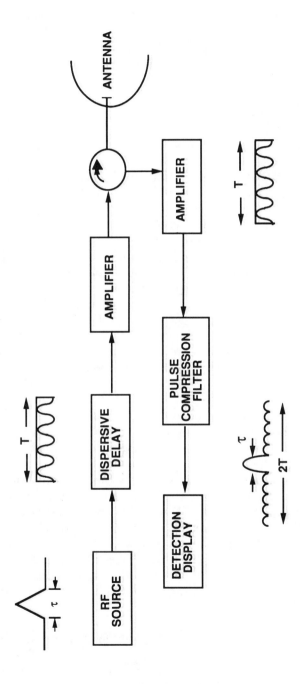

Figure 9.2 Pulse compression processing in a radar system.

peak power and a coded pulse of greater duration and then compressing the echo pulse.

A radar system that incorporates pulse compression processing rather than a simple pulse system to achieve high range resolution generally provides, among others, the following advantages:

1. Improved detection performance for a given peak power;
2. Mutual interference reduction;
3. Increased system operational flexibility.

Probability of detection increases with SNR, which in turn is proportional to average power. Average power, in turn, is the product of peak power, transmit pulse length, and PRF. Pulse compression allows the increase of transmit pulse length without any concomitant spoiling of range resolution. Thus, pulse compression allows the system designer to trade pulse length for peak power or PRF. The trade-off between pulse length and peak power is usually the driving item because of the realities and state of the art in component capability. In general, operation of transmitters at high peak powers leads to problems with insulation, arcing, and generation of X-rays. Special techniques permit generation of pulses shorter than 1 ns, but it is difficult to achieve high enough pulse energy in such systems for long-range radar operation. Furthermore, the advent of practical solid-state transmitters, which are characteristically low-powered, has greatly underscored the utility of reducing peak power requirements by means of pulse compression. In addition, the probability of intercepting a radar transmission is generally directly proportional to its peak power.

Mutual interference among radars can become severe in a dense, active electromagnetic environment. Pulse compression radars can be designed to reduce mutual interference by equipping each radar with a different modulation code and matched filter.

Finally, if the radar signal processor is designed so that various codes and compression ratios can be effected, a system using a transmitter operating at a fixed PRF, peak power, and duty cycle can still provide variable range resolution capability. This flexibility in range resolution may be exploited to incorporate target detection, acquisition, track, and identification modes in a single radar system.

Concomitant with these potential advantages are complications associated with relying on pulse compression as opposed to the transmission of a simple pulse of the requisite duration and power to achieve the aims of a particular radar system. Among these complications are:

1. The requirement for a power amplifier to bring the pulse compression signal to useful levels;
2. Increased system processing requirements due to requisite compression and side-lobe suppression processing;

3. The need for fine control of the waveform parameters that carry the code information;

4. A broadening of the range blind zones;

5. Waveform sensitivity to doppler shift.

The first three items may be viewed as an increase in the radar-processor complexity and sophistication requirement, that is, system cost, size, and maintenance. The fourth issue is discussed in Chapter 12. The last is the result of the interaction of the coding scheme and the doppler induced on the returns from moving targets, which can work to degrade the performance of the pulse compression scheme. This last item is discussed in much greater detail in Section 9.7.

The precise nature and extent of the potential advantages and restrictions just discussed are, in fact, heavily dependent on the system's mission requirements and the particular pulse compression technique used.

Techniques based on frequency modulation and techniques based on phase modulation of the transmit waveform are well developed and have seen widespread application in modern pulse-doppler radar systems. These techniques are discussed in detail in the following sections.

9.2 FREQUENCY MODULATION TECHNIQUES

A radar's carrier frequency may be modulated to increase the bandwidth of the radar transmission and allow pulse compression on receive. Linear frequency modulation (LFM or *chirp*) is the oldest and best developed of all the pulse compression techniques, having first been proposed in the late 1940s [4].

Frequency stepping, which entails changing the frequency of the transmit signal discretely on a pulse-by-pulse basis, with or without chirping the frequency during each pulse, is a more recently developed technique which has been spurred by the recent advances in digital technology [5]. However, frequency stepping is primarily a multiple-pulse technique that requires relatively long look times and can be severely degraded in the presence of target or platform motion. Thus, frequency stepping is generally used for achieving extremely high range resolutions in instrumentation and experimental scenarios rather than for airborne pulse-doppler applications. We therefore restrict the following discussions to the LFM waveforms.

9.2.1 Linear Frequency Modulation

The LFM (or chirp) waveform consists of a rectangular transmit pulse of duration $T = t_2 - t_1$, as shown in Figure 9.3(a). The carrier frequency f is swept linearly (chirped) over the pulse length by an amount Δf, as depicted in Figure 9.3(a).

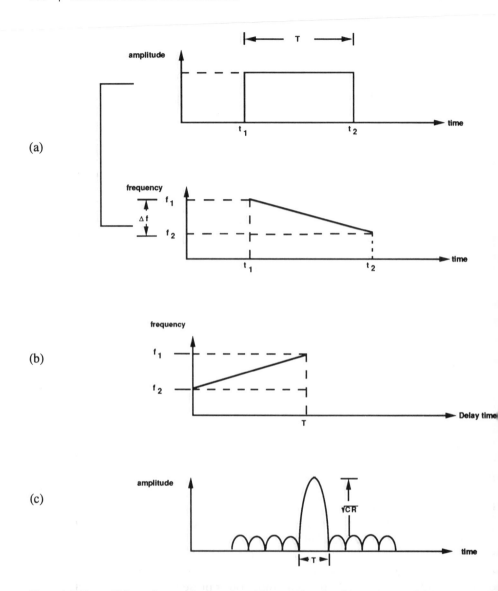

Figure 9.3 Linear FM waveform and processing: (a) transmit pulse; (b) receiver; and (c) compressed pulse.

A pulse compression filter in the radar receiver is matched to the transmitted waveform so that the received signal experiences a frequency-dependent time delay. As shown in Figure 9.3(b), the higher frequency components of the received signal experience correspondingly longer delay times than the lower frequency components at the output of the dispersive delay line. The receiver is constructed so that the delay is proportional to frequency, where the proportionality factor is the negative of the slope (in time frequency) of the transmitted waveform; that is, the receiver constitutes a filter matched to the transmitted waveform. The output signal from the pulse compression filter may be characterized by an envelope having a higher amplitude and a narrower pulse length than the transmitted envelope, as is conceptually depicted in Figure 9.3(c). In fact, the shape of the compressed waveform can be characterized as a $\sin(x)/x$ function when the time-bandwidth product, $T\Delta f$, is large [3]. In particular, the peak side lobe of such a waveform is 13 dB below the system's peak (compressed) response. It can be shown [6] that the compressed (4-dB) pulse width is given by

$$\tau = 1/\Delta f \qquad (9.6)$$

and thus the pulse compression ratio CR is given by

$$CR = T/\tau = T\Delta f \qquad (9.7)$$

Since Δf is the bandwidth B of this signal, the product $T\Delta f$, as noted earlier, is also defined to be the system's time-bandwidth product. The range resolution of this pulse compression radar is given by

$$\delta_r = c/(2\Delta f) = c\tau/2 \qquad (9.8)$$

For example, a pulse compression radar using a 1-μs pulse could have a 50-MHz chirp bandwidth resulting in a range resolution of $\delta_r = 3.0 \times 10^8/(2 \times 50 \times 10^6) = 3$m. This resolution represents a fiftyfold improvement over the resolution of the uncompressed 1-μs pulse from the same radar. Note that the compressed pulse length, and hence the range resolution, is independent of the transmitted pulse length; both are a function only of the frequency excursion Δf, as can be seen from (9.6).

An important feature of linear FM that makes it most suitable among the compression codes for many applications is its relative insensitivity to degradation in response to doppler-shifted signals. Although a more general description of this insensitivity must be postponed until the introduction of the ambiguity diagram in Section 9.6, a discussion of the phenomenon is presented here because linear FM itself can be derived as a second-order approximation to the ideal, completely doppler-invariant waveform [2, p. 427].

Let the solid line in Figure 9.4 represent the time-frequency return from a stationary target. Now assume a moving target at the same range, but with a radial velocity v sufficient to impart a doppler shift of f_d Hz on the return signal. Its return is represented by the dotted line parallel to, but shifted up from, the stationary return in frequency. Since the compression filter is matched to the stationary return, only that portion of the shifted return falling in the frequency region between f_1

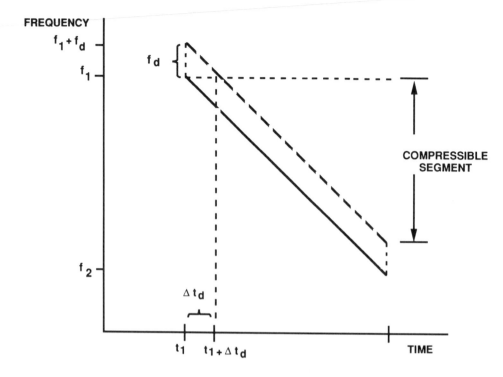

Figure 9.4 The range-doppler coupling of linear FM.

and f_2 will be effectively compressed. The effect of this doppler-shifted return passing through the compression filter will thus be twofold: (1) a fractional loss in output power, given by $\Delta t_d/T$, due to the foreshortened compressible segment, and (2) a delay in compression of

$$\Delta \tau_d = (T/\Delta f) f_d \qquad (9.9)$$

seconds, which corresponds to a decrease in apparent range of $\Delta t_d/\tau$ range bins. For high pulse compression ratios, $T \gg \tau$. It follows that $(\Delta t_d/T) \ll \Delta t_d/\tau)$. Thus, the moving target's return may be compressed almost fully (small $\Delta t_d/T$), yet there may be a significant shift in the target's apparent range (large $\Delta t_d/\tau$). This range-velocity coupling may be highly desirable for applications in which detection is the primary goal, while it may present a problem for applications in which accurate range or velocity measurements are required. It should be noted, however, that these range-velocity ambiguities may be resolved by using chirped waveforms with different time-frequency slopes just as range ambiguities can be resolved using pulse trains with different PRFs.

9.2.2 Stretch

LFM, as described above, uses a passive receive filter. As such, it may be implemented over an unlimited number of range bins and is thus an excellent implementation for a surveillance mode. Another technique appropriate for achieving high time-bandwidth products is through the transmission of an LFM waveform while implementing the compression in the receiver via an active correlational process. This implementation is often referred to as *Stretch* processing after [7].

Active correlation consists of multiplying returns by a replica of the transmitted waveform, filtering to extract the unmodulated waveform envelope, and then integrating the resulting difference-frequency product across the pulse width to complete the matched-filtering operation. This multiply-and-integrate function must in general be performed for each range gate instrumented by the radar system. The range at which the process is applied is determined by the time delay between transmission and the beginning of the active correlation process. The size of the range gate is determined by the time duration of the transmit waveform.

LFM supports very efficient implementation of active correlation, as illustrated in Figure 9.5. The receiver output is mixed with an LFM signal whose slope and time extent match those of the transmitted waveform. Hence, active-correlation multiplication is conducted at RF followed by lowpass filtering to extract the difference-frequency terms. Thereafter, the signal is split into I and Q components and digitized for further processing.

The return from each range bin within the selected range gate thus corresponds to a pulsed sinusoid at the output of the active difference mixer. The later the return, the higher the residual frequency. The frequency of the tone corresponding to the ith range bin from the beginning of the range gate is given by $\tau_i k$, where τ_i is the range delay from the first range bin to the ith range bin and k is the frequency slope coefficient (that is, $k = B/T$, where B is the bandwidth and T is the transmit pulse duration) of the waveform. Pulse compression is completed by performing a spectral analysis of the difference-frequency output to transform the pulsed tones into corresponding frequency resolution cells. In practice, this spectral analysis is performed by digitizing the difference-frequency output and processing it through an FFT.

LFM active correlation reduces computational throughput requirements compared with other digital compression techniques by performing multiplication within the receiver so that the signal processor must only implement integration (spectral analysis) over a set of samples of a comparatively long time duration. Active correlation and FFT computation are performed just once for each specified range window and each FFT output corresponds to the return from a specific range bin within the specified range window.

Stretch pulse compression gain is nominally $CR = T \times B$, less the side-lobe suppression weighting and any system losses, just as in an LFM implementation.

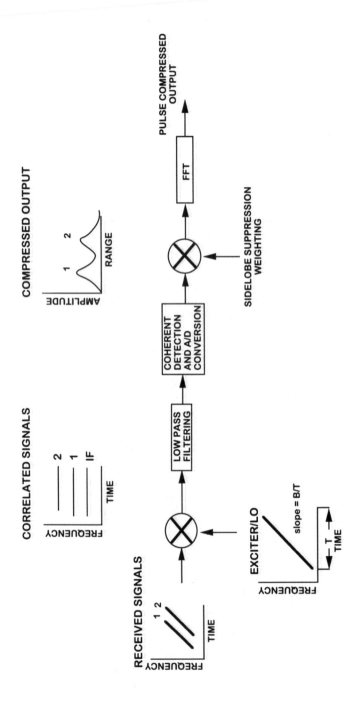

Figure 9.5 Stretch pulse compression

Active-correlation LFM pulse compression is commonly used for both real-time and postprocessing wideband pulse compression requirements. The relatively simple implementation of Stretch pulse compression for a small number of range gates makes it an ideal choice for tracking waveforms and range-limited surveillance modes. A common application in airborne pulse-doppler radars is in a high-range-resolution-while-search mode, where the Stretch waveform and processing may be used to provide range information [8]. A number of significant extensions of the LFM active-correlation process have been suggested and implemented as well. For example, specialized techniques have been developed to accommodate the range-walk correction required in SAR processing [9]. Advantages of Stretch pulse compression over conventional LFM processing include ease of implementation and waveform flexibility due to the simple hardware mechanization that can use programmable digital hardware. Furthermore, Stretch pulse compression systems can be designed with the ability to predistort the transmit and receive ramps to correct for system errors such as frequency nonlinearities and frequency-dependent system losses.

The Stretch technique can be implemented as an augmentation to a high-range-resolution-while-search mode in an airborne pulse-doppler radar in the following way. After detection in a range-ambiguous High PRF mode, the radar can be made to generate two chirps of differing slopes and to Stretch process the received signals. The two different slopes allow unwrapping of the range-doppler ambiguity, and the processing of multiple range gates allows full range coverage. An example illustrating range coverage and resolution follows. While the example discusses a simple (say, up-) chirp, one may assume a second (say, down-) chirp and identical processing for resolving the resulting range-doppler ambiguity.

For a system using an LFM transmit pulse of duration T and a delay of time ΔT_0 in the generation of the receiver waveform replica, the resulting enhanced range-resolution map will begin at range $R = c \times \Delta T_0/2$, and the range extent of the map will be $R = c \times T/2$. Thus, for $\Delta T_0 = 160$ μs and $T = 80$ μs, the radar map will begin at the range $R = (3 \times 10^8) \times (160 \times 10^{-6}/2) = 240 \times 10^2 = 24$ km, and the range extent of the radar map will be $R = (3 \times 10^8) \times (80 \times 10^{-6}/2) = 120 \times 10^2 = 12$ km. If the bandwidth B of the transmitted LFM waveform is 80 kHz, then the range resolution will be $\delta r = c/(2B) = 3 \times 10^8/(2 \times 80 \times 10^3) = 1.88$ km. By generating an active reference chirp in the receiver every 80 μs, for, say, eight contiguous times, one may achieve full range coverage from 24 km (approximately 15 mi) out to 120 km (75 mi).

9.2.3 Frequency Stepping

As described in the previous sections, LFM and Stretch use continuous modulation of the carrier frequency to encode the transmit pulse for subsequent compression. When discrete modulation of the carrier frequency is used to encode the transmission, the technique is termed *frequency stepping*. In the most common form of frequency stepping, the frequency is changed on a pulse-by-pulse basis in an interpulse, sampled version of LFM. In airborne pulse-doppler radars, the single pulse per frequency is often replaced by a pulse burst per frequency to allow doppler processing at each frequency prior to range pulse compression. Such a waveform, like Stretch, is an excellent candidate for forming high-range-resolution profiles of targets with an airborne pulse-doppler radar.

A single-pulse-per-frequency-stepped FM waveform is depicted in Figure 9.6, where:

T = transmit subpulse duration;
τ = effective compressed pulse duration;
N = number of subpulses;
F_0 = frequency of the first pulse;
Δf_s = frequency step between pulses;
F_k = $F_0 + (k - 1)\Delta f_s$ = frequency of the kth pulse;
ΔF = $N \times \Delta f_s = B$ = total bandwidth of the transmission.

At each moment, the radar transmitter and receiver are operating in a narrowband, $B_{pulse} = 1/T$. At the ith pulse in the process, the receiver extracts the I and Q samples from the narrowband reception at frequency F_{i-1} for each implemented range gate. The ensemble of I,Q samples is then passed through an inverse FFT to provide a range-resolved map of the instrumented range gate. Since this process requires separate processing of each instrumented range gate, stepped-frequency waveforms are most appropriate for either track modes or for surveillance modes where the number of range gates to be processed is small.

Since the ensemble of returns is collected over a bandwidth B, the resulting resolution one can achieve is given by

$$\delta r = c/(2B) = c/(2\Delta F) = c/(2N\Delta f_s)$$

and it turns out that there is a necessary relationship between T and Δf_s. For targets contained wholly within the resolution of one subpulse (that is, wholly within one radar range gate, $c/(2T)$), we must design the waveform so that $\Delta f_s \leq 1/T$; for otherwise we will have undersampled the frequency response of the target [10]. That is, $\Delta f_s = 1/T$ represents critical sampling of the process. In practice, especially for targets whose extent is greater than one range gate of the system, designs are typically constructed with $1/2T \leq \Delta f_s \leq 1/T$, yielding an oversampling of the process to resolve ambiguities introduced by the IF filtering of the extended range map [10].

Assume for now that we choose critical sampling, $\Delta f_s = 1/T$. Then the achievable range resolution is given by

$$\delta r = c/(2B) = c/(2\Delta F) = c/(2N/T) = (cT/2)/N$$

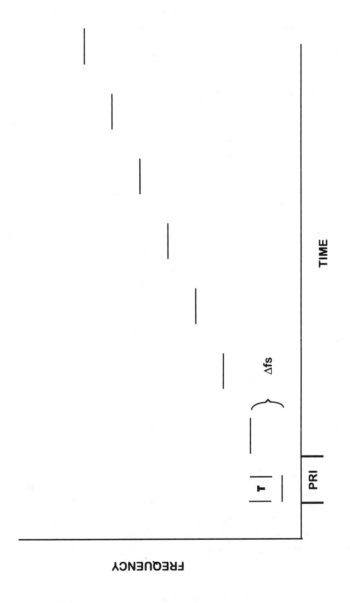

Figure 9.6 Stepped-frequency transmit waveform.

That is, the achievable resolution of the waveform is one-Nth that of the transmit pulse width. That is, the process divides the radar's range gate, as determined by its pulse width, into N resolution cells. It also follows that the pulse compression ratio of the wideband transmission is given by

$$CR = NT \times B = NT \times N/T = N^2$$

where one factor of N is contributed by the N pulses on target and the second factor of N is contributed by the total frequency excursion of the waveform.

Returning to the relationship $\Delta f_s = 1/kT$, $1 \le k \le 2$, we see that the rate at which the frequency is stepped is bounded by the reciprocal of a multiple of the chosen radar transmit pulse width. Smaller step sizes mean longer step sequences and therefore longer target dwells to achieve a given total bandwidth (range resolution). A way to break the dependency between transmit pulse width and step size is to use some form of intrapulse pulse compression on the transmit pulse. In doing so, the step size restriction is then made a function of the compressed rather than the transmit pulse width, thereby allowing both long transmit pulses and large frequency step sizes.

In practice, for airborne pulse-doppler systems, stepped-frequency waveforms are usually implemented in a track mode with pulse bursts at each frequency. The pulse burst is doppler-processed to yield the doppler gate of interest, and only the I,Q samples from this gate are retained for high-resolution (stepped-frequency) processing. In this way, the computational load for formation of high-range-resolution maps of targets is kept reasonable and the velocity of the target can be tracked with high accuracy to mitigate the range-doppler ambiguities that can affect the resulting high-range-resolution profile.

9.3 PHASE MODULATION TECHNIQUES

A transmitted radar pulse of duration T may also be coded for subsequent compression by dividing it into N subpulses, each of duration τ, and by coding these subpulses in terms of the phase of the carrier. The binary phase (biphase) codes can be represented simply by pluses and minuses, where a plus subpulse designation represents no phase shift and a minus subpulse designation represents a carrier phase shift of π rad. Polyphase codes are more complex and allow for any of M phase shifts on a subpulse basis, where M is called the *order of the code* and the possible phase states are

$$\phi_i = (2\pi/M)\ i, \text{ for } i = 1, \ldots, M \tag{9.10}$$

The process of coding a transmit pulse in phase and then using this modulation to effect pulse compression on receive is most easily exhibited by an example based on biphase coding of the pulse and compression through a tapped delay line matched filter. Figure 9.7 represents transmission of a 13-element biphase ($M = 2$) code, where the abscissa represents time, the ordinate represents amplitude, + represents no phase change in the carrier, and—represents a carrier phase shift of π rad. The autocorrelation, or *compressed output* in response to a point target, may be found by passing this coded waveform through its matched filter, given as a tapped delay line in Figure 9.7(b). The autocorrelation is computed by passing the code through the filter from right to left and shifting one subpulse (τ) for each computation. After each shift, the code and filter values that occupy the same (time) positions are multiplied, and the results of all these multiplications are added. This value is output, the code is shifted one more subpulse to the left, and the process is repeated. Thus, after 12 shifts of the code through the filter, we obtain

code	+	+	+	+	+	−	−	+	+	−	+	−	+	
x filter	+	+	+	+	+	−	−	+	+	−	+	−	+	
= sum	1	1	1	1	−1	1	−1	1	−1	−1	−1	−1		= 0

where we take the product of two like signs to be 1 and the product of two different signs to be −1. Similarly, after 13 shifts,

code	+	+	+	+	+	−	−	+	+	−	+	−	+	
x filter	+	+	+	+	+	−	−	+	+	−	+	−	+	
= sum	1	1	1	1	1	1	1	1	1	1	1	1	1	= 13

which is the main peak of the compression function. Figure 9.7(c) gives a plot of all the values so obtained. The total system time response for a point target echo is on the order of $2T$, where T is the transmit pulse duration; the compressed (3-dB) pulse length is on the order of τ, where τ is the subpulse length; and the effective amplitude of the peak response is 13 times that of the amplitude of the inserted coded pulse. The response between $t = -\tau$ and $t = \tau$ is the compressed signal. The responses for $|t| > \tau$ represent the range side lobes generated by the system. In general, given a transmit pulse of duration T coded in N subpulses each of duration τ, matched filtering will result in a compressed pulse that has an effective peak amplitude N times the input pulse and a resolving capability equivalent to that of a simple pulse system that uses a pulse of width τ. The pulse compression ratio is therefore

$$CR = N = T/\tau = B \times T \tag{9.11}$$

where, as usual, $B = 1/\tau$.

Although recent work in the field has generated renewed interest in polyphase coding schemes for pulse compression, most fielded systems that use phase coding use biphase coding schemes. We will therefore focus on biphase techniques here and invite the reader to pursue independent study in the areas of Frank

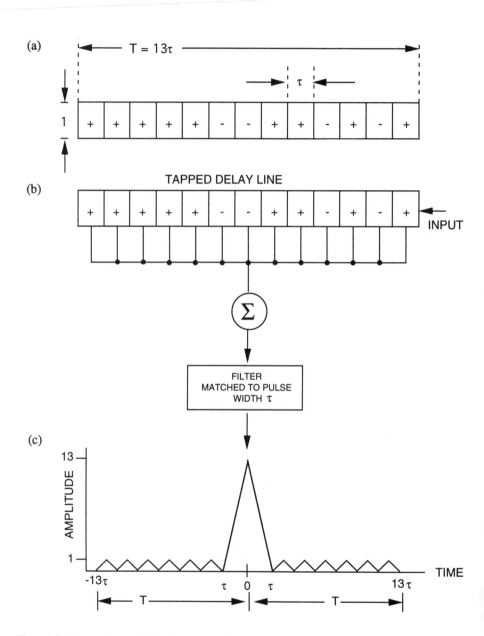

Figure 9.7 Binary phase-coded pulse compression.

Codes [11], Welti side-lobe canceling Codes [12], and the so-called NRL Codes [13]. A general overview of these codes is also available in [14] and [5, Chap. 12]. In the remainder of this section, we discuss in detail some of the particularly useful binary phase codes: the Barker, near-perfect, combined Barker, pseudorandom, and Golay codes.

9.3.1 Barker Codes

Barker codes are the binary phase codes with the property that the peak side lobes of their autocorrelation functions are all equal to $1/N$ in magnitude, where N is the code length, and the output signal voltage (i.e., the voltage of the maximum output) is normalized to 1. All known Barker codes along with the ISL and PSL (see Section 9.1) of their matched systems are given in Table 9.1. Extensive searches for Barker codes of length greater than 13 have been conducted by various researchers, but none have been found. Odd-length Barker codes of length greater than 13 do not exist [15]. Even-length Barker codes of length greater than 13 and less than several thousand do not exist [16]. In fact, those listed in Table 9.1 are thought to be the only Barker codes possible.

The great attractions of the Barker codes are that (1) their side-lobe structures contain the minimum energy that is theoretically possible,a and (2) this energy is uniformly distributed among the side lobes. By virtue of these properties, Barker codes are sometimes called *perfect codes.*

The search for longer perfect codes (to achieve higher pulse compression ratios) having so far failed, it is natural to seek out longer near-perfect or "good" codes—codes whose peak side lobes are the best that can be attained for a given code length. Table 9.2 provides a listing of the minimum peak side lobes that can be achieved with matched filtering for code lengths 7 through 34. Each listing also

Table 9.1
Known Barker Codes

Code Length	Code Elements	PSL (dB)	ISL (dB)
1	+	—	—
2	+ −, + +	−6.0	−3.0
3	+ + −, + − +	−9.5	−6.5
4	+ + − +, + + + −	−12.0	−6.0
5	+ + + − +	−14.0	−8.0
7	+ + + − − + −	−16.9	−9.1
11	+ + + − − − + − − + −	−20.8	−10.8
13	+ + + + + − − + + − + − +	−22.3	−11.5

Table 9.2
Summary of the Near-Perfect Codes of Lengths 7 through 34 [17]

Length	Number	Peak Side Lobe	ISL (dB)	Sample Code
7	1	1	−9.12	0100111
8	16	2	−6.02	10010111
9	20	2	−5.28	011010111
10	10	2	−5.85	0101100111
11	1	1	−10.83	01001000111
12	32	2	−8.57	100110101111
13	1	1	−11.49	1010110011111
14	18	2	−7.12	01010010000011
15	26	2	−6.89	001100000101011
16	20	2	−6.60	0110100001110111
17	8	2	−6.55	00111011101001011
18	4	2	−8.12	011001000011110101
19	2	2	−6.88	1011011101110001111
20	6	2	−7.21	01010001100000011011
21	6	2	−8.12	101101011101110000011
22	756	3	−7.93	0011100110110101011111
23	1,021	3	−7.50	01110001111110101001001
24	1,716	3	−9.03	011001001010111111100011
25	2	2	−8.51	1001001010100000011100111
26	484	3	−8.76	10001110000000101011011001
27	774	3	−9.93	010010110111011101110000111
28	4	2	−8.94	1000111100010001000100101101
29	561	3	−8.31	10110010010101000000011100111
30	172	3	−8.82	100011000101010010010000001111
31	502	3	−8.56	0101010010010011000110000001111
32	844	3	−8.52	00011101100011010011101111110011
33	278	3	−9.30	011001100101010100101100001111111
34	102	3	−9.49	1100110011111111100001101001010101

includes one example of such a code (where a 0 is used to represent −1 for convenience), the number of such codes, and the best ISL that can be achieved for each length. It should be noted that these results were achieved empirically by, essentially, an exhaustive search technique, the only approach known that can lead to such absolute results [17].

9.3.2 Combined Barker Codes

One scheme to generate codes longer than 13 bits is the method of forming combined Barker codes using the known Barker codes. For example, in order to devise a system with a 20:1 pulse compression ratio, we may use either the

5×4 Barker code or the 4×5 Barker code. The 5×4 Barker code consists of the 5-bit Barker code, each bit of which is the 4-bit Barker code. Thus, the 5×4 combined Barker code is the 20-bit code

$$| + + - + | + + - + | + + - + | - - + - | + + - + |$$

where the 4-bit Barker code $+ + - +$ is modulated by

$$| \quad + \quad | \quad + \quad | \quad + \quad | \quad - \quad | \quad + \quad |$$

which is the 5-bit Barker code.

Combined codes consisting of any number of individual codes can be defined analogously. The filter associated with a combined code is a combination of the filters matched to the individual codes. The individual codes (and corresponding filters) are called the *subcodes* (or subsystems or components) of the full code (system). The matched filter for a combined code may be implemented directly as a tapped delay line whose impulse response is the time inverse of the code, or as a combination of subcode matched filters. Figure 9.8 is an example of a combined matched filter for the 5×4 combined Barker code. The first stage of the filter (on the right) is simply the matched filter to the inner (4-bit) code. The second stage represents a filter matched to the 5-bit Barker code except that the active elements are spaced four taps apart. This filter is equivalent to the 20-bit tapped delay line matched filter (their impulse responses are identical); however, the number of active arithmetic elements (plus or minus multiplications) in the combined filter is 9, the sum of the subcode lengths, and the number in the 20-bit filter is 20, the product of the subcode lengths. These results generalize to codes that are the combination of any number of subcodes.

There are various advantages to creating high-pulse-compression-ratio systems using combined Barker codes and filters:

1. By combining the subcodes in various ways, one can generate numerous codes of various lengths appropriate for different modes in a single radar system (such as surveillance, track, and identification).
2. As indicated above, the number of arithmetic elements required for processing is the sum of the subcode lengths when implementing a combined filter as opposed to the product when implementing the matched filter as a single tapped delay line.
3. The analytic procedure for deriving the tap weights of a length n ISL-optimized filter (as described in Section 9.4.2) requires the solution of a system of n linear equations in n unknowns. As n grows large, the solution becomes difficult. These side-lobe reduction techniques may be applied to a long combined code by using a combination of filters, each optimized for the corresponding subcode [18].
4. The side-lobe characteristics of a combined system are easily and inexpensively studied, since knowledge of the side-lobe characteristics of its components is

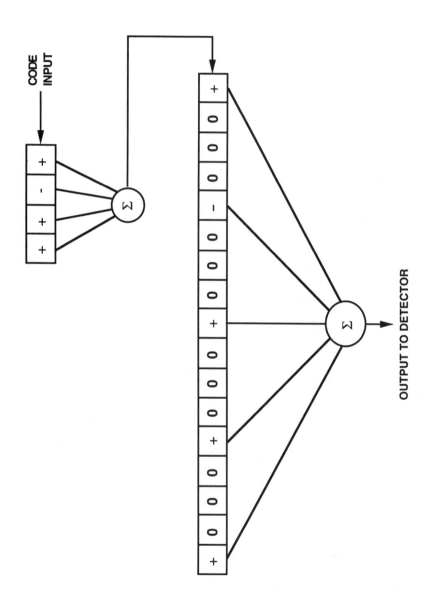

Figure 9.8 Combined match filter for 5 × 4 combined Barker code.

essentially all we need to know. In particular, the PSL of a system is approximately the PSL of its weakest component, the ISL is approximately the root sum square of the subsystems' ISLs, and the LPG is approximately the sum of the individual LPGs, though the ISL and LPG show some sensitivity to order [18].

9.3.3 Pseudorandom Codes

Another set of long binary phase codes that are relatively easy to generate, have good side-lobe properties, and can be changed algorithmically are the so-called *pseudorandom sequences* (PN codes). The PN codes of primary interest are the ones of maximal length (maximal-length binary shift register sequences, maximal-length sequences, or simply *m-sequences*). Because of the properties just mentioned, as well as their spread spectrum characteristics, these sequences have proved popular for some radar and very many communications applications [19].

These sequences can be generated by initializing, at any nonzero state, a binary shift register with feedback connections such as the one depicted in Figure 9.9, clocking the system to circulate the bits, and picking off the appropriate outputs. The result of this process is a sequence of length $2^n - 1$, where n is the number of shift registers employed. In order for the output to be of maximal length and, therefore, a nonrepeating $2^n - 1$ sequence, the feedback paths must correspond to the nonzero coefficients of an irreducible, primitive polynomial modulo 2 of

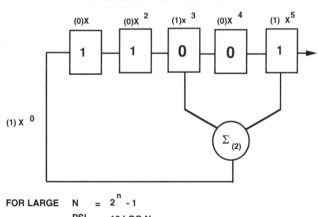

PSEUDORANDOM CODES

FOR LARGE N $= 2^n - 1$
PSL $=$ 10 LOG N
ISL $=$ RELATIVELY HIGH
LPG $=$ 0 (FOR MATCHED FILTER IMPLEMENTATION)

Figure 9.9 Maximal-length binary shift register with initial state.

degree n. An example of such a polynomial of degree 5 is $(1) + (0) x + (0) x^2 + (1) x^3 + (0) x^4 + (1) x^5$. The five-stage shift register that corresponds to this polynomial is given in Figure 9.9. Note that the constant 1 term corresponds to the feedback from the adder to the first stage in the register. The 1 coefficients of the x^3 and x^5 terms correspond to the feedback paths from the third and fifth registers to the adder. In general, the shift registers may be initialized in any nonzero state to generate an m-sequence. An initial state of 0, 1, 0, 0, 0 is shown. See also Table 9.3, borrowed from Nathanson [5, p. 465]. For each possible degree (number of shift registers employed) between one and eight, the shift

Table 9.3
Maximum-Length Pseudorandom Codes and Their Properties

Degree (Number of States) and Length	Polynomial Octal	Lowest Peak Side-Lobe Amplitude	Initial** Conditions, Decimal	Lowest rms Side-Lobe Amplitude	Initial Conditions, Decimal
1(1)	003*	0	1	0.0	1
2(3)	007*	−1	1.2	0.707	1.2
3(7)	013*	−1	6	0.707	6
4(15)	0.23*	−3	1, 2, 6, 8, 10, 11, 12	1.39	2.8
5(31)	0.45*	−4	5, 6, 26, 29	1.89	6.25
			(9 conditions)	1.74	31
			2, 16, 20, 26	1.96	6
6(63)	103*	−6	1, 3, 7, 10, 26, 32, 45, 54	2.62	32
			(9 conditions)	2.81	35
			(9 conditions)	2.38	7
7(127)	203*	−9	1.54	4.03	109
	211*	−9	9	3.90	38
	235	−9	49	4.09	12
	247	−9	104	4.23	24,104
	253	−10	54	4.17	36
	277	−10	14, 20,73	4.15	50
	313	−9	99	4.04	113
	357	−9	15, 50, 78, 90	4.18	122
8(255)	435	−13	67	5.97	135
	453	−14	(20 conditions)	5.98	254
	455	−14	124, 190, 236	6.10	246
	515	−14	54	6.08	218
	537	−13	90	5.91	
90	543	−14	(10 conditions)	6.02	197
	607	−14	(6 conditions)	6.02	15
	717	−14	124, 249	5.92	156

*Only single Mod-2 adder required.
**Mirror images not shown.

registers that yield m-sequences with the best peak and rms (ISL) side-lobe levels are given. The octal numbers of column 2 can be translated to binary numbers to obtain the polynomial coefficients appropriate for defining the feedback connections. The binary representations of the decimal integers of column 4 represent the initial states of the shift register that yield the minimum peak side-lobe amplitudes cataloged in column 3. Column 5 gives the minimum rms side-lobe levels for the codes, and column 6 gives the decimal equivalents of the binary initial conditions necessary to achieve those levels. Note that shift registers for mirror images of the defined codes are not included in the table. Thus, the total number of these "best" codes is twice those explicitly defined.

By referring to a list of irreducible polynomials (see [20], for example), one can construct the appropriate feedback connections to generate any m-sequence of degree 1 through degree 34 (output length $2^{34} - 1$).

For large $N = 2^n - 1$, the peak side-lobe out of the filter matched to such a code when the signal is normalized to 1 is approximately $N^{-0.5}$ in voltage. The actual value varies with the particular sequence. For example, with $N = 127$, the PSL varies between -18 and -19.8 dB as opposed to the -21 dB predicted by the above rule of thumb. As N increases, the rule of thumb approximation improves.

Especially interesting, perhaps, is the fact that in response to a continuous, periodic flow of one of these codes through its matched filter, the output is a periodic peak response of N (in voltage) and a flat range side-lobe response of -1. That is, for a periodic code response, the pseudorandom codes are perfect.

The fact that PSL is inversely proportional to the square root of code length makes m-sequences appropriate codes for pulsed-radar applications in which few closely spaced targets are expected in the field of view, such as the imaging of space-based objects; however, the relatively high ISL of these output waveforms make them unsuitable for high-target-density and extended-clutter situations. Furthermore, either a full tapped delay line or a bank of shift register generators, one for each code bit, must be employed in the receiver to compress at all ranges. These considerations are some of the primary reasons that m-sequences have been more popular in communications and CW radar applications than in pulse-radar applications.

9.4 GOLAY SIDE-LOBE CANCELING CODES

The classes of Welti and Golay codes are side-lobe-canceling codes.[12,21] They are sets of pairs of codes with the property that the sum of the autocorrelations of the two codes in a pair sum to twice the voltage of a single autocorrelation at the peak and to zero elsewhere. The Welti codes form a large, general set of polyphase codes that have this property. The Golay codes form the subset of these side-lobe canceling codes that are binary. Figure 9.10 exhibits a pair of Golay codes, their autocorrelations, and the zero side-lobe sum of their autocorrelations. As can be

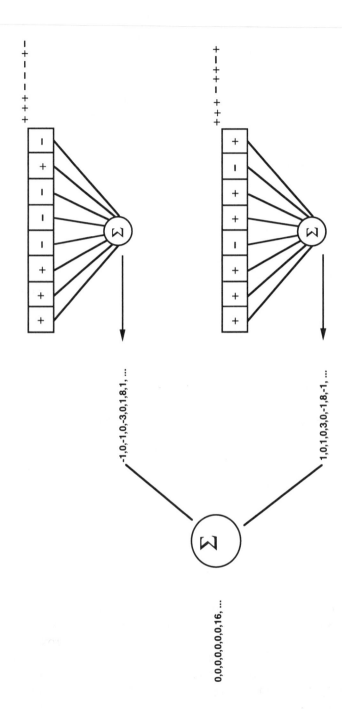

Figure 9.10 Golay side-lobe-canceling code pair of length 8.

seen from the figure, the key to the side-lobe canceling property of Golay code pairs is that the range side lobes of one are equal in amplitude and opposite in sign to the side lobes of the other.

Although these codes may seem to represent the ideal solution to the side-lobe suppression problem, the difficulties involved with their implementation as well as their sensitivities to echo fluctuations and doppler shift make them less than ideal for most applications.

9.5 RANGE SIDE LOBES AND WEIGHTING

As mentioned in Section 9.1, every method of pulse compression leads to the generation of range side lobes. Since the side lobes from any range bin appear as targets in adjacent range bins, the suppression of side lobes is critical in applications where there are high target densities, targets of varying reflectivity, or extended clutter. For example, given two targets whose RCS differs by 15 dB, the first side lobes due to the larger target will cause a false alarm in a linear FM system designed to detect the smaller target unless side-lobe suppression is employed. (Recall that the first side lobe appears approximately 13 dB down from the peak in the matched-filter implementation of linear FM.)

In general, side-lobe suppression is achieved by tapering the matched-filter response by weighting the transmitted waveform, the matched filter, or both in either frequency or amplitude. The weighting that is employed should ideally be applied to both the transmitted waveform and the matched filter so as to retain the matched-filter characteristics and thus avoid a loss of SNR (LPG); however, this is often impractical. Therefore, weighting is usually applied only to the matched filter, and the ensuing LPG is accepted as a necessary system loss.

The following subsections describe various mismatch-filter weighting functions used for side-lobe suppression. The discussion is divided into functions for FM waveforms and functions for discrete phase-coded waveforms.

9.5.1 Side-Lobe Suppression for FM Waveforms

The same illumination functions used in antenna design to reduce spatial side lobes can also be applied to the frequency domain to reduce the time side lobes in pulse compression. A comparison of several types of spectral weighting functions for an LFM signal with a rectangular spectrum is shown in Table 9.4. The Dolph-Chebyshev weighting theoretically results in all side lobes being equal; however, it is physically unrealizable. The Taylor weighting is a practical approximation to the Dolph-Chebyshev. The Taylor weighting with $\bar{n} = 6$ means that the peaks of the first five (or $\bar{n} - 1$) side lobes are designed to be equal; the side lobes then fall off at 6 dB per octave. Weighting the received-signal spectrum to lower the side lobes increases the main-lobe width and reduces the peak SNR as compared with

Table 9.4
Weighting Function Data

Weighting Function	PSL (dB)	Pulse Widening	Mismatch Loss (dB)	Far Side-Lobe Fall-off Rate
Dolph-Chebyshev	−40.0	1.35	—	1
Taylor, $\bar{n} = 6$	−40.0	1.41	−1.2	$1/t$
$k + (1 - k) \cos^n$				
Hamming ($k = 0.08$, $n = 2$)	−42.8	1.47	−1.34	$1/t$
Cosine-squared ($k = 0$, $n = 2$)	−32.2	1.62	−1.76	$1/t^3$
Cosine-cubed ($k = 0$, $n = 3$)	−39.1	1.87	−2.38	$1/t^4$
$n = 1$, $k = 0.04$	−23.0	1.31	−0.82	$1/t$
$n = 2$, $k = 0.16$	−34.0	1.41	−1.01	$1/t$
$n = 3$, $k = 0.02$	−40.8	1.79	−2.23	$1/t$

Source: [5].

unweighted pulse compression. These effects, cataloged in columns 3 and 4 of Table 9.4, are due to the filter not being matched to the received waveform. For example, reducing the side lobes to a level of −42.8 dB with the Hamming weighting results in a loss in peak SNR (LPG) of 1.34 dB and a broadening of resolution by 47%.

The nonlinear-FM, constant-amplitude waveform provides a compressed waveform with low time side lobes at the output of its matched filter without the LPG that is incurred with the linear-FM waveform and mismatched filter. The nonlinear variation of frequency with time can be made approximately equivalent to amplitude-weighting the transmitted signal spectrum in terms of side-lobe reduction, yet it allows maintaining the rectangular pulse shape that is desired for efficient transmitter operation. However, this method of weighting also corrupts the doppler invariance characteristics inherent in LFM.

9.5.2 Side-Lobe Suppression for Phase-Modulated Waveforms

Mismatched filtering for side-lobe suppression is often required with biphase-coded waveforms as well. Various weighting functions for this purpose have been studied and suggested [5, p. 489]. The technique discussed here is an optimal method for reducing the integrated side-lobe level of the code response through the compression filter.

Optimal ISL suppression filters of any length may be derived analytically for any discrete phase-coded waveform [18,22]. Given a code of length M and a desired mismatched filter length N, one can derive the expression for the total energy in the side-lobe structure of the output waveform in terms of the unknown coefficients of the optimal filter. Optimization is achieved by setting all the partial derivatives of the expression with respect to the unknown coefficient equal to zero. The result is a system of N linear equations in N unknowns whose solution is the set of optimal coefficients. The filters so derived have the property that they minimize the ISL for a given mismatch length. Of importance is the fact that in so doing, they drive down the PSL rather quickly as well.

Table 9.5 gives the performance of particular ISL-optimized filters derived for the six nontrivial Barker codes. These levels were computed assuming that the filter weights and processor arithmetic were quantized to 5 bits. Note that significant side-lobe reduction is achieved with relatively little loss in processing gain. As an example, the 31-bit filter for the 13-bit code affords approximately 15-dB improvement in PSL and 13-dB improvement in ISL over the matched filter case (Table 9.1), while inducing only a 0.2-dB LPG. Figure 9.11 provides a good visual comparison of the compressed waveforms that result from the filtering of the 13-bit Barker code through its matched filter (Figure 9.11(a)) and through the 31-bit optimal mismatched filter (Figure 9.11(b)). In both cases, the outputs are given in decibels.

As a final comment, we note that the matrix inversion required to derive the requisite ISL-optimized filter coefficients becomes unmanageable for large compression ratios. However, if one forms a large code by combining Barker codes, then an effective side-lobe suppression filter may be formed by combining subcode ISL-optimized filters, as indicated in Section 9.3.2.

9.6 THE RADAR AMBIGUITY FUNCTION

The autocorrelation function of the radar's transmitted waveform represents the time (range) response of the radar receiver's matched filter in the presence of a point target that has zero radial velocity with respect to the radar. In pulse-doppler radar applications, there is almost always an undetermined velocity between the radar and the targets of interest. The mathematical functions called *time-frequency autocorrelation functions,* or *ambiguity functions,* were first introduced by Woodward [23] to allow representation of the time response of a signal processor when a target has significant radial velocity with respect to the radar [14]. The ambiguity function, its basic properties, and specific ambiguity function types are discussed briefly [2] as well.

The output from a filter $v(t)$ at time T_R in response to a transmit waveform $u(t)$ that has been doppler shifted f_d Hz is given by

$$X_{uv}(T_r, f_d) = \int_{-\infty}^{\infty} u(t)\, v^*(t + T_R)\, \exp(j2\pi f_d t)\, dt \qquad (9.12)$$

where:

Table 9.5
ISL-Optimized Barker Systems

Barker Code Length	Filter Length	Filter Taps	PSL, dB	ISL, dB	LPG, dB
3	15	0.029, 0.048, 0.088, 0.143, 0.236, 0.381, 0.618, 1, −0.618, 0.381, −0.236, 0.143, −0.088, 0.048, −0.029	−37.6	−32.5	1.26
4	18	0.082, 0.129, 0.043, −0.198, −0.391, −0.239, 0.360, 1.882, −0.480, 0.262, −0.141, 0.077, −0.042, 0.023, −0.012, 0.007, −0.004	−30.1	−28.2	1.65
5	17	0.078, 0.044, −0.101, −0.264, −0.292, 0.044, 0.716, 1.233, 0.764, −1.233, 0.716, −0.044, −0.292, 0.264, −0.101, −0.044, 0.078	−35.5	−28.9	0.61
7	29	0.073, 0.084, 0.101, 0.107, 0.143, 0.198, 0.268, 0.282, 0.223, 0.187, 0.335, 0.724, 1, 0.676, −0.287, −0.676, 1, −0.724, 0.335, −0.187, 0.223, −0.282, 0.268, −0.198, 0.143, −0.107, 0.101, −0.084, 0.073	−29.9	−21.2	1.25
11	31	0.126, −0.042, 0.188, −0.125, 0.131, −0.208, 0.188, −0.291, 0.126, −0.256, 0.758, 0.417, 0.737, −0.993, −0.684, −1, 0.684, −0.993, −0.737, 0.417, −0.758, −0.256, 0.−0.126, −0.291, 0.188, −0.208, −0.131, −0.125, −0.188, −0.042, −0.126	−27.2	−17.1	0.71
13	31	−0.115, −0.124, −0.048, −0.040, −0.115, −0.093, −0.045, −0.146, 0.025, 0.618, 0.842, 0.680, 0.839, 0.638, −0.812, −1, 0.812, 0.638, −0.839, 0.680, −0.842, 0.618, −0.025, −0.146, 0.045, −0.093, 0.115, −0.040, 0.048, −0.124, 0.115	−33.4	−25.4	0.20

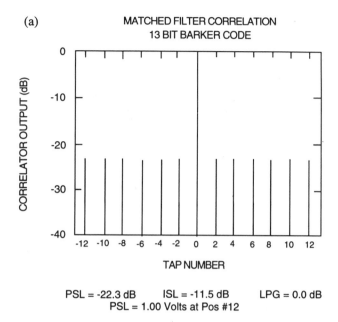

(a) MATCHED FILTER CORRELATION
13 BIT BARKER CODE

PSL = -22.3 dB   ISL = -11.5 dB   LPG = 0.0 dB
PSL = 1.00 Volts at Pos #12

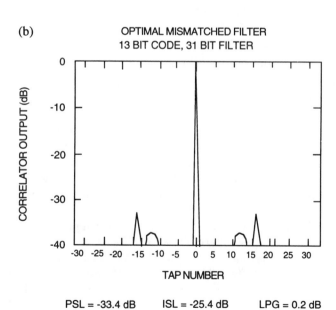

(b) OPTIMAL MISMATCHED FILTER
13 BIT CODE, 31 BIT FILTER

PSL = -33.4 dB   ISL = -25.4 dB   LPG = 0.2 dB
PSL = 0.27 Volts at Pos #16

Figure 9.11 Outputs from 13-bit Barker with (a) matched and (b) mismatched filtering.

v^* = complex conjugate of v.

The peak response occurs when $T_R = f_d = 0$.

If $u \equiv v$ (i.e., the matched-filter case) and we let

$$\Psi(T_R, f_d) = \|X_{uu}(T_R, f_d)\|^2$$

we obtain

$$\Psi(T_R, f_d) = \left\|\int_{-\infty}^{\infty} u(t)\, u^*(t + T_R)\, \exp(j2\pi f_d t)\, dt\right\|^2 \tag{9.13}$$

The function Ψ is called the ambiguity function of u, and its plot in time-frequency-amplitude space is the ambiguity diagram of u. The function $\Psi(T_R, f_d)$ has the following properties:

1. The peak response of Ψ occurs at $T_R = 0$, $f_d = 0$ and, in power, is given by $(2E)^2$, where E = received energy.
2. The diagram is symmetric about $T_R = -f_d$.
3. When there is no doppler, the ambiguity function is the autocorrelation function of the transmit waveform $u(t)$.
4. $|\Psi(0, f_d)|^2 = |\int u^2(t)\, \exp(j2\pi f_d t)\, dt|^2$; that is, the frequency profile at $T_R = 0$ is proportional to the spectrum of $u^2(t)$.
5. The volume under the entire ambiguity surface is independent of $u(t)$ and is equal to $(2E)^2$.

In terms of ramifications for radar waveform design, from property 5 above one, can deduce perhaps the single most important lesson to be learned from ambiguity function analysis: the total energy in the response over the entire frequency-range plane of a returned signal of fixed energy is likewise fixed; thus, attempts to reduce the energy content in any particular region must lead to increased energy content in other regions.

Transmit waveforms with different ambiguity diagrams are desirable for different applications. For example, if precise measurements of both range and doppler of an echo source are required, then the ideal waveform would have an ambiguity diagram consisting of a single peak at the origin sufficiently thin in both dimensions to achieve the desired resolutions. An idealized ambiguity function of this sort, generally called a *thumbtack ambiguity*, is depicted in Figure 9.12(a) (adapted from [2, p. 413]). This diagram represents the general response of a biphase-coded 13-μs pulse, whose bandwidth (chipping rate) is 1 MHz. Note that the maximum value of Ψ is $(2E)^2$ and that since the total volume under Ψ must be $(2E)^2$, the thinner the peak (i.e., the more stringent the resolution requirements), the more the energy must be spread throughout the remainder of the response.

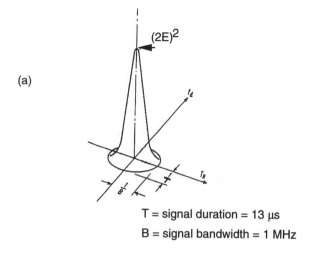

(a)

T = signal duration = 13 μs

B = signal bandwidth = 1 MHz

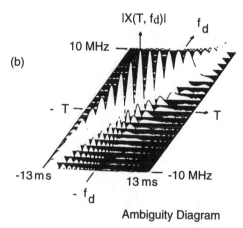

(b)

Ambiguity Diagram

Figure 9.12 (a) Thumbtack ambiguity function and (b) an ambiguity diagram for a linear FM pulse of 13-μs duration and 1-MHz bandwidth.

The result of requiring both fine range and velocity resolutions is increased processor requirements. Generally, one doppler filter must be implemented for each of the ambiguity functions (with, say, 3-dB overlaps) required to cover the entire doppler range of interest, and for each doppler filter, sufficient memory is required to store the values for each range bin of interest (and this number is directly proportional to the range coverage required and inversely proportional to the range resolution required). These numbers can be quite formidable for the coverages often required of pulse-doppler radars.

For certain applications, it may in fact be desirable to accept some ambiguity in target location in range and doppler in order to reduce processor complexity.

An example of one such waveform is given in Figure 9.12(b) adapted from [24, p. 134]. Here, the "knife edge" surface slanting across the range-doppler plane represents the ambiguity diagram of a linear FM pulse of 13-μs duration and 1-MHz bandwidth. Note that the responses from targets separated by 1 μs in range and 76.9 kHz in doppler will be almost identical, and thus they will be unresolvable; however, full range and velocity coverage can be achieved with relatively few doppler channels per range bin.

Figure 9.13 represents the ambiguity diagram of a 13-bit Barker code. The abscissa represents range (time) in range bins; the ordinate represents per-pulse-doppler shift, sampled every 45 degrees, and the height represents amplitude. Note that at 0 doppler, the code compresses perfectly through its matched filter to give the classic 13-bit Barker compression. Since the received code is more seriously mismatched due to imposed doppler, the compression gets less effective. Figure 9.14 gives the compression for a received pulse with a 180-degree phase shift over the pulse. At this point, the target in range gate 13 is no longer visible, and, instead, large range side lobes, which could easily cause false detections, have appeared at range gates 8, 17, and 18.

Figure 9.15 represents the ambiguity diagram of a linear FM waveform assumed to have pulse length and bandwidth equal to those of the Barker code exhibited

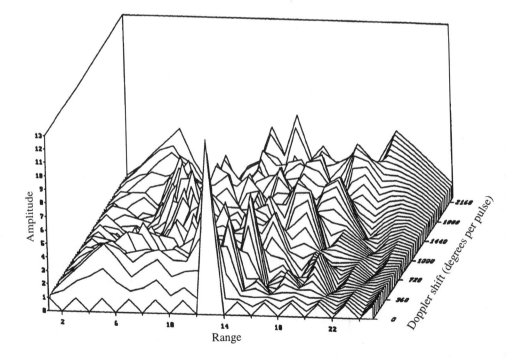

Figure 9.13 13-bit Barker ambiguity diagram.

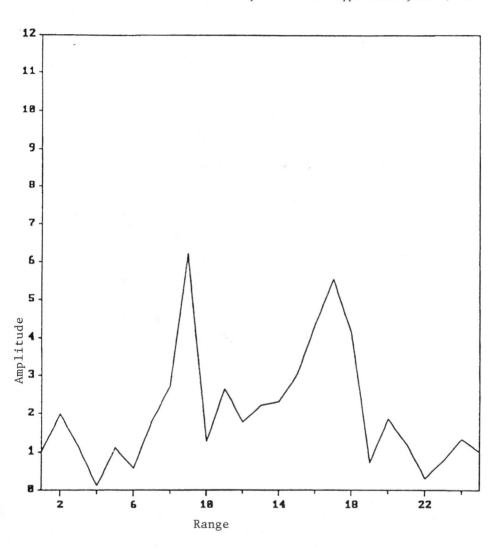

Figure 9.14 13-bit Barker ambiguity diagram 180-degree phase shift over pulse.

in Figure 9.13. Again, compression is perfect for those returns with 0 doppler shift; however, in this case, the compression returns to near perfect levels periodically with increasing doppler. The principal difference between the compression at 0 doppler and that at 720 degrees doppler is the shifting of the peak one full range bin, as can be seen in Figure 9.16, which represents the compression at a 720-degree shift over the pulse. As an aside, we note that the compressed waveform at 0 doppler shift does not appear to be a $\sin(x)/x$ function. This is due to the small (13:1) pulse compression ratio (time-bandwidth product) being modeled.

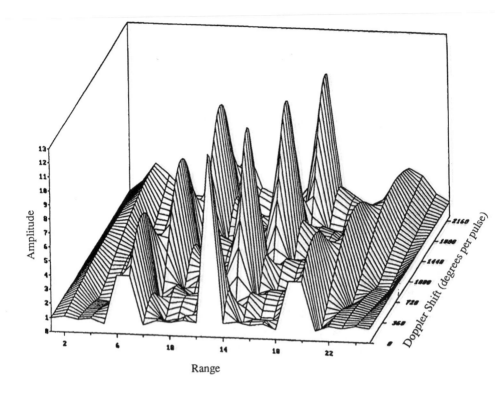

Figure 9.15 Linear FM ambiguity diagram.

The preceding two waterfall plots provide a reasonable first order analysis of compression as a function of code type (Figure 9.13 for biphase codes and Figure 9.15 for linear FM codes) and total phase shift per pulse. Figure 9.17 gives the relationship between radial velocity and degrees of (doppler) phase shift per pulse for a 9.5-GHz, 13-μs pulse. Since this shift is directly proportional to carrier frequency, pulse length, and radial velocity, this figure permits extrapolation to scenarios dealing with any carrier frequency, any pulse length, and all radial velocities.

Thus, there is no single ideal waveform for all situations, but particular waveforms are well suited for particular applications. With this in mind, we present a discussion of several commonly used waveforms and compare them in terms of their ambiguity functions.

The waveforms selected for comparison are:

1. A simple 13-μs pulse;
2. A linear FM pulse with 1-MHz sweep and 13-μs duration;
3. A 13-bit Barker phase code pulse of 13-μs duration.

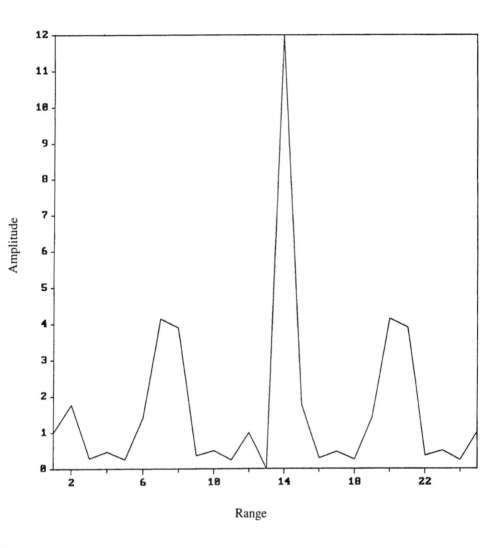

Figure 9.16 Linear FM ambiguity diagram 720-degree phase shift over pulse.

Figure 9.18 compares the ambiguity functions of each of these three waveforms by superimposing the 3-dB contours of the ambiguity functions for each waveform. Note that all three waveforms have the same doppler resolution near $T_R = 0$ because all three waveforms have the same basic time envelope—a 13-μs rectangular pulse. The doppler resolution is equal to the reciprocal of this rectangular pulse envelope, which, for this example, is 76.9 kHz. The time or range resolution is directly proportional to the transmitted bandwidth. The 1-MHz bandwidth of the 13-bit Barker and linear FM coded pulses provide range resolution of approximately

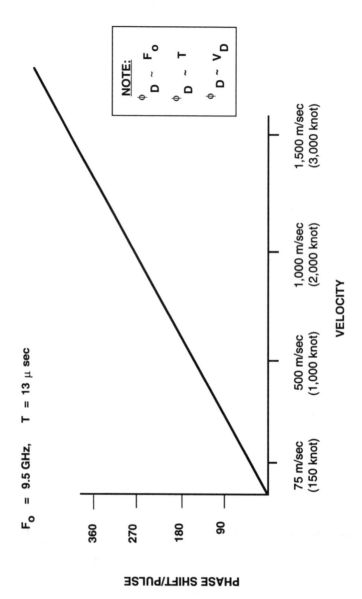

Figure 9.17 Phase shift per pulse versus velocity.

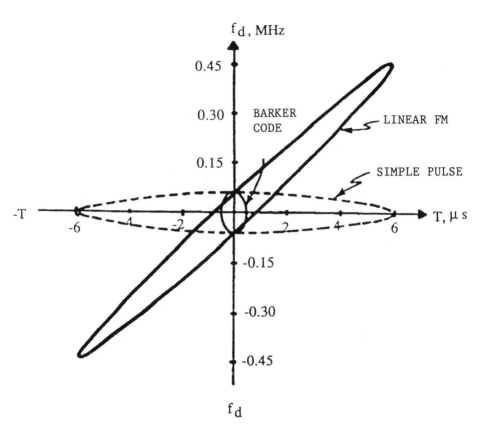

Figure 9.18 Ambiguity function 3-dB contours for simple pulse, linear FM, and 13-bit Barker code. (Adapted from [24, p.136])

150m; the 76.9-kHz bandwidth of the simple pulse provides range resolution of approximately 1,950m. The FM pulse experiences a doppler-range cross coupling, as evidenced by the skewing of its 3-dB contour. For the actual value of range or doppler to be determined, the other would have to be known a priori, or at least two waveforms with different chirps would have to be used. Thus, three quite different response functions can be produced with three waveforms having the same time duration and amplitude.

9.7 SUMMARY

In general applications, radar waveforms may be modulated in phase or frequency to increase the bandwidth of the transmitted pulse. This enhanced bandwidth may then be used by matched (or mismatched) filtering on receive to increase the range resolution of the radar system. This general technique, called *pulse compression*, is often used in modern radar systems to concomitantly maintain a required range resolution while increasing pulse energy and resulting average power. Each technique for effecting pulse compression has a unique set of characteristics depending on both the underlying theory and the state of the art of the technology involved. Waveform choice and design thus continue to pose challenging problems to the radar system engineer. Implementation of pulse compression in a radar system requires a stable signal source for generating the transmitted carrier, a method for coding the transmitted carrier, and implementation of the appropriate signal processing in the receiver to compress backscattered radar signals. All of these factors translate into increased radar system complexity, cost, and chance for failure. The paragraphs that follow summarize some of the most salient properties of simple pulsed, frequency-modulated, and phase-coded waveforms with respect to their utility and use in radar systems.

Simple pulse radars are the least complex, least expensive, and easiest to maintain of the radar systems. Simple pulse waveforms are used in older generation radars with magnetron transmitters, in radars where cost is a primary constraint, and in systems where range resolution and detection requirements can both be met using a simple pulse.

Of all pulse compression techniques, linear frequency modulation is the oldest and most well developed. It is used to improve detection performance while maintaining range resolution. High-time-bandwidth products and fine range resolution are possible. It is particularly useful for detection of moving target, since it can provide broad doppler coverage even with long-duration transmit waveforms. A simple linear FM waveform results in a range-doppler ambiguity function that does not allow for precise determination of either unless the other is in extracted some other way.

Phase coding techniques may also be employed to improve detection performance (average power) with no impact on range resolution. Biphase coding of the carrier is the best developed of these techniques. Extremely high pulse compression ratios can be achieved, though resolution for system functions such as target recognition is limited compared with frequency modulation techniques because of instantaneous bandwidth limitations in current digital technology. The discrete nature of the coding makes these waveforms easy to generate and particularly flexible. They may be implemented with a bank of doppler filters to give precise range and doppler information and can be used in conjunction with MTI to improve subclutter visibility.

In practice, modern pulse-doppler radar systems tend to use pulse compression in both the low- and medium-PRF modes. Duty cycle constraints and the high average powers developed obviate the use of pulse compression in high-PRF mode.

Typically, biphase coded waveforms are used in conjunction with doppler corrections and doppler filter banks to provide range-doppler maps of all detections. For example, the F-15 AN/APG-63 radar uses a 13-bit Barker-coded transmit waveform in its medium-PRF mode for enhanced detection. The incoming I and Q data are doppler-compensated before compression to keep the incoming signals within the proper ambiguity region for good compression. Thereafter, a full set of range-doppler bins are generated to provide an accurate display of target positions and relative speeds, and blind zones are minimized by the use of multiple PRFs [25].

As we have seen in Section 9.4.2, the received waveforms may be processed either with a matched filter to optimize SNR or with mismatched filtering techniques to optimize side-lobe suppression. Furthermore, by using combined Barker, maximal-length PN, or near-perfect codes and mismatched filtering on receive, we could develop higher pulse compression ratios, and therefore achieve greater detection ranges as long as we respect the limits of the system duty cycle and blind zone constraints.

References

[1] Cook, C. E., and M. Bernfield, *Radar Signals*, New York: Academic Press, 1967.

[2] Skolnik, M. I., *Introduction to Radar Systems*, 2nd edition, New York: McGraw-Hill, 1980.

[3] Rihaczek, A. W., *Principles of High-Resolution Radar*, New York: McGraw-Hill, 1969.

[4] Klauder, J. R., A. C. Price, S. Darlington, and W. J. Albersheim, "The Theory and Design of Chirp Radars," *BSTJ*, Vol. 39, No. 4, July 1960.

[5] Nathanson, F. E., J. P. Riley, and M. N. Cohen, Chapter 13 in *Radar Design and Principles*, 2nd edition, New York: McGraw-Hill, 1991.

[6] Cook, E., "Pulse Compression, Key to More Efficient Radar Transmission," *Proc. IRE*, Vol. 48, No. 3, March 1960, p. 310.

[7] Caputi, W. J., "Stretch: A Time Transformation Technique," *IEEE Trans.*, Vol. AES-7, No. 2, March 1971, pp. 269–278.

[8] Stimson, G. W., Chaps. 14 and 27 in *Introduction to Airborne Radar*, Hughes Aircraft Company, El Segundo, CA, 1983.

[9] Curlander, J. C., and R. N. McDonough, Chaps. 9 and 10 in *Synthetic Aperture Radar Systems and Signal Processing*, New York: John Wiley & Sons, 1991.

[10] Wehner, D. R., Chap. 5 in *High Resolution Radar*, 2nd edition, Boston: Artech House, 1995.

[11] Frank, R. L., "Polyphase Codes With Good Nonperiodic Correlation Properties," *IEEE Trans. Inf. Theory*, January 1963.

[12] Welti, G. R., "Quaternary Codes for Pulsed Radar," *IRE Trans. Inf. Theory*, June 1960.

[13] Lewis, B. L., F. F. Kretschmer, Jr., and W. W. Shelton, *Aspects of Radar Signal Processing*, Norwood, MA: Artech House, 1986.

[14] Eaves, J. L., and E. K. Reedy, Chap. 15 in *Principles of Modern Radar*, New York: Van Nostrand Reinhold Company, 1987.

[15] Turyn, R., "Optimal Codes Study," Sylvania Electronics Systems Final Technical Report, Air Research Development Command Contract No. AF19(604)-5473, January 1960.

[16] Turyn, R., "On Barker Codes of Even Length," *PROC. IEEE*, Vol. 51, September 1963, p. 1256.

[17] Cohen, M. N., and P. E. Cohen, "Near-Perfect Codes and Optimal Filtering for Suppression of Their Range Side Lobes," *Proc. 18th European Microwave Conference*, Stockholm, Sweden, September 1988.

[18] Cohen, M. N., "Waveform Design Study," Final Technical Report for Standard Elektrik Lorenz, Mannheim, West Germany, January 1989.

[19] Dixon, R. C., *Spread Spectrum Systems*, New York: John Wiley & Sons, 1976.

[20] Peterson, E. W., and E. J. Weldon, Jr., Appendix C in *Error Correcting Codes*, 2nd edition, Cambridge, MA: MIT Press, 1972.

[21] Golay, M. J. E., "Complementary Series," *IRE Trans. Inf. Theory*, April 1961.

[22] Ackroyd, M. H., and F. Ghani, "Optimum Mismatched Filters for Side-Lobe Suppression," *IEEE Trans. Aerospace and Electronic Systems*, Vol. AES-9, March 1973, pp. 214–310.

[23] Woodward, P. M., *Probability and Information Theory, With Applications to Radar*, New York: McGraw-Hill, 1953.

[24] Brookner, E., *Radar Technology*, Norwood, MA: Artech House, 1982.

[25] "PSP Computer Program Development Specification," Hughes Aircraft Company Technical Report, Document No. DS31325-147, Volume 1, Revision B, October 1985.

CHAPTER 10

Synthetic Aperture Processing

Mark A. Richards

High-resolution ground mapping has become an important mode for airborne fire control radars. Examples in which finely detailed images are needed include terrain-matching navigation, surveillance mapping, and ground target classification and identification. The map resolution required for airborne radar applications typically ranges from 35m down to as little as 1m or 2m [1], sometimes less. Range resolution this fine is relatively easy to achieve using pulse compression. Comparable cross-range[1] resolution, however, is not possible in conventional operation. For example, a radar operating at 10 GHz with a 1m antenna has a beamwidth of about 30 mrad, giving cross-range resolution of about 300m at 10 km. Furthermore, unlike range resolution, cross-range resolution degrades with range.

SAR operation provides a solution to the problem of attaining good cross-range resolution. SAR techniques achieve constant cross-range resolution commensurate with range resolution by using sophisticated signal processing to exploit the aircraft's motion relative to the scene being imaged. Figure 10.1 is an example of the quality obtainable in SAR imagery by the early 1990s. Collected by the Hughes Aircraft Company's ASARS-2 radar, this image obtains a resolution of 6m at ranges of tens of kilometers.

The cost of SAR lies in the complex signal processor required and the associated demands on the precision of other system parameters. In some cases, SAR imposes some operational limitations as well. As we shall see, the signal processing loads can be traded against resolution requirements to create a menu of SAR modes, from unfocused doppler beam sharpening (DBS) to spotlight SAR.

---

[1]By *cross-range* resolution we mean resolution measured in the ground (Earth) plane, in a direction orthogonal to the ground range direction. Cross range and ground range give the dimensions of an SAR pixel. Cross-range resolution is not the same as along-track resolution. This difference and our allowance for nonlevel flight account for additional geometric factors in the equations in this chapter, which are not found in some other texts.

Figure 10.1 ASARS-2 image of a portion of the Los Angeles area. Marina del Rey and the runways of Los Angeles International airport are clearly visible. The image resolution is 6m. (Photograph courtesy of Hughes Aircraft Company. Used with permission.)

In this chapter we describe SAR basics. We will consider the focused strip-mapping and spotlight modes in the greatest depth, but will also touch on the lower resolution unfocused and DBS modes. Interestingly, spotlight SAR is a radar equivalent of computerized tomography, or "CAT scanning." Historically, SAR processing was first done optically. It is now done almost exclusively digitally, and we will consider only digital processing. Summaries of optical SAR processing are available in [2–4].

10.1 THE SYNTHETIC APERTURE CONCEPT

To understand the need for SAR processing and the fundamental concept of its operation, we must first consider the cross-range resolution of a real-aperture radar and consider typical values for an airborne multimode radar system. The 3-dB beamwidth of a conventional antenna of dimension D operating at a wavelength λ is nominally

$$\theta = \frac{\lambda}{D} \tag{10.1}$$

Aperture weighting for side-lobe control may increase this somewhat, but the functional form remains the same. The corresponding cross-range resolution at a distance R_0 is

$$\delta_{cr} = 2R_0 \tan\left(\frac{\theta}{2}\right) \cong R_0\theta = \frac{R_0\lambda}{D} \tag{10.2}$$

The approximation is valid to within 1% for θ less than 0.34 rad, just under 20 degrees. We see that cross-range resolution is proportional to range. Equation (10.2) indicates that, for a given operating frequency, we must increase the aperture size D to improve cross-range resolution. When improvements of one or two orders of magnitude are required to match range resolution, the antenna becomes unrealistically large. For example, at 10 GHz a resolution of 5m at 10 km requires an antenna 60m wide.

A 60m antenna is too big to carry, but for the moment assume we have it anyway. Furthermore, assume it takes the form of a linear phased array. Instead of providing all of the elements needed to populate a large array (and having to carry it around), a SAR multiplexes a single element by moving it from one array element location to the next in sequence. Thus, SAR samples the desired large aperture in both space and time using the relatively small real aperture as a single array "element." Figure 10.2 illustrates this basic SAR concept. A narrow effective beam can be formed by correctly phasing and combining the successively collected returns from a set of wide-beam measurements, just as with any phased array.

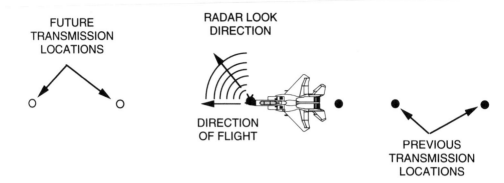

Figure 10.2 The synthetic-aperture concept: the effect of a large array is obtained by moving the real antenna from one element location to another, transmitting at each.

The SAR can collect echoes from a particular scatterer so long as it remains in the field of view of the real aperture. The maximum effective aperture size, D_{SAR}, can be computed with the aid of the simplified geometry in Figure 10.3. A point scatterer P at nominal range R_0 is illuminated by a real-aperture beam with 3-dB beamwidth θ at a squint angle ψ from forward looking. (We will generalize to three-dimensional geometry in Section 10.2.) The distance along the line a-a' within the real beam is the distance the aircraft will travel while P is illuminated. The maximum effective aperture size is this length projected orthogonal to the boresight (i.e., along the line a'-a''). For a pencil beam antenna with $\theta < 20$ degrees, this is approximately [5]

$$D_{SAR} \cong R_0 \theta \qquad (10.3)$$

Although (10.3) gives the largest possible synthetic aperture size, there is a subtlety that must be accounted for before the resolution can be computed. In a

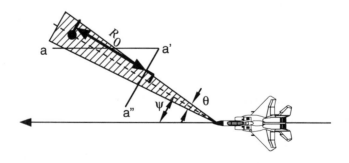

Figure 10.3 Basic geometry of squint mode SAR.

physical array, all elements transmit at the same time. The target is illuminated with and reradiates a single composite field having some particular phase. Because the path length back to each element varies, there is a relative phase shift between the echoes received at different elements, and the spread in the echo phases limits the angle over which the responses add essentially in phase. This phenomenon is the reason the array is directive (i.e., has a narrow gain pattern). In the SAR system, transmission and reception are carried out independently at each element location. Consequently, the target reradiates different phases to each element corresponding to the different path lengths traveled by the incoming pulses. The result is that all relative phase shifts between elements when the echo is received are doubled, giving the effect of a physical array of twice the size of the synthetic aperture [1]. Thus, in SAR the 3-dB beamwidth becomes

$$\theta_{SAR} = \frac{\lambda}{2D_{SAR}} = \frac{\lambda}{2R_0\theta} \tag{10.4}$$

Using (10.1) for the real-aperture beamwidth θ and substituting the resulting synthetic-antenna beamwidth θ_{SAR} into (10.2) then gives the ideal SAR cross-range resolution δ_{cr} as

$$\delta_{cr} = \frac{D}{2} \tag{10.5}$$

There are several remarkable implications of (10.5):

1. The finest achievable cross-range resolution is a constant depending only on the real aperture size. It does not depend on the squint angle.
2. δ_{cr} does not vary with range, because the effective synthetic-aperture size increases with range, as shown in (10.3).
3. δ_{cr} does not depend on wavelength.
4. δ_{cr} improves as the real aperture gets smaller, since the wider real beamwidth keeps each scatterer in view longer.

These classic properties show why SAR is ideal for ground mapping.

10.2 FOCUSED FORWARD-SQUINTED STRIP-MAPPING SAR CHARACTERISTICS

In forward-squinted strip mapping, the real antenna beam stares ahead and to the side of the aircraft. The aircraft is usually flying at approximately constant velocity during SAR operation, and is usually, though not necessarily, level. Consequently, a strip of terrain is illuminated for SAR mapping. SAR processing is carried out in real time, with the image being displayed on a cockpit monitor.

10.2.1 Geometry

The squint mode geometry is shown in more detail in Figure 10.4. The aircraft flies at velocity v and a descent angle ξ from horizontal. The boresighted point scatterer P in the y-z (earth) plane is at a depression angle δ from horizontal and squint angle ψ from forward-looking. R_0 is the nominal range to P at time $t_0 = 0$, when P is on boresight. An important derived parameter is the total cant angle ϕ between the aircraft velocity vector and the boresight vector, which is related to ξ, δ, and ψ according to [5]

$$\cos \phi = \sin \xi \sin \delta + \cos \xi \cos \delta \cos \psi \qquad (10.6)$$

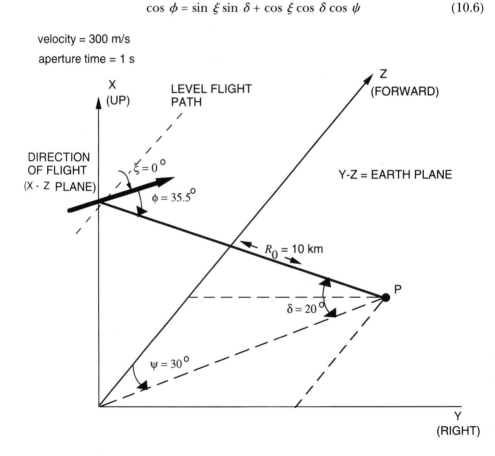

Figure 10.4 Geometry of squint mode SAR for nonlevel flight.

For forward-squinted radars, $\cos \phi$ is always positive.

The illuminated area at any given instant is defined by the projection of the real-antenna 3-dB pattern onto the terrain. For a flat surface and a pencil beam antenna, this is the ellipse with down-range and cross-range axes of approximately $R_0 \theta / \sin \delta$ and $R_0 \theta$, respectively. Of course, this illumination ellipse will normally be divided into range cells $c\tau_c / (2 \cos \delta)$ meters, long where τ_c is the compressed pulse length. If the center of this pattern passes over the scatterer P, it will be illuminated for [5]

$$T_{max} = \frac{R_0 \theta}{v \gamma} \text{ seconds} \tag{10.7}$$

where γ is the geometric factor

$$\gamma = \cos \xi \sqrt{\sin^2 \psi + \cos^2 \psi \sin^2 \delta} \tag{10.8}$$

T_{max} is therefore the maximum aperture time. In airborne systems, T_{max} is typically on the order of 1 to 10 sec, but it can be significantly greater or smaller.

The slant range from the aircraft to P can easily be computed from the coordinates of each. So long as $vT_{max}/R_0 \ll 1$, the result is approximately [5]

$$R(t) \cong R_0 \left[1 - \cos\phi\left(\frac{vt}{R_0}\right) + \frac{1}{2}\left(\frac{vt}{R_0}\right)^2 \sin^2\phi \right] \tag{10.9}$$

Thus, the range varies quadratically with time.

10.2.2 Resolution vs. Aperture Time

The cross-range resolution achievable according to (10.5) is often better than required. In this case, a smaller synthetic aperture can be used by simply processing data over a shorter collection time than the maximum of $R_0\theta/v\gamma$. The data collection time is called the *aperture time* and is denoted T_a. The effective aperture size is the distance flown in the ground plane (*y-z* plane in Figure 10.3) and orthogonal to the nominal line of sight during the aperture time, which is $vT_a \cos \xi \sin \psi$. Using this value for D_{SAR} in (10.2) and (10.4) gives a more general expression for cross-range resolution [5]:

$$\delta_{cr} = \frac{\lambda R_0}{2vT_a \cos \xi \sin\psi} \tag{10.10}$$

Equation (10.10) shows that resolution improves as the amount of data processed (i.e., the aperture time) increases. Furthermore, obtaining cross-range resolution that is independent of range requires that we increase the aperture time linearly with range; that is, more data must be processed to resolve more distant targets to the same degree as near-in targets.

Equation (10.10) also shows that, for a given range and aperture time, the squint-mapping SAR cross-range ground resolution degrades as the flight path departs from level or the boresight vector approaches the velocity vector (i.e., forward-looking). This effect can be compensated for by increasing the aperture time T_a, but there are practical limits to that approach. Consequently, airborne SAR mapping is usually restricted to squint angles of 10 degrees or more. If we set $T_a = T_{max}$ in (10.10), we obtain the ideal lower bound on cross-range resolution for the squint SAR. It is

$$\delta_{cr} = \frac{D}{2}\sqrt{1 + \text{ctn}^2 \psi \sin^2 \delta} \qquad (10.11)$$

In conventional side-looking geometry ($\psi = 90$ degrees, $\delta = \xi = 0$ degrees), (10.11) reduces to the classical result $\delta_{cr} = D/2$ found by basic arguments in Section 10.1.

10.2.3 Point Scatterer Signature

Up to the point where the complex image is finally passed through a detector to convert it to pixel data suitable for display, the SAR signal processing system is nominally linear. Consequently, we can understand the behavior of the SAR system, including such characteristics as resolution, side lobes, and the response to complex multiscatterer scenes, by considering the response to a single isolated point scatterer. In the nomenclature of linear systems analysis, the point scatterer response is the SAR system *impulse response.*

For any pulse transmitted when the isolated point scatterer is not within the radar's field of view, that is, the main beam of the physical aperture, we will assume the echo is effectively zero, since it is greatly attenuated by being both transmitted and received through the antenna side lobes. For any pulse transmitted when the point scatterer is within the field of view, the received signal will be a replica of the transmitted signal, delayed by $2R(t)/c$, where $R(t)$ is the range to the scatterer as a function of time and was given in (10.9). Assume a pulse of bandwidth B Hz and duration T sec. The Nyquist sampling rate in fast time is then B samples/s, so the echo from the point scatterer will occupy BT consecutive samples in the fast time dimension for that particular pulse. These samples will extend from a delay of $2R(t)/c$ to $2R(t)/c + T$.

The quadratic variation of range with time means that echoes from successive pulses occur at different ranges, a phenomenon known as range migration. Range migration consists of two components, range walk and range curvature. Range walk is the variation in the middle term (linear in t) of (10.9) over the aperture time, while range curvature is the variation in the third (quadratic in t) term.

In this squinted mode, the maximum and minimum values of the linear term of (10.9) occur at $t = +T_a/2$ and $t = -T_a/2$, respectively. Their difference is defined as the range walk:

$$\Delta R_w = vT_a \cos \phi \qquad (10.12)$$

Range walk is significant if the range resolution δ_r is less than ΔR_w. Figure 10.5 illustrates range walk by showing a notional range/cross-range signature of the point scatterer both before and after pulse compression. The two-dimensional matrix is the same slow time/fast time matrix of radar echo samples described in Chapter 8 and Figure 8.2. Each range line for fixed cross-range corresponds to the returns from a single PRI; successive cross-range positions are the returns from successive PRIs. As an example, if $\psi = 30$ degrees, $\delta = 20$ degrees, $\xi = 0$ degrees, $v = 300$ m/s, and $T_a = 1$ sec, then $\phi = 35.5$ degrees and range walk will occur if $\delta_r < 244$m, certainly the case in any high-resolution mode. Thus, for radars squinted forward significantly, a frequent case with airborne fire control radars, range walk can be quite large and the echo from a fixed scatterer can move through many range bins as the scatterer moves through the beam. Also, note that range walk is the same for all scatterers; that is, it is independent of range, as shown by (10.12).

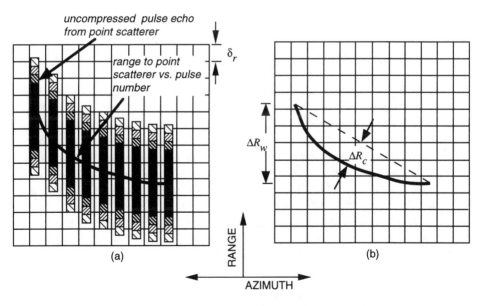

Figure 10.5 Range/cross-range response to a point scatterer, showing range curvature and range walk: (a) before pulse compression and (b) after pulse compression.

Similarly, range curvature is the difference between the minimum and maximum values of the quadratic term of (10.9) over the aperture time, which occur at $t = 0$ and $t = \pm T_a/2$; this is

$$\Delta R_c = \frac{(v T_a \sin \phi)^2}{8 R_0} \tag{10.13}$$

For the same parameters and a range of 10 km, ΔR_c is 0.38m, probably less than a resolution cell. Range curvature is often modest (a few range bins) and sometimes insignificant in airborne squint mapping, in contrast to the large range walk. Another difference is that range curvature is not independent of range. Rather, it varies as R_0^{-1} for fixed aperture time,[2] so that scatterers at short ranges exhibit greater range curvature than scatterers at longer ranges.

Note that both range walk and range curvature can be computed using only parameters that are known to the radar system: own-ship velocity and attitude, aperture time, nominal range, and antenna pointing angles. Thus, it is possible for the aircraft data processor to estimate the range migration parameters. The extent of the signature in range before pulse (range) compression will be BT resolution cells, where BT is the waveform time-bandwidth product. It is also very important to note that the maximum cross-range spread of the signature varies with range, a phenomenon also shown in exaggerated fashion in Figure 10.6. Consequently, the point scatterer signature is not shift-invariant.

10.2.4 Frequency Domain Interpretation of Linear FM—Deramp SAR Data

If a linear FM waveform is used, then a special demodulation technique called *deramp processing* can be applied to obtain a direct measurement of the Fourier transform of the scene reflectivity in the range dimension [6–8].[3] Consider a linear FM pulse with center frequency f_0, bandwidth B Hz, and duration τ sec. Assume that the range extent of the region being imaged is ΔR meters, and can be completely overlapped with the pulse (i.e., $\Delta R < c\tau/2$). In a conventional quadrature demodulator, the received pulse is, in effect, mixed with a complex exponential at the radar center frequency, as shown in Figure 10.7(a). In deramp processing, the pulse is mixed instead with a replica of the transmitted pulse, as shown in Figure 10.7(b). It can be shown [8] that under these circumstances, common in spotlight SAR, the output $g(t)$ of the demodulator is, to a good approximation, equal to the

---

[2]However, if we want the cross-range resolution to be independent of range, then T_a has to be proportional to range. Range curvature is then directly, rather than inversely, proportional to R because of the longer observation times for more distant scatterers.

[3]Deramp processing is essentially identical to so-called stretch processing used in some wideband surveillance radars, and to the processing used in some frequency-modulated continuous wave (FMCW) radars.

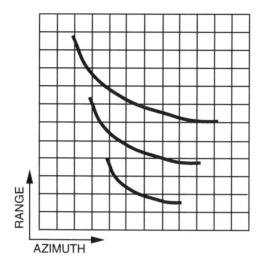

Figure 10.6 Range/cross-range response (exaggerated) to point scatterers at different ranges. In a focused mode where aperture time is proportional to range, both range walk and range curvature increase linearly with range.

Fourier transform of the range profile $i(x)$, which describes the scene reflectivity in the range dimension over the frequency interval $[f_0 - B/2, f_0 + B/2]$.

If we now consider a two-dimensional scene $i(x, y)$, then the demodulator output is in fact a portion of the Fourier transform of the *projection* of $i(x, y)$ onto the radar line of sight, which is at a squint angle ψ with respect to the axes (see Figure 10.4) [7,8]. The projection-slice theorem of Fourier analysis then tells us that the demodulator output is an estimate of slice through the two-dimensional Fourier transform over the range $[f_0 - B/2, f_0 + B/2]$ at an angle in Fourier space equal to the radar squint angle ψ [9]. Thus, for the geometry shown in Figure 10.4, using a linear FM waveform and deramp processing results in a data array that represents samples of the Fourier transform $I(\omega_x, \omega_y)$ of $i(x, y)$ on the polar grid shown in Figure 10.8. The clear implication is that an estimate of the reflectivity $i(x, y)$, that is, an image of the scene, could be obtained by computing the inverse Fourier transform of the collected data.

10.2.5 Doppler Bandwidth

Equation (10.9) showed that the range to a given scatterer varies quadratically with time. Since the phase φ of the echo from a scatterer is just $4\pi/\lambda$ times the range, it follows that the phase also varies quadratically. Thus, if we consider the slow time series of pulse returns from a single scatterer, the relative motion between the radar platform and the scatterer induces a linear FM chirp in the returned signal.

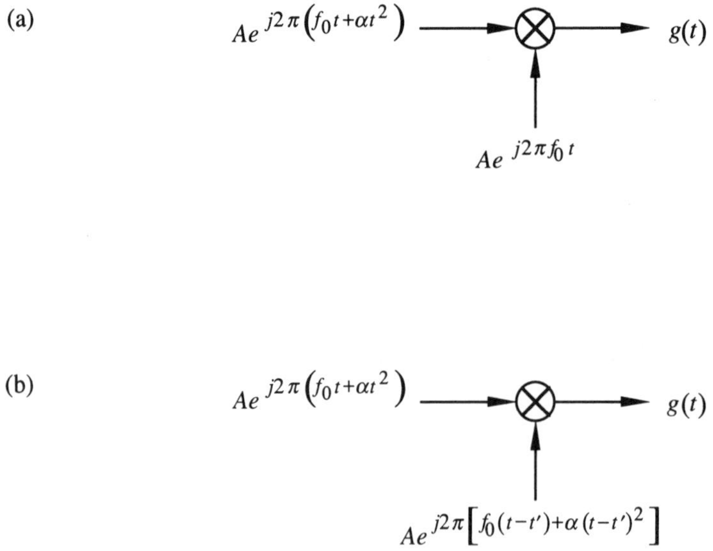

(a)

$$Ae^{j2\pi\left(f_0 t + \alpha t^2\right)} \longrightarrow \otimes \longrightarrow g(t)$$

$$Ae^{j2\pi f_0 t}$$

(b)

$$Ae^{j2\pi\left(f_0 t + \alpha t^2\right)} \longrightarrow \otimes \longrightarrow g(t)$$

$$Ae^{j2\pi\left[f_0(t-t') + \alpha(t-t')^2\right]}$$

Figure 10.7 Complex representations of conventional and deramp demodulators: (a) conventional quadrature (I/Q) demodulator using complex sinusoid as the reference function and (b) deramp demodulator using a replica of the transmitted linear FM waveform as the reference function.

The radar's PRI samples this chirp. By considering the relation between aperture time and motion-induced chirp bandwidth, we can determine the signal bandwidth that must be processed in slow time to achieve a given cross-range resolution.

Instantaneous frequency f_D is given by $(-1/2\pi)\ d\varphi/dt$. Applying this definition to (10.9) and evaluating the change in f_D over an aperture time gives the doppler bandwidth of the returns from any given scatterer:

$$W = \frac{2v^2 T_a \sin^2\phi}{\lambda R_0} \text{ Hz} \tag{10.14}$$

We can then combine (10.14) and (10.10) to get an expression for the doppler bandwidth required for a given cross-range resolution:

$$W = \frac{v \sin^2\phi}{\cos \xi \sin \psi} \frac{1}{\delta_{cr}} \cong \frac{v \sin \psi}{\delta_{cr}} \tag{10.15}$$

where the approximation is valid for the usual case of level flight ($\xi = 0$ degrees) and a shallow incidence angle ($\delta \cong 0$ degrees). This result shows that the doppler

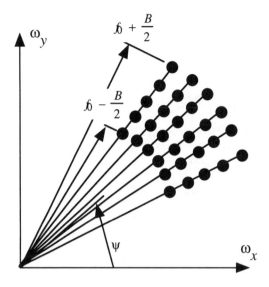

Figure 10.8 Polar format Fourier-domain data collected by spotlight SAR system using linear FM waveform and deramp processing.

bandwidth that must be processed increases as the inverse of the desired cross-range resolution.

The maximum doppler bandwidth occurs when $T_a = T_{max}$ and is

$$W_{max} = \frac{2v\theta \sin^2\phi}{\lambda\gamma} \cong \frac{2v\theta}{\lambda} \sin\phi \qquad (10.16)$$

where again the approximation holds for $\xi = 0$ degrees and $\delta \cong 0$ degrees. Equation (10.16) is the maximum doppler bandwidth that must be processed to obtain the ideal resolution of $D/2$. It is also the difference in absolute doppler shift for scatterers at the leading and trailing edges of the radar beam. For the parameters used in the example of Section 10.2.3, this is about 1,162 Hz.

These results suggest that the echoes from scatterers at the same range but slightly different aspect angles (though still within the main beam) can be separated based on their slightly different doppler shifts. This idea can be followed to again obtain (10.5); see, for example, [10].

10.2.6 Signal-to-Noise Ratio and Pulse Repetition Frequency

We now consider the impact of synthetic-aperture operation on the radar system SNR (S/N) and PRF. One standard form of the range equation is

$$S/N = \frac{P_p G^2 \lambda^2 \sigma n_i}{(4\pi)^3 R_0^4 kT_0 B F} \qquad (10.17)$$

where the new symbols are:

P_p = peak transmitted power;
G = antenna gain;
n_i = number of coherently integrated pulses;
σ = radar cross section of the imaged object;
k = Boltzmann constant;
B = receiver bandwidth;
T_0 = receiver temperature;
F = receiver noise figure.

Let $P_p = P_{av}/\tau_u PRF$, where P_{av} is the average power and τ_u is the uncompressed pulse length. Assume $B = Q/\tau_u$, where Q is the time-bandwidth product of the waveform. Since SAR is coherent, $n_i = T_a PRF$, and we can in turn use (10.10) to express T_a in terms of the cross-range resolution δ_{cr}. The antenna gain can be expressed in terms of the horizontal and vertical beamwidths θ_h and θ_v as $G \cong 4\pi\eta/\theta_h\theta_v$, where η is the antenna efficiency factor. In turn, we can use (10.1) to write $\theta_h = \lambda/D_h$ and $\theta_v = \lambda/D_v$. The RCS $\sigma = \sigma° \delta_r \delta_{cr}$, where σ degrees is the reflectivity of the imaged patch. These substitutions can be used to reduce (10.17) to

$$S/N = \frac{P_{av}\eta^2 D_h^2 D_v^2 \delta_r \sigma°}{8\pi R_0^3 Q\lambda v k T_0 F \cos \xi \sin \psi} \qquad (10.18)$$

This equation shows that the SNR increases with increasing average power, real-aperture size, or frequency, but decreases as velocity increases or range resolution becomes finer. Also, the SNR decreases in squint modes ($\psi \neq 90$ degrees) compared to side-looking modes ($\psi = 90$ degrees). Note that the SNR does not depend on the cross-range resolution.

Equation (10.18) is for an arbitrary aperture time. If we maximize resolution by maximizing the aperture time, we can use (10.11) to relate δ_{cr} and D_h. For simplicity, we now restrict ourselves to the side-looking case and use instead (10.5). Then $\delta_{cr} = D_h/2$ and (10.18) becomes

$$S/N = \frac{P_{av}\eta^2 D_v^2 \delta_r \delta_{cr}^2 \sigma°}{2\pi R_0^3 Q\lambda v k T_0 F} \qquad (10.19)$$

Equation (10.19) shows that, in the case of maximum aperture time, the observed SNR now does depend on the cross-range resolution, and in fact decreases as its square.

The PRF is limited by the doppler bandwidth and limits the maximum unambiguous range. To satisfy the Nyquist sampling criterion, the PRF must be at least equal to the maximum doppler bandwidth W_{max} of (10.15) [11]. At the same time, the maximum unambiguous slant range is

$$R_{ua} = \frac{c}{2PRF} \qquad (10.20)$$

W_{max} is usually quite low, perhaps a few hundred hertz in forward-squinted operation to a few kilohertz in side-looking operation. Consequently, SAR is a low-PRF mode.

10.3 OTHER SYNTHETIC-APERTURE MODES

10.3.1 Doppler Beam Sharpening and Unfocused SAR

DBS and unfocused SAR are two simplified versions of synthetic-aperture processing that trade reduced computational requirements for reduced (but still better than real-beam) resolution. In unfocused SAR, range walk is corrected but range curvature is not. Consequently, a residual quadratic variation in range, and hence in phase shift, remains when the data are correlated in cross-range. The defocusing caused by this phase variation is controlled by limiting aperture time to a value that keeps the range curvature less than some specified value that depends on the amount of degradation in point target focusing quality that can be tolerated [12]. A typical standard is $\Delta R_c < \lambda/8$, corresponding to an uncorrected SAR two-way phase shift of 90 degrees. The aperture time required to give this curvature can be found from (10.9) or (10.14) to be

$$T_a = \frac{\sqrt{\lambda R_0}}{v \sin \phi} \qquad (10.21)$$

Thus, T_a still increases with range, but not as rapidly as in the focused case. The corresponding unfocused resolution becomes [5]

$$\delta_{cr} = \frac{\sin \phi}{2 \cos \xi \sin \psi} \sqrt{\lambda R_0} \qquad (10.22)$$

which varies as the square root of range.

In focused SAR, the aperture time is proportional to range. In unfocused SAR, it is proportional to the square root of range. DBS is a still coarser form of SAR in which the aperture time is constant for all ranges. In DBS, the antenna usually actively and continuously scans a field of view to one or both sides of the aircraft. T_a is determined by the real-antenna beamwidth and scan rate. Since T_a

is normally less than the unfocused aperture time, processing is unfocused. The cross-range resolution is given by (10.10) with T_a constant. We see that cross-range resolution in a DBS mode is proportional to range, just as in a real-beam radar. However, the resolution will be finer than the real-beam resolution. If we denote the real-beam resolution of (10.2) as δ_{rb} and the resolution in the DBS mode as δ_{DBS}, we can define the DBS *beam sharpening ratio*:

$$BSR = \frac{\delta_{rb}}{\delta_{DBS}} = \frac{2vT_a \cos \xi \sin \psi}{D} \qquad (10.23)$$

Beam sharpening ratios on the order of 10 are common in practice.

Figure 10.9 shows the relative resolution of real aperture radar, DBS radar with a beam sharpening ratio of 10, unfocused SAR, and focused SAR using the same parameters as those described in Section 10.2.2. For that case, the unfocused aperture time becomes about 100 ms, giving a resolution of about 10m at 10 km.

10.3.2 Spotlight SAR

Unfocused SAR and DBS use shorter aperture times and achieve poorer resolution than focused strip-mapping SAR. Aperture time and resolution may also be *increased*

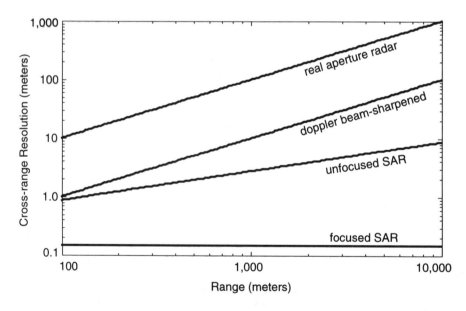

Figure 10.9 Comparison of ideal maximum cross-range resolution for real aperture, DBS with BSR = 10, unfocused SAR, and focused SAR operation. A 0.3m antenna operated at 10 GHz is assumed.

beyond the focused strip-map case. This is done by "spotlighting" the area to be imaged (i.e., scanning the antenna so as to keep it trained on the same ground patch as the aircraft flies by). This technique allows the aperture time to be increased substantially.

The scenario is sketched in Figure 10.10. The spotlight mapper "flies around" the target region. Consequently, it is common to express the resolution in terms of the angle of rotation during integration, shown as β in the simplified geometry of the figure. The resolution is then [6]

$$\delta_{cr} = \frac{\lambda}{2 \sin \beta} \cong \frac{\lambda}{2\beta} \qquad (10.24)$$

with β expressed in radians.[4] The approximation applies when β is small, as is normally the case. The greater performance of spotlighting is gained at the expense of area coverage. Spotlighting also aggravates range migration problems, since a region is illuminated longer than in other modes.

Figure 10.11 is an example of a high-resolution spotlight SAR image. Figure 10.12 compares the type of imagery obtainable from the same radar operating in different modes. The SAR images in these figures were obtained using the

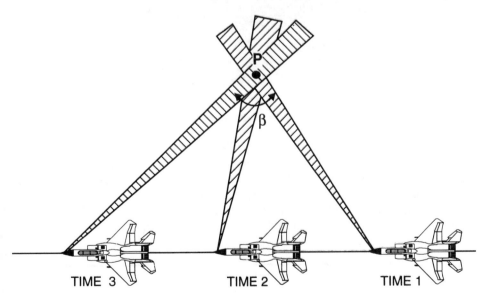

Figure 10.10 Illustration of spotlight SAR mapping, in which the antenna is steered to remain on a target area, increasing the aperture time.

---

[4]This formulation can be applied to focused strip mapping also. In that case, β equals the real-aperture beamwidth $\theta = \lambda/D$, again confirming the ideal resolution of $D/2$.

APG-76 radar developed by Westinghouse Norden Systems. The same scene of oil tanks in a coastal area is shown in all four parts of the figure. The upper left image is a real-beam map. The upper right is a DBS image, the lower right is a 2.5m resolution SAR image, and the lower left is a spotlight SAR image. (The maps are not to the same scale. The DBS map covers the largest area, followed by the real beam, SAR, and spotlight maps.) The improvement in image quality from real beam to DBS, to SAR, and then to spotlight SAR is obvious.

Figure 10.11 Spotlight mode SAR image of the Pentagon. (*Source:* [13]. Reprinted with permission.)

Interestingly, spotlight mapping is recognized as a radar equivalent of the computer-aided tomography (CAT) common in medical diagnosis [14]. The analogy is closest when the aircraft either flies a trajectory that keeps the nominal range to the mapped area constant, or uses motion compensation procedures that achieve an equivalent result (see Section 10.4.2). In this case, polar format spotlight processing [6] can be used. In comparison to CAT, spotlight SAR uses very narrow viewing angle ranges and very narrowband signals.

10.4 IMPLEMENTATION OF FOCUSED SAR

10.4.1 SAR Matched Filtering

The ideal SAR signal processor is a filter matched to the range/cross-range signature of the point scatterer response. The output of this matched filter would have a

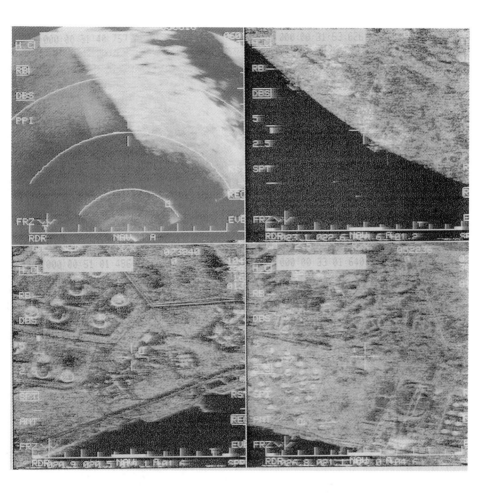

Figure 10.12 Comparison of radar mapping modes. The image is of a cluster of oil storage tanks in a coastal area. Upper left: real-beam image. Upper right: DBS image. Lower right: SAR image with 2.5m resolution. Lower left: spotlight SAR image. (Courtesy of Westinghouse Norden Systems. Used with permission.)

strong peak at the range/cross-range coordinates of the scatterer P. If we denote the sampled return of Figure 10.4 as $s(n, m)$, where n is the range bin and m the cross-range bin, then the desired operation is the two-dimensional correlation

$$i(n, m) = \sum_k \sum_l s(k, l) h^*(k + n, l + m; n, m) \qquad (10.25)$$

where $i(n, m)$ is the estimate of the complex image at the processor output and $h^*(k, l; n, m)$ is the conjugate of the signature of a point scatterer signature located at coordinates (n, m); we shall call h^* the *reference function* or *reference kernel*. Note that in general we have a different reference kernel for each output sample in recognition of the shift-varying echo signature of a point target. Consequently, the signal processor must either maintain a library of reference kernels as a function of range and azimuth position within the image, or must compute the reference kernel on the fly.

Equation (10.25) represents a process of focusing and integrating. Multiplying a complex sample of s by a sample of h^* rotates the echo phase (and weights its amplitude); the summation then integrates the phase-corrected echo samples. When h^* is centered on the true coordinates (n, m) of the scatterer, all of the focused phases will coincide, and the integration will give a maximum response.

10.4.2 Basic Processing Algorithms

Figure 10.5(a) illustrated the usual range and cross-range format for collecting raw SAR data in a two-dimensional memory within a digital processor, and we have seen that a two-dimensional correlation with the reference function h^* is the desired processing. Furthermore, recall that the point scatterer signature, and therefore h^*, is shift-varying. Figure 10.13 shows the overall structure of a generic SAR processor that can perform the required computations. This figure is applicable to arbitrary SAR waveforms; we will consider the special case of spotlight SAR and polar format processing in Section 10.4.7.

In airborne systems, the range profile of a point scatterer is the same at each cross-range position; it is merely shifted closer or farther by range walk and curvature. Consequently, the first step is standard range dimension pulse compression within each PRI. This converts the signature of each scatterer from a form dispersed over BT range cells like that of Figure 10.5(a) to the range-compressed version of Figure 10.5(b).

In general there still remains, however, a two-dimensional correlation, usually called cross-range or azimuth compression, which is the heart of SAR processing. Cross-range compression for airborne digital processors is generally performed in the following three steps: (1) cross-range presum, (2) range walk correction, and (3) cross-range correlation. Recall that range walk may be large, while range curvature will usually be slight.

Cross-range presumming is a technique for reducing the data rates in the SAR processor. It is possible when the maximum doppler bandwidth of (10.15) is greater than needed for the desired resolution as given in (10.16), which is frequently the case. Equivalently, less than the maximum aperture time is required.

Regardless of the doppler bandwidth required to achieve the desired resolution, the radar PRF must be set to at least the maximum doppler bandwidth in order to avoid aliasing the raw data [11]. However, if this is more than required for the desired resolution, but all processing is done at the PRF rate, the processor

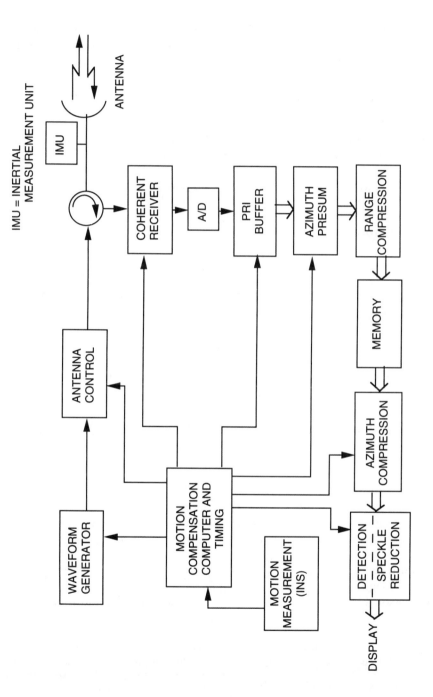

Figure 10.13 Typical SAR system structure (*After:* [6]).

design is stressed needlessly. If less than the full bandwidth is needed, standard digital decimation techniques [15] can be used to lower the sampling rate. These techniques basically involve lowpass filtering of the data in the slow time (cross-range) dimension, followed by discarding excess samples.

Figure 10.14 illustrates the effect of cross-range presumming on the doppler spectrum. The spectrum of the raw data repeats itself at the PRF as shown in Figure 10.14(a). If only, for example, one-quarter of the doppler bandwidth is to be processed, then a digital lowpass filter is applied to the data, giving the spectrum in Figure 10.14(b). The sampling rate can then be reduced to one-quarter of the

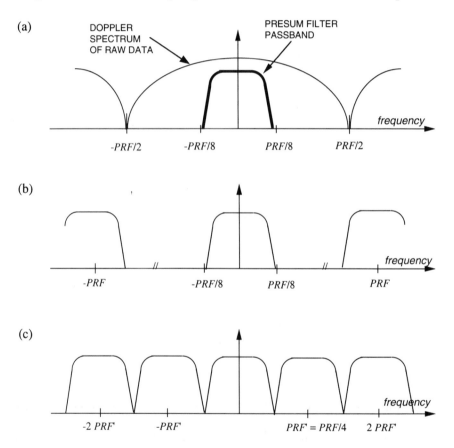

Figure 10.14 Doppler spectra when cross-range presumming is used: (a) doppler spectrum at radar input—only the portion of the full bandwidth within the presum filter passband is needed for resolution, but the PRF must accommodate the full bandwidth; (b) spectrum after cross-range presum filtering; and (c) spectrum after filtering and sampling rate reduction—no information has been lost in the decimation.

PRF by simply discarding three out of every four samples, giving the spectrum of Figure 10.14(c) and reducing the data rates in the rest of the processor by a factor of four. Symmetric FIR filters are usually used, since they can be designed to have exactly linear phase response (crucial for subsequent focusing), are easy to design, and need be operated themselves only at the lowered data rate. The technique can be extended, with some complications, to reducing the data rate by any rational factor [15].

Cross-range presumming can be performed before or after range compression. If performed first, it will reduce the total range compression workload by reducing the number of cross-range lines; however, it will also increase the signal dynamic range required in the pulse compressor. For a digital pulse compressor, the reduction in computation rate at the cost of a longer digital word length may be a desirable trade-off, and cross-range presumming may be done prior to pulse compression. In analog pulse compression (e.g., surface acoustic wave for linear FM waveforms), speed is not an issue, but dynamic range is limited. Consequently, pulse compression will often precede presumming.

In some image formation algorithms, a correlation or filtering operation is done in the cross-range dimension. If presumming, another cross-range filtering operation, is also required, the two can sometimes be combined. In this case, cross-range processing should be completed first so that down-range processing is not performed on extraneous range lines which will be removed by the presumming decimation.

Range walk correction is implemented by using inertial navigation system (INS) and antenna pointing data to compute the range walk at each cross-range position relative to a reference cross-range line. One technique for correcting range walk is then to simply shift the range data for each cross-range position by an appropriate number of range cells to remove the range walk, as shown in Figure 10.15. (Note that quadratic range curvature is still present.) A more sophisticated implementation of this idea uses a digital interpolating filter on each range line to correct for fractional-bin offsets.

Range walk correction by data shifting and interpolation is most appropriate for strip-mapping SAR systems. Another technique applicable primarily to spotlight SARs uses the computed range walk to vary the time delay from the time of pulse transmission to the time when the A/D converter at the receiver output is triggered to begin sampling. The delay is chosen such that the echo from some reference point, generally in the middle of the imaged scene, always occurs in the same range bin. For example, if the aircraft is getting nearer to the target area, the time delay before the A/D converter begins sampling is reduced on each successive pulse. This technique eliminates the computation required for range walk correction at the cost of more complicated receiver control.

After presumming and range walk correction, the SAR data are ready for cross-range correlation. Because the range curvature that remains can sometimes exceed one range bin, cross-range correlation can in fact be a two-dimensional

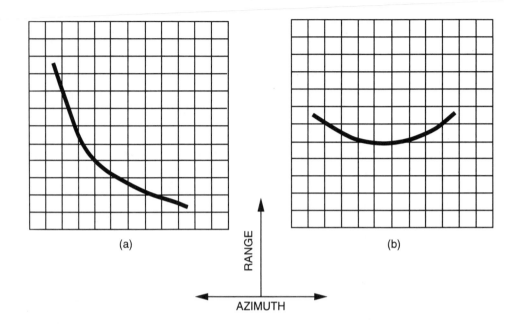

Figure 10.15 Effect of range walk removal on the range-compressed point scatterer signature: (a) signature before range walk removal and (b) signature after range walk removal. Range curvature is still present.

correlation. There are many approaches to performing the cross-range compression. We will briefly outline four:

1. Direct matched-filter correlation;
2. Fast correlation;
3. Deramp and spectral analysis;
4. Polar format processing.

They differ in the assumptions made about the waveform and demodulation scheme used and their ability to compensate for range curvature. Additional details and descriptions of the other methods can be found in [6–8,16,17] for a variety of algorithms.

10.4.3 Direct Implementation of Matched-Filter SAR Processing

If a programmable signal processor is used for the SAR cross-range compression, (10.25) is implemented directly in software on data stored in a two-dimensional memory array. Alternatively, Figure 10.16 shows the structure of a straightforward

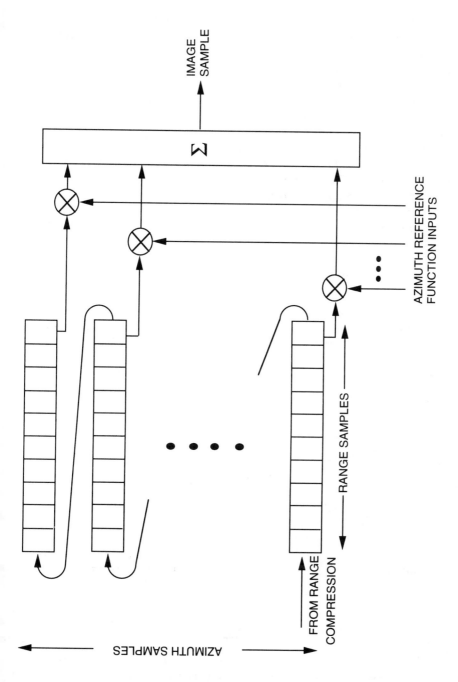

Figure 10.16 Structure of direct matched-filter cross-range correlator (*After:* [4]).

hardware implementation of a direct matched-filter correlation processor. Range-compressed data continuously fill a cross-range delay line equal to the swath length (range extent of the imaged data). Samples from each cross-range line are taken in parallel, weighted by the reference function corresponding to the cross-range response of a point scatterer, and summed to complete the correlation for a single range/cross-range position in the output map. Again, a library of reference functions is maintained and cycled through as the cross-range compression is carried out for different range positions.

Figure 10.16 shows some of the cross-range samples extracted from different taps in their delay lines. This would be done if range curvature exceeded a resolution cell; the delay lines would be tapped in a pattern mimicking the range curvature. A more accurate system would do a finer curvature correction by interpolating samples from several taps from each range line to account for fractional-bin curvatures, as shown for a single delay line in Figure 10.17.

10.4.4 Fast Correlation Cross-Range Compression

The fast correlation method simply substitutes FFT correlation techniques for the spatial domain direct correlation approach of (10.25). In general, a two-dimensional FFT of the full range/cross-range map cannot be used to accomplish the two-dimensional correlation because of the shift-varying reference function. This problem is dealt with in different ways, depending on the degree of range curvature.

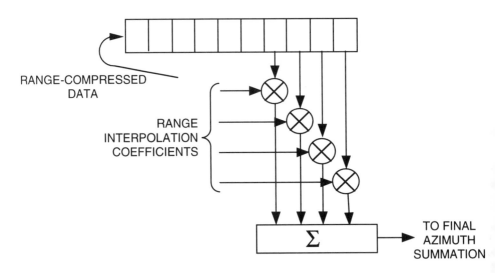

Figure 10.17 Illustration of range interpolation for fine curvature correction in brute-force cross-range correlation. Only one cross-range line from Figure 10.16 is shown.

If the range curvature is less than one range bin, the two-dimensional cross-range correlation reduces to a series of one-dimensional cross-range-only correlations. While the reference function still varies with range, each one-dimensional correlation is shift-invariant by itself and may be computed with standard FFT methods. If the range curvature is greater than a range bin, then a technique called *hybrid correlation* [18] may be used to break the two-dimensional reference function into several one-dimensional cross-range-only components. Individual cross-range-only correlations are performed with each of the component reference functions using FFT methods. The outputs are then shifted and summed to get the total response. Thus, the cross-range compression is performed with frequency-domain techniques in the cross-range direction and spatial-domain techniques in range. Hence the name *hybrid correlation*.

10.4.5 Cross-Range Compression by Deramp and Spectral Analysis

In Section 10.2.4 we pointed out that if the transmitted waveform is a linear FM chirp and deramp processing is used in the receiver, then the time series at the receiver output is an estimate of the Fourier transform of the reflectivity in the range dimension. Furthermore, we showed that the range and thus the phase of the echoes from any given scatterer from one pulse to the next vary quadratically. Consequently, the *cross-range waveform* obtained by considering echoes from a fixed scatterer over multiple pulses (the slow-time dimension of the data matrix) is also a linear FM chirp. The down-range chirp is directly generated by the transmitted radar waveform, while the cross-range chirp is induced by the relative motion between the radar and scatterer.

Because the point target response in cross-range is also a linear FM signal, we can apply deramp processing in that dimension to perform azimuth compression in a manner very similar to the down-range processing. Details are given in [7]. The basic requirement is that the data be multiplied by a complex exponential in cross-range that matches the expected chirp rate for some reference range, typically in the middle of the imaged swath. After compensation, each scatterer contributes a constant frequency term of the form

$$s(n_0, m) = k \exp\left[-j\frac{4\pi v \Delta_{az}}{\lambda R_0} mT\right] \tag{10.26}$$

where T is the PRI, n_0 is the fixed range bin (fast time) index, and m is the pulse number (slow time) index. Spectral analysis in the cross-range dimension, typically with a FFT, produces a map of the scatterer locations in cross-range for that range bin.

Note that the frequency of the demodulator output, and therefore the cross-range scale factor, is a function of the nominal range. It may not be necessary in practice to recompute the cross-range scale factor for every range bin if the variation is small enough; one scale factor can be used over several range bins. The number of range bins over which a single scale factor can be used without significant loss of focusing quality is called the *depth of focus* and is described next.

10.4.6 Depth of Focus

We have noted that the cross-range correlation reference function is range-dependent, so that in principle a different correlator reference kernel or FFT scale factor is needed for each range bin. In practice, a single reference function or scale factor can usually serve over a number of range bins and sometimes the entire imaged swath without significant degradation in the quality of the image. The range extent over which this can be done is the depth of focus.

Once range walk is compensated for, it is range curvature that determines the form of the reference kernel. Thus, depth of focus is determined by finding the change in range R_0, which will cause a specified change in range curvature over the aperture time. The "tolerable" variation is usually taken, somewhat arbitrarily, as $\lambda/4$, corresponding to 90 degrees of phase shift. Equation (10.13) gave a formula for range curvature. Differentiating (10.13) with respect to range gives the change in curvature as nominal range changes; setting that quantity equal to $\lambda/4$ gives the depth of focus as [7,19]

$$DOF = \frac{2\lambda R_0^2}{(vT_a \sin \phi)^2} \tag{10.27}$$

For the example parameters we have been using, the depth of focus is about 308m with a 1-sec aperture time, increasing to 4.94 km if the aperture time is limited to 0.25 sec. Similar computations can be carried out in the cross-range and height dimensions [20]. If the imaged swath extends, say, 2 km in range, then seven distinct reference kernels would be required to maintain good focus quality across the full range extent if a 1-sec aperture time is used. On the other hand, if lower resolution are required such that a 0.25-sec aperture time is adequate, a single reference kernel would suffice for the entire range extent. In this case, the two-dimensional shift-varying correlation would become shift-invariant, and two-dimensional FFT techniques could be used to implement direct matched-filter cross-range compression.

10.4.7 Polar Format Processing for Spotlight SAR

In Section 10.2.4 we saw that when a linear FM waveform is used in conjunction with deramp processing on receive, the two-dimensional fast time/slow time data array can be interpreted as samples of the Fourier transform $I(\omega_x, \omega_y)$ of the complex image $i(x, y)$ on a polar format grid, as sketched in Figure 10.8. An inverse Fourier transform, using the FFT, of $I(\omega_x, \omega_y)$ should result in an estimate of the image $i(x, y)$.

Unfortunately, the FFT implicitly assumes that the Fourier domain data is sampled on a *rectangular* grid. The nonrectangular nature of the spotlight data becomes more significant as the resolution is increased in either range (increasing the ratio between the upper and lower frequency limits of the sampled portion of Fourier space) or cross-range (increasing the angular spread of the Fourier data). Using a standard inverse FFT on the data without compensating for the nonrectangular grid results in a loss of resolution and gain and especially an increase in side lobes in the point target response.

A common solution to this problem is to interpolate the data from a polar to a rectangular grid prior to applying the FFT, resulting in the spotlight mode SAR processing approach diagrammed in Figure 10.18. One possible approach to this interpolation is illustrated in Figure 10.19. In Figure 10.19(a), the data are first interpolated along the spatial frequency radii (fast time dimension) to give samples that fall on equally spaced parallel lines in the cross-range spatial frequency dimension. As shown in Figure 10.19(b), the data are next interpolated in the cross-range (slow time) dimension to give a final rectangular grid. The FFT can then be applied to generate the complex image.

Dudgeon and Mersereau [9] describe one algorithm for polar-to-rectangular interpolation. The interpolation is typically done with low-order FIR digital filters. Munson et al. [21] describe the performance of several alternative interpolating filters.

In some cases, after cross-range interpolation the data samples are evenly spaced in range but merely offset from one range line to the next. If the range sample spacing is as desired, then only shifting of the range samples is required. In this case, the shifting can be combined with the range FFT processing by multiplying the data by an appropriate linear phase function prior to the FFT, eliminating the need for explicit range interpolation.

10.5 INTERFEROMETRIC THREE-DIMENSIONAL SAR

One of the newest developments in SAR is the ability to do high-resolution imaging in three dimensions using interferometric techniques, commonly called IFSAR. The basic approach employs two complex SAR images of a scene formed using physically displaced apertures. The two apertures are typically physically separate receive apertures on a single aircraft antenna structure, usually with a common transmit aperture. However, IFSAR has also been demonstrated using images collected from a conventional single-aperture system on multiple passes.

Figure 10.20 illustrates the case in which the two apertures are separated horizontally by a baseline distance B and are at an altitude above the reference

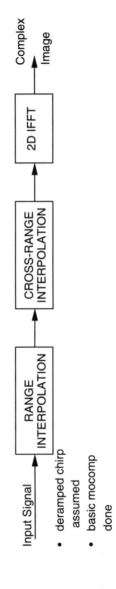

Figure 10.18 Processing flow for polar format image formation algorithm.

(a)

(b)

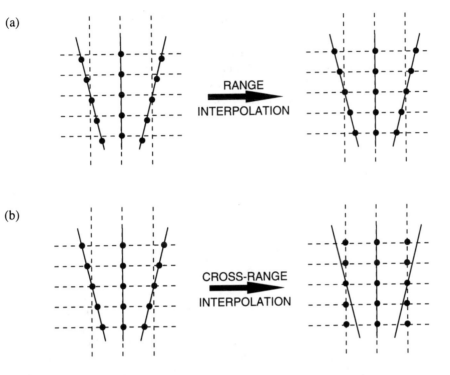

Figure 10.19 One method of interpolation from a polar grid to a rectangular grid so that the two-dimensional FFT can be used for image formation: (a) interpolation in fast time (radial) dimension to straight lines in cross-range and (b) interpolation in slow time dimension to final rectangular grid.

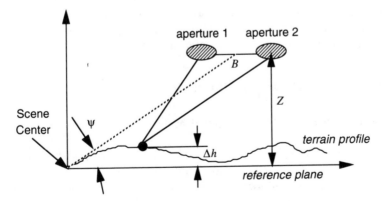

Figure 10.20 Geometry of interferometric SAR height estimation using horizontally displaced apertures.

plane of Z. The reference plane is arbitrary; it is shown here as a ground reference, but could equally well be the plane connecting the two apertures, for example, with some minor modifications to the final expressions. If we denote the two conventionally formed, complex, two-dimensional SAR images of the scene as $s_1(n, m)$ and $s_2(n, m)$, then the first step in estimating the height distribution of scatterers is to form the so-called *interferogram*:

$$if(n, m) = s_1^*(n, m) \, s_2(n, m) \qquad (10.28)$$

where the asterisk denotes complex conjugation. Note that the phase of each pixel of $if(n, m)$ will be the difference between the phases of the corresponding pixels of $s_1(n, m)$ and $s_2(n, m)$. The estimated height Δz of each pixel relative to the reference plane is then [7]

$$\Delta z(n, m) \cong - \left[\frac{Z \cot \psi}{4 \pi B \sin \psi} \right] \arg if(n, m)\} \qquad (10.29)$$

Figure 10.21 is a striking IFSAR image of the downtown Washington, D.C., area. In this image, lighter shades represent higher elevations. Many Washington landmarks are clearly visible: the Potomac river, the Pentagon, the Washington Mall, the runways of National Airport, and others. This image was made with data collected by the NASA Jet Propulsion Laboratory's TOPSAR C-band interferometric imaging radar.

IFSAR is an immature technology at this writing, with many details beyond those described here. Care must be taken in aligning the two individual complex SAR images. The height measurement can be highly ambiguous, requiring two-dimensional phase unwrapping of the interferogram. The SNR must be high for good phase estimates; thus, IFSAR performance degrades in low-reflectivity areas. Researchers are actively experimenting with approaches to filtering and smoothing the interferogram to improve the height image quality without excessive loss of resolution. For additional details and references, see [7].

10.6 OTHER ASPECTS OF SAR OPERATION

10.6.1 Phase Errors and Motion Compensation

We have implicitly assumed so far that the aircraft is flying a constant velocity path, typically straight and level at constant speed. In practice, this is never quite true. Atmospheric turbulence, minor maneuvers and course corrections, vibration, gimbal transients, and similar effects all cause accelerations. These deviations manifest themselves as phase errors in the processing, since the algorithms typically assume

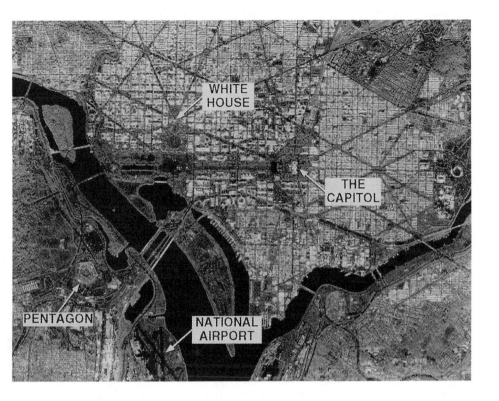

Figure 10.21 Interferometric SAR height image of Washington, D.C. Lighter shades of gray correspond to higher elevations.

constant-velocity flight when computing focusing phases. These phase errors, in turn, degrade the effective SAR antenna pattern [22].

Specifically, suppose the transmitted waveform is of the form $A \exp[j2\pi ft]$ and the range to some scatterer is R on a particular pulse. Then the received signal is

$$y(t) = A \exp\left[j2\pi f\left(t - \frac{2R}{c}\right)\right] \qquad (10.30)$$

Now suppose that the radar platform is, for whatever reason, displaced from the intended path by a distance ϵ; then the received signal will instead be

$$\tilde{y}(t) = \tilde{A} \exp\left[j2\pi f\left(t - \frac{2R - 2\epsilon}{c}\right)\right] = \frac{\tilde{A}}{A} \exp\left[-j4\pi\frac{\epsilon}{\lambda}\right] y(t) \qquad (10.31)$$

The amplitude term in (10.30) will be very nearly unity, so the primary effect of motion errors is a phase rotation to the data. The job of motion compensation is therefore to estimate ϵ and correct the data by a simple counterphase rotation:

$$\hat{y}(t) = \exp\left[+j4\pi\frac{\epsilon}{\lambda}\right]\tilde{y}(t) \cong y(t) \qquad (10.32)$$

The difficult part, of course, is estimating ϵ.

Phase errors are frequently categorized as either low- or high-frequency [23]. Low-frequency errors are those that repeat on a time scale greater than the aperture time, which is typically anywhere from 1 to 10 sec, but may be significantly more or less. Thus, any periodicity of low-frequency errors is not apparent during formation of an image. High-frequency errors are subdivided into deterministic and noiselike errors. In the former, periodic structure is evident during the aperture time. Low-frequency errors produce net phase tapers across the aperture, affecting the synthetic pattern main lobe and side lobes. High-frequency errors affect primarily the side lobes. Table 10.1 describes the major effect of the most important phase errors.

Phase errors are minimized by using data from the aircraft's INS to track the deviations from a constant-velocity flight path and compute the additional phase corrections necessary to compensate for them. Some newer systems mount an additional strap-down inertial measurement unit (IMU) onto the antenna structure [23,24]. The IMU accounts more accurately for motions of the antenna relative to the airframe; these relative motions include not only intentional scanning, but also airframe flexure. Inertial antenna mounts can also be used to stabilize the antenna as much as possible. Figure 10.22 is a block diagram of a typical modern motion compensation instrumentation package. This system combines inputs from an antenna-mounted IMU, the platform INS, radar-derived doppler velocity

Table 10.1
Effects of Various Phase Errors on the Synthetic-Aperture Pattern

Error Type	Dominant Effect
Low-Frequency	
Linear	Main-lobe pointing error
Quadratic	Main-lobe broadening
Cubic	Main-lobe asymmetry and pointing error
High-Frequency	
Deterministic	Discrete "paired echo" high side lobes
Random	Increased side-lobe level

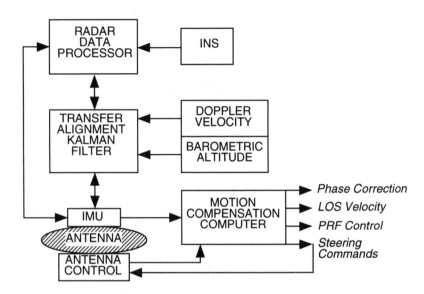

Figure 10.22 Generic motion compensation instrumentation package.

estimates, and an auxiliary barometric altimeter. Other systems might also employ data from a Global Positioning System (GPS) receiver. In this system, the INS provides an initial attitude reference to the IMU. IMU data are sent to the radar data processor (RDP), which corrects for the lever arm displacement between the INS and IMU. The difference in INS and corrected IMU position estimates is input to the Kalman *transfer alignment* filter, whose output is fed back to the IMU to update its state. The updated IMU state is fed to the motion compensation computer, which uses these data to compute the position error ϵ and the associated phase correction. The motion compensation computer typically also computes antenna steering commands to stabilize the line of sight, timing commands needed to compensate for range walk as discussed in Section 10.4.2, and PRI adjustments to maintain constant cross-range sampling intervals.

Of course, motion compensation is not perfect either. An uncorrected low-frequency quadratic phase error of 114 degrees will spread the main lobe of the synthetic pattern by 10% [1]; this phase error corresponds to about one-third of a wavelength, a mere 0.4 in at 10 GHz!

Starting from a tolerable quadratic phase error specification, such as 114 degrees, a motion compensation error budget can be developed for the various INS/IMU errors. Table 10.2 is an example of such an analysis [23]. For the example parameters we have used with an aperture time of 0.2 sec, gravitational acceleration $g = 10.8$ m/s^2, and a line-of-sight acceleration A of 3 m/s^2, we see that cross-range velocity errors dominate the budget; the total cross-range velocity error must be less than about 4 m/s. However, the aperture time in this example is short. Since

Table 10.2
Motion Compensation Error Budget

| Error Source | Formula | Value | | Root Sum Square (RSS) Total | Quadratic Phase Error (rad) |
		INS	IMU		
Cross-range velocity error, v_e (m/s)	$\dfrac{\pi v v_e T_a^2}{\lambda R_0}$	4	1	4.12	0.52
Boresight accelerometer bias, A_b (μg)	$\dfrac{\pi A_b T_a^2}{2\lambda}$	700	800	1,063	0.022
Platform tilt, α (mrad)	$\dfrac{\pi \alpha a T_a^2}{2\lambda}$	2	1	2.24	0.046
Boresight accelerometer scale factor error, s (parts per million)	$\dfrac{\pi s A T_a^2}{2\lambda}$	1,500	1,500	2,121	0.013
Total RSS quadratic phase error					0.53

Source: [23].

the phase error contributions all increase as the square of the aperture time, the error margins can be much tighter in many situations [23]. Also, at longer ranges the other error effects become more significant as velocity errors become smaller. Hepburn et al. [24] give another example of a motion compensation error budget.

Two factors mitigate somewhat the sensitivity of SAR to motion compensation errors. First, only motion compensation errors over the aperture time are of any consequence. Buildups in position estimation error over longer time scales are of no consequence to any one image. Second, only *changes* in position error over the aperture time affect the final image. A constant bias in position results in a constant phase error on every data sample. The only effect is a complex constant multiplying the complex image, and this constant is removed by the detector when the final image is formed. Thus, the motion compensation system need only stay within the error budget over intervals of T_a sec; long-term biases and drift are unimportant to image formation.[5]

---

[5]Long-term drift may be quite important, however, to image interpretation. Targets or land features detected in an SAR image may need to be precisely geolocated for targeting or for surveying purposes. It is then essential to have accurate information on the absolute position of the radar platform.

10.6.2 Clutter Locking and Autofocus

Motion compensation techniques account for deviations from the assumed nominal velocity and position during the aperture time. Doppler tracking of the clutter spectrum is often used to ensure that the nominal velocity itself is correct. If an incorrect nominal velocity is used in the SAR processing, the cross-range correlation reference kernel will be incorrect in both duration and centroid. The processed data will then be defocused and displaced.

Clutter locking is a simple doppler tracking technique used in many SAR systems [19]. It measures the doppler spectrum centroid, which should be equal to the radial velocity along the boresight. Any difference is used to drive a simple feedback loop to correct the assumed doppler.

Autofocusing attempts to estimate remaining phase errors present in the data after motion compensation and compensate for them to reduce image blurring. A number of techniques have been suggested; see [7] for a good overview. Two of the most common are the map drift method and the phase gradient method. The map drift algorithm is designed specifically to correct quadratic phase errors. It divides the SAR data into two halves, corresponding to the first half of the aperture, and the second half, and forms a lower resolution image from each [7,19]. If quadratic phase error is present, peaks in the images will be shifted away from their correct locations, but in opposite directions. Cross-correlating the two images produces a peak whose offset from the origin is directly proportional to the amount of quadratic phase error left. It is then relatively simple to correct the data for this quadratic error and form the final full-resolution image from the complete aperture of data. The map drift algorithm must usually be applied iteratively to get the best results. Extensions to the map drift algorithm that divide the data into more than two subapertures are capable of estimating higher order phase error terms.

The phase gradient algorithm [7,25] is a newer approach that is not limited to a particular order of phase error and appears to offer excellent, robust performance. The basic approach of the algorithm is to estimate the first derivative of the cross-range phase error from a partially compressed SAR image. Details of the procedure for estimating this derivative are beyond the scope of this chapter, and in fact are still areas of active research; see [7,25]. The phase error derivative is then integrated to get the actual phase error function, and a correction is applied to the data. As with map drift, the phase gradient algorithm is usually applied iteratively to get the best results. Figure 10.23 illustrates the results of a simulation in which three iterations of the phase gradient autofocus algorithm were sufficient to restore the badly blurred image of Figure 10.23(a) to the quality of Figure 10.23(b).

10.6.3 Speckle Reduction

Like any coherent imaging system, SAR produces images contaminated by *speckle,* an aptly named multiplicative noise. Speckle is the natural result of the coherent

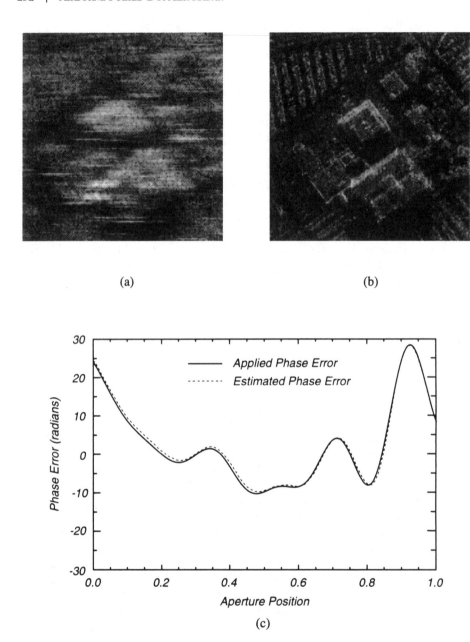

(a)   (b)

(c)

Figure 10.23 Autofocus using the phase gradient algorithm. A high-order phase error of several tens of radians was introduced into an image: (a) image obtained by spotlight SAR processing without autofocus; (b) image obtained after three iterations of the phase gradient algorithm; and (c) comparison of the actual phase error function and the estimate generated by the phase gradient algorithm. (*Source:* [13]. Reprinted with permission.)

combination of echoes from many different scatterers to form an image pixel. If the amplitude distribution is Gaussian (ensured by the law of large numbers when many scattering centers are involved) and the phase distribution is uniform, then the pixel amplitude will be Rayleigh distributed [19]. Thus, pixels representing areas with the same RCS (mean echo amplitude) can give rise to different pixel amplitudes. These variations are not due to thermal, quantization, or other noise sources, but are nonetheless considered noise because of their effect on image quality.

Speckle is reduced by noncoherent integration of multiple uncorrelated images, or *looks*, of the same scene. This process reduces the pixel variance, reducing the amplitude variations among pixels representing the same RCS. Uncorrelated looks can be obtained using transmitter frequency or polarization agility. Since the maximum doppler bandwidth often exceeds that required to meet resolution goals, the most common approach is to divide the doppler band into multiple subbands, each wide enough to form an image of the proper resolution. Images are calculated for each band and combined. Typically, 4 to 10 looks might be used for speckle reduction. Figure 10.24 demonstrates the image enhancement obtained by integration of 10 looks. Another method simply averages adjacent pixels of a high-resolution image to form one lower resolution pixel as illustrated in Figure 10.25. In some cases, the maximum doppler bandwidth may be small enough so that a trade-off between resolution and speckle reduction becomes necessary.

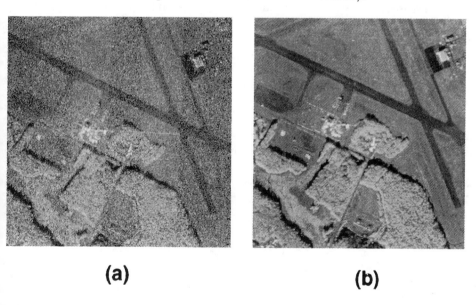

(a)　　　　　　　　　　　　　　　**(b)**

Figure 10.24 Integration of independent looks to reduce image speckle: (a) single look image and (b) image obtained by integration of 10 looks. (*Source:* Westinghouse Norden Systems. Used with permission.)

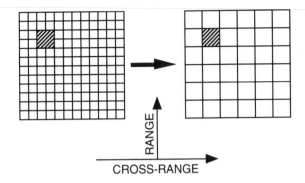

Figure 10.25 Averaging adjacent pixels to reduce image speckle. Speckle noise power is reduced, but the resolution is also reduced by a factor of two in each dimension.

10.7 SUMMARY

SAR is a technique for getting the angular resolution of a very large array antenna by multiplexing a single small antenna in time and space. SAR modes in airborne fire control radar provide high-resolution sensing to support mapping and target imaging applications. Squint-focused strip map, spotlight, and DBS modes are most commonly used, with spotlight becoming increasingly dominant.

SAR requires a highly coherent radar, a sophisticated digital signal processor with a large memory and high-speed computational unit, and tight integration with inertial motion measurement systems. An extensive menu of processing algorithms exists; the choice depends on particulars, such as the degree of range curvature or available processing power. Adaptive control algorithms such as clutter locking and autofocus are required for best performance. Recent years have seen great progress in the understanding of SAR imaging and the development of very high quality imaging algorithms. Ideas have been borrowed extensively from the fields of seismic processing and medical imaging. New algorithms such as the range migration algorithm and the chirp scaling algorithm have been developed which better account for range curvature than traditional approaches [7]. Areas of rapid current development include three-dimensional imaging and imaging with maneuvering platforms and targets.

References

[1] Stimson, G. W., *Introduction to Airborne Radar*, Hughes Aircraft Co., El Segundo, CA, 1983.

[2] Eaves, J. L., and E. K. Reedy, eds., *Principles of Modern Radar*, New York: Van Nostrand Reinhold, 1987.

[3] Fitch, J. P, *Synthetic Aperture Radar*, New York: Springer-Verlag, 1988.

[4] Ausherman, D. A., "Digital vs. Optical Techniques in Synthetic Aperture Radar (SAR) Data Processing," *Optical Engr.*, Vol. 19, No. 2, Mar./Apr. 1980, pp. 157–167.

[5] Lunsford, G. H., and A. H. Green, Jr., "Modeling of Synthetic Aperture Radar for Nonconventional Geometry," *Proc. IEEE 1985 Intl. Radar Conf.*, May 1985, pp. 366–371.

[6] Ausherman, D. A., et al., "Developments in Radar Imaging," *IEEE Trans. Aerospace and Electronic Systems*, Vol. AES-20, No. 4, July 1984, pp. 363–400.

[7] Carrara, W. G., R. S. Goodman, and R. M. Majewski, *Spotlight Synthetic Aperture Radar*, Norwood, MA: Artech House, 1995.

[8] Munson, D. C., Jr., and R. L. Visentin, "A Signal Processing View of Strip-Mapping Synthetic Aperture Radar," *IEEE Trans. Acoustics, Speech, and Signal Processing*, Vol. 37, No. 12, December 1989, pp. 2131–2147.

[9] Dudgeon, D. E., and R. M. Mersereau, *Multidimensional Digital Signal Processing*, Englewood Cliffs, NJ: Prentice-Hall, 1984.

[10] Levanon, N., *Radar Principles*, New York: John Wiley & Sons, 1988.

[11] Oppenheim, A. V., and R. W. Schafer, *Digital Signal Processing*, Englewood Cliffs, NJ: Prentice-Hall, 1975.

[12] Richards, M. A., "Nonlinear Effects in Fourier Transform Processing," Chapter 6 in *Coherent Radar Performance*, J. A. Scheer and J. L. Kurtz, eds., Norwood, MA: Artech House, 1993.

[13] Jakowatz, C. V., et al., *Spotlight Mode Synthetic Aperture Radar*, Boston, MA: Kluwer Academic Publishers, 1996.

[14] Munson, D. C., J. D. O'Brien, and W. K. Jenkins, "A Tomographic Formulation of Spotlight-Mode Synthetic Aperture Radar," *Proc. IEEE*, Vol. 71, No. 8, August 1983, pp. 917–925.

[15] Lim, J. S., and A. V. Oppenheim, eds., Chap. 3 in *Advanced Topics in Signal Processing*, Englewood Cliffs, NJ: Prentice-Hall, 1988.

[16] Bamler, R., "A Comparison of Range-Doppler and Wavenumber Domain SAR Focusing Algorithms," *IEEE Trans. Geoscience and Remote Sensing*, Vol. 30, No. 4, July 1992, pp. 706–713.

[17] Giglio, D. A., ed., *Algorithms for Synthetic Aperture Radar Imagery*, Proc. SPIE, Vol. 2230, April 1994.

[18] Wu, C., K. Y. Liu, and M. Jin, "Modeling and a Correlation Algorithm for Spaceborne SAR Signals," *IEEE Trans. Aerospace and Electronic Systems*, Vol. AES-18, No. 5, September 1982, pp. 563–574.

[19] Elachi, C., *Spaceborne Radar Remote Sensing: Applications and Techniques*, New York: IEEE Press, 1988.

[20] Robinson, P. N., "Depth of Field for SAR With Aircraft Acceleration," *IEEE Trans. Aerospace and Electronic Systems*, Vol. AES-20, No. 5, September 1984, pp. 603–616.

[21] Munson, D. C., Jr., et al., "A Comparison of Algorithms for Polar-to-Cartesian Interpolation in Spotlight Mode SAR," *Proc. IEEE Intl. Conf. on Acoustics, Speech, and Signal Processing*, 1985, pp. 1364–1367.

[22] Kirk, J. C., Jr., "Motion Compensation for Synthetic Aperture Radar," *IEEE Trans. Aerospace and Electronic Systems*, Vol. AES-11, No. 3, May 1975, pp. 338–348.

[23] Kennedy, T. A., "The Design of SAR Motion Compensation Systems Incorporating Strapdown Inertial Measurement Units," *Proc. IEEE 1988 Natl. Radar Conf.*, April 1988, pp. 74–78.

[24] Hepburn, J. S. A., et al., "Motion Compensation for High Resolution Spotlight SAR," *Record of IEEE Position Location and Navigation Symposium (PLANS 84)*, November 1984, pp. 59–65.

[25] Wahl, D. E., P. H. Eichel, D. C. Ghiglia, and C. V. Jakowatz, Jr., "Phase Gradient Autofocus—A Robust Tool for High Resolution SAR Phase Correction," *IEEE Trans. Aerospace and Electronic Systems*, Vol. AES-30, No. 5, July 1994, pp. 827–835.

CHAPTER 11

Medium-PRF Detectability and Range Resolving

Guy V. Morris

The target signal return must satisfy two criteria to be detected:

1. Its signal strength must exceed the minimum signal strength needed for detection.
2. Its range and doppler frequency must be in the "clear region" of the range-doppler space provided by the PRF.

The target must be detected on at least two PRFs to resolve the range ambiguity. Generally radars are designed to require detection on three PRFs to ensure proper range ambiguity resolution even if the returns from two different ranges are received simultaneously. Targets detected on fewer than three PRFs are not displayed to the operator. Therefore, the actual detection criterion becomes detection on at least three PRFs.

The total antenna dwell time must be subdivided to provide a coherent processing interval at each PRF. Each time the PRF is changed, fill pulses must be transmitted that are not used in coherent processing. Using multiple PRFs is less energy-efficient than using a single coherent processing interval equal to the antenna dwell time. Therefore, the total number of PRFs used should be minimized subject to meeting the other requirements described in this chapter.

This chapter discusses the general principles of:

1. Range-doppler blind zones;
2. Number of clear PRFs;
3. Effect of duty cycle;

4. Pulse compression;

5. Range resolving.

The detailed method of calculation of the set of medium PRFs is reserved for Chapter 12.

11.1 RANGE-DOPPLER BLIND ZONES

Figures 11.1 and 11.2 show the blind zones for two PRFs. The unambiguous range R_u is inversely proportional to the PRF. The blind region in the range dimension

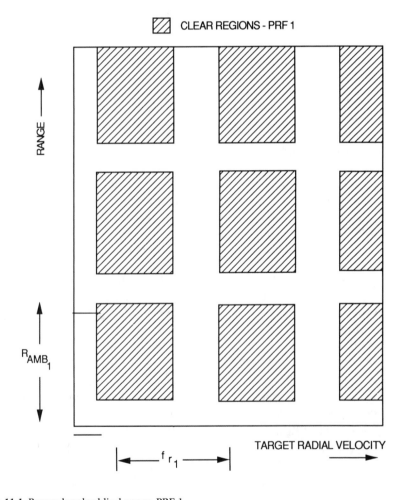

Figure 11.1 Range-doppler blind zones: PRF 1.

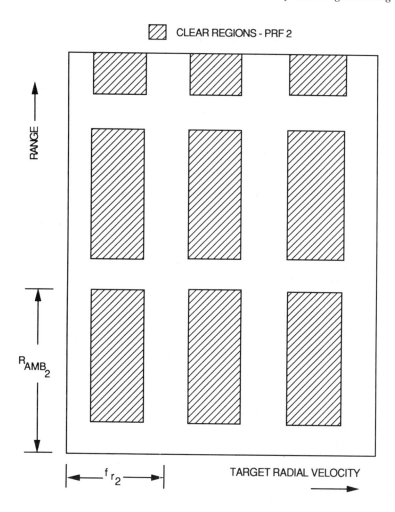

Figure 11.2 Range-doppler blind zones: PRF 2.

corresponds to the amount of time that the receiver is turned off during transmission. The unambiguous doppler frequency f_u is equal to the PRF. The doppler blind region shown is the area in which main-lobe clutter is expected. (Main-lobe clutter is not necessarily present, say, in look-up scenarios. However, since the main application of medium-PRF radar is a look-down scenario against a low-flying target, the existence of main-lobe clutter will be assumed.) The spectral width of the main-lobe clutter signal varies with the speed of the aircraft and with the radar pointing angle. One simplification often made in radar design is to determine the maximum width of main-lobe clutter expected and to notch out that portion of the spectrum. Also, medium-PRF signal processors often translate the spectrum to center main-lobe clutter at zero frequency. Figures 11.1 and 11.2 show the main-lobe clutter notch centered at zero frequency.

Target returns falling within the shaded area within R_u and f_u are detected unambiguously in both range and doppler. It is easy to show that the total unambiguous range-doppler space is a constant, specifically:

$$R_u f_u = \frac{c}{2} \tag{11.1}$$

Equation (6.1) is an alternative form of (11.1) expressed in velocity units.

Target detections are possible within all the shaded areas of Figure 11.1 or 11.2, but will be ambiguous in range or doppler or both. One criterion for choosing a set of medium PRFs is to ensure that all targets within the range-doppler space of interest fall in the clear region on at least three PRFs.

11.2 EFFECTS OF THE NUMBER OF CLEAR PRFs

The probability of detection P_d is given by

$$P_d(N_c) = \sum_{m=M}^{N_c} \left(\frac{N_c!}{m!\,(N_c - m)!} \right) P_{dm}^m (1 - P_{dm})^{N_c - m} \tag{11.2}$$

where:

P_{dm} = probability of detection on each of the m PRFs;
M = number of PRFs required for detection;
N_c = number of clear PRFs.

The probability of false alarm P_{fa} can be computed using

$$P_{fa} = \sum_{m=M}^{N_t} \left(\frac{N_t!}{m!\,(N_t - m)!} \right) P_{fam}^m (1 - P_{fam})^{N_t - m} \tag{11.3}$$

where:

N_t = total number of PRFs;
P_{fam} = probability of false alarm on each of the m PRFs.

Figure 11.3 shows $P_d(N_c)$ as a function of the number of clear PRFs N_c, assuming the detections and false alarms on each PRF are independent and of equal probability. The specific parameters used are:

N_t = 8;
N_c = 3;
P_{dm} = 0.5;
P_{fam} = 0.001.

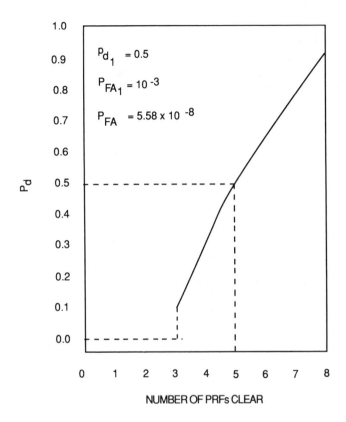

Figure 11.3 Probability of detection versus number of clear PRFs.

Note that the target must be clear on five of the PRFs to ensure that $P_d(N_c)$ exceeds P_{dm}. Note also that the resultant false alarm probability P_{fa} is much lower than P_{fam}. The position of the target return in the range-doppler map of each PRF is deterministic; that is, the range-doppler location is given by (1.10) and (1.11). One can see intuitively that false alarms on the various PRFs must fall in a certain pattern to be interpreted erroneously as a target. Therefore, the detection thresholds on the individual PRFs may be set to a lower value, which increases P_{fam}, than would be acceptable in a system that uses single-PRF, single-detection criteria. Lower detection thresholds increase $P_d(N_c)$.

11.3 THE EFFECT OF THE DUTY CYCLE

Many airborne radar transmitters are operated with the final power amplifier saturated. When the peak power is thus constrained, it is desirable to operate a radar at the maximum transmitter duty cycle, because maximizing the transmitter average power maximizes the signal-to-noise and signal-to-jamming ratios. (If the transmitter is capable of maintaining the average power by increasing the peak power when the duty factor is reduced, then the effects described in this section are not present.) When the set of medium PRFs is chosen to ensure that the target is detectable in doppler frequency, the ratio between the highest and the lowest PRF may be as much as 2:1. If a constant transmitter pulse width is used, then the average power and SNR also vary by 2:1.

The effect of duty cycle variation on the probability of detection can be illustrated by considering the following example. One form of the radar range equation developed in [1] applicable to medium-PRF radar is

$$\left(\frac{S}{N}\right) = \frac{P_t G^2 \lambda^2 \sigma_t T_d d_u}{(4\pi)^3 R^4 k T_e L_s} \tag{11.4}$$

where d_u is the transmitter duty cycle. The definition of the other terms and the values used in this example are shown in Table 11.1. The probability of detection versus duty cycle shown in Figure 11.4 was derived using the values of (S/N) versus duty cycle computed using (11.4) and the detection curves of Chapter 14 for a nonfluctuating target. Note that an increase of duty cycle from 0.01 to 0.02 results in an increase in P_{dm} from approximately 0.2 to 0.8. There is less performance variation versus duty cycle for realistic (that is, fluctuating) targets. If a similar analysis is performed using the Swerling 1 detection curves of Chapter 15, then the increase in P_{dm} is from approximately 0.3 to 0.5.

In many radars that use an FFT for doppler processing, the number of pulses integrated is constant and the dwell time is not. The following substitution can be made in (11.4):

$$T_d d_u = \left(\frac{N_p}{\mathrm{PRF}}\right)(\tau_t \mathrm{PRF}) = N_p \tau_t \tag{11.5}$$

Table 11.1
Radar Parameters Used in Example

Parameter	Symbol	Value	Unit
Peak transmitter power	P_t	10.0	kW
Antenna gain	G	30.0	dB
Wavelength	λ	0.031	m
Range	R	27.8	km
Noise energy	kT_e	−178.0	dBm/Hz
Target cross section	σ_t	5.0	m^2
System losses	L_s	14.0	dB
Dwell time	T_d	6.0	ms

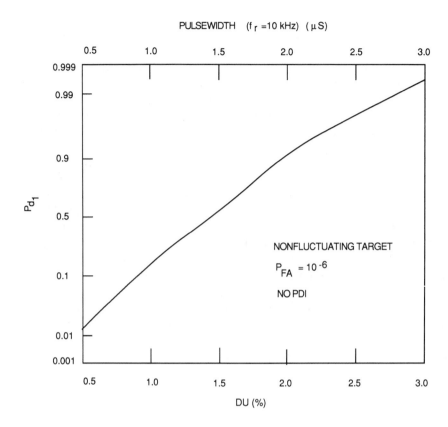

Figure 11.4 Probability of detection versus transmitter duty cycle.

where:

N_p = number of pulses;

τ_t = transmitter pulse width.

In either case, constant N_p or constant T_d, P_{dm} may be maximized for each PRF by changing the transmitter pulse width to maintain a constant duty cycle. The receiver video bandwidth should also be changed correspondingly to maintain the bandwidth match. If the time between range gates is kept equal to the transmit pulse width, then the total number of receive range gates is a constant for all PRFs. The difference in range-gate spacing is an additional complexity that must be accommodated in the range-resolving circuits. Wider transmit pulses cause each range blind zone to be wider, but the blind zones are spaced further apart because the PRF is lower. The total extent of the range blind zone regions is constant for a constant duty cycle. The clear region duty cycle d_r can approach

$$d_r = 1 - d_u \tag{11.6}$$

11.4 PULSE COMPRESSION

Pulse compression is a technique that can be used to increase average power without sacrificing range resolution. Pulse compression is discussed in some detail in Chapter 9 and will only be mentioned here to focus attention on the implications of applying the technique to medium-PRF radar. In multimode mode radars, the high-PRF mode requirements often result in the selection of a transmitter tube with high duty cycle capability (e.g., 0.5) but relatively low peak power. When used in the medium-PRF mode with duty cycles on the order of 0.02, the average power is inadequate and pulse compression is used.

Because of the increased transmitter duty cycle, pulse compression increases the range blind zones. Consider the example depicted in Figure 11.5. Three PRFs and a two-out-of-three detection criteria are used. When a constant transmitter pulse width of 1.9 μs is used with each of the three PRFs, 0.985 of the range region is clear. If a 13 to 1 Barker pulse compression code is used, resulting in a transmitter pulse width of 24.7 μs, then the clear region drops to 0.7225.

Some radars use a transmitter tube that is capable of an 8- to 10-dB increase in peak power, relative to its high-PRF output power, when the tube is operated at the lower duty factors of medium-PRF modes. Pulse compression is usually not used, since the increase in peak power provides an increase in performance approximately equal to 13-bit Barker pulse compression.

11.5 RANGE RESOLVING

The nature of range ambiguities will be explained using a two-PRF example before considering the three-PRF case, which is the most common implementation. T_t in Figure 11.6 is the true (unambiguous) target round-trip propagation time. T_{a1} and T_{a2} are the apparent (ambiguous) propagation times at PRF_1 and PRF_2, respectively. If we make the following definitions:

$$T_1 = \frac{1}{PRF_1} = \frac{1}{N \cdot PRF_0} \tag{11.7}$$

$$T_2 = \frac{1}{PRF_2} = \frac{1}{(N+1) \cdot PRF_0} \tag{11.8}$$

$$PRF_0 = \frac{c}{2R_{us}} = \frac{1}{T_0} \tag{11.9}$$

where:
R_{us} = desired system unambiguous range after range resolving;
N = an integer.

T_t can be expressed as a function of T_1 and T_2 as follows:

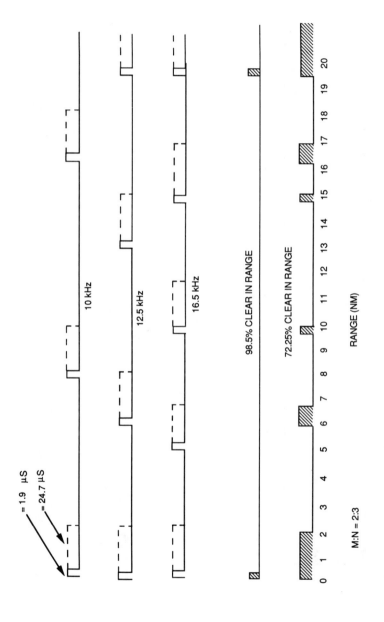

Figure 11.5 Pulse compression effect on blind zones.

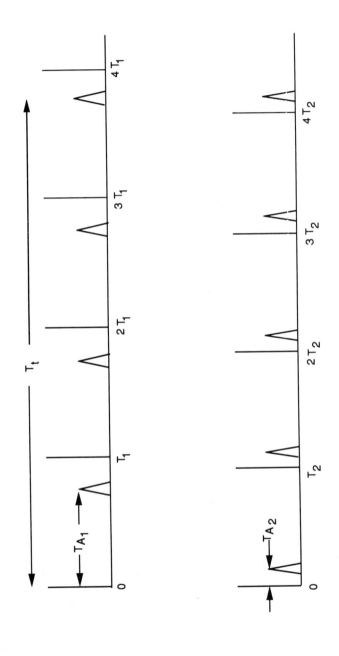

Figure 11.6 Range resolution with two PRFs.

$$T_t = nT_1 + T_{a1} \qquad\qquad n = 0, 1, 2, \ldots \qquad\qquad (11.10a)$$

$$T_t = (m + n)T_2 + T_{a2} \qquad\qquad m = 0, 1 \qquad\qquad (11.10b)$$

Equation set (11.10) is useful for resolving the range ambiguity for

$$T_t \leq nT_1 = (n + 1)T_2 \qquad\qquad (11.11)$$

The range that causes $nT_1 = (n + 1)T_2$ is called the coambiguous range. The 12.5- and 16.5-kHz PRFs of Figure 11.5 graphically illustrate the concept of a *coambiguous range*.

Ideally, (10.6) through (10.9) could be solved simultaneously to yield an expression for T_t in terms of the measurements T_{a1} and T_{a2} and the known quantities PRF_0 and N. In the practical case, the delay measurements are corrupted by interference, quantization errors, and noise. The measurement errors may result in different values of T_t for integer values of n and m when used in equation set (10.9). A value of T_t that minimizes the difference is then chosen. A simple algorithm for determining T_t is as follows: (1) compute trial values of Tt using equation set (10.9) for various values of m and n; (2) ideally, the true value of T_t is the one that is produced by both equations of the set that is identically the same. Practically, some disagreement between the two answers is accepted, perhaps plus or minus one range cell.

This simple algorithm, extended to use three PRFs, will now be described using Figure 11.7. Three returns labeled A, B, and C are observed on each of the three PRFs. Obviously, the apparent range position of target A is invariant with PRF and thus must be unambiguous. Figure 11.7 shows a map of the detections as a function of the apparent range-gate position for each of the three PRFs. For simplicity, the range-gate width is constant for all PRFs. Note that there is a different number of range gates for each PRF because of the different interpulse interval. Figure 11.8 is constructed by repeating the hit pattern for each of the PRFs. This process is functionally equivalent to computing the trial values of T_t described earlier. The true ranges of the targets correspond to the ranges in Figure 11.8, in which hits are observed in all three PRFs.

Three PRFs are needed to ensure that two targets will be properly resolved. (In the previous example, three targets were accurately resolved, which was the fortuitous result of the target returns staying in the same ordered position. In general, four PRFs are required to resolve three targets without generating a false target.) The false targets or "ghosts" that can be generated by two targets in a two-PRF resolver are illustrated in Figure 11.9. Two different symbols for the returns of the two targets are used for clarity. The ghosts occur when coincidence occurs in the range resolver between the return from target 1 on one PRF and the return from target 2 on the other PRF.

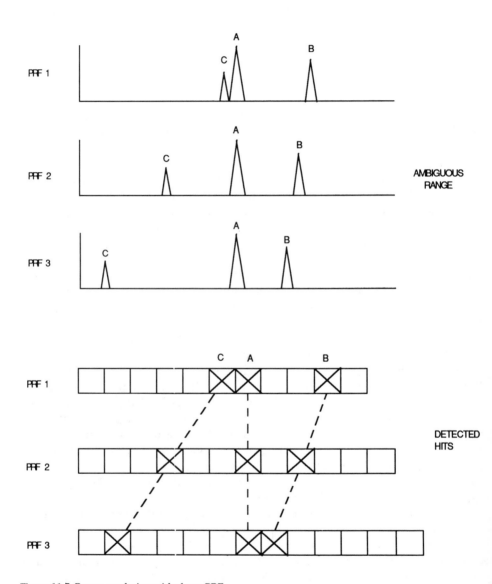

Figure 11.7 Range resolution with three PRFs.

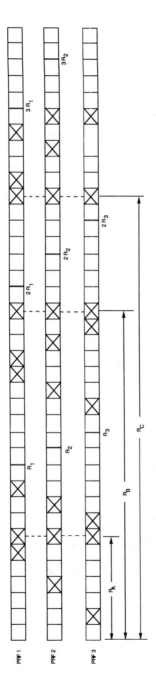

Figure 11.8 Hit pattern repeated.

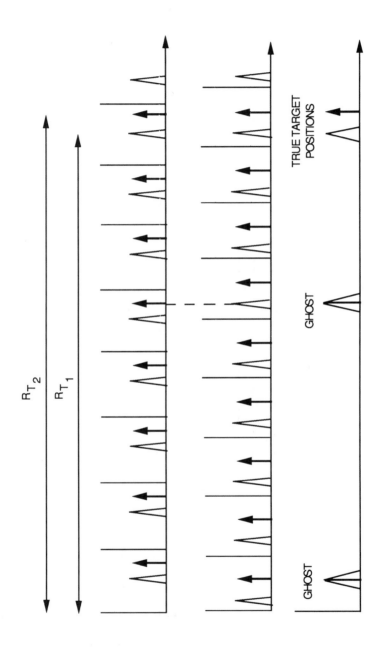

Figure 11.9 Ghosting.

Reference

[1] Hovanessian, S. A., *Radar System Design and Analysis*, Norwood, MA: Artech House, 1984, p. 95.

CHAPTER 12

Selection of the Medium PRFs

Guy V. Morris

The probability of detection P_d, computed using (11.2), is actually $P_d|C$; that is, the probability of detection given that the target return is in a clear range-doppler region. The global probability that a target will be detected during an antenna beam dwell P_{dg} is

$$P_{dg} = (P_d|C)P_c \qquad (12.1)$$

where P_c is the probability that the target is in a clear range-doppler region.

This chapter deals with choosing a set of PRFs that simultaneously achieve the required P_c and are satisfactory for range resolving. The topics include:

1. The major-minor PRF selection method;
2. The $M{:}N$ PRF selection method;
3. The use of blind zone charts to confirm PRF choice;
4. Summary of trade-offs in PRF selection.

The major-minor technique was the first one used in the development of medium-PRF modes, and radars of this type are still in service. It has been largely displaced by the $M{:}N$ method for radars that use a mechanically scanned antenna. Radars that use an electronically scanned antenna can efficiently use a PRF control strategy based on the major-minor technique; therefore, the major-minor method will continue to be used.

12.1 MAJOR-MINOR PRF SELECTION METHOD

The major-minor method can be summarized as follows:

1. Choose major PRFs to cover the doppler frequencies of interest. A minimum of two PRFs is required; usually, three are used.

2. Choose two minor PRFs associated with each major PRF to provide range resolution. The range resolver associates the detections on the minor PRFs with the respective major PRF only.

The method will be explained by describing the overall considerations and by presenting a representative example.

12.1.1 Selection of the First Major PRF

Several factors influence the choice of the PRF, including:

1. Ensuring that an acceptably high percentage of the spectral region between PRF lines is free of main-lobe clutter (or the main-lobe rejection filter);
2. Ensuring that the duty cycle (i.e., the average power) is sufficient to achieve the required probability of detection P_d;
3. Equipment considerations such as the maximum allowable duty cycle;
4. Other considerations that affect the choice of pulse width, such as range resolution, visibility of targets at short ranges, and size of the clutter cell.

Each of these factors is discussed below.

The width of the main-lobe clutter can be estimated using (2.7). For this example, assume that the maximum main-lobe clutter width at the maximum antenna scan angle is 2.125 kHz. Each PRF must be greater than 8.5 kHz for the PRF to provide a P_c greater than 0.75.

The minimum range requirement of 1,500 ft (i.e., the range at which the return from the target arrives when the receiver is off) corresponds to a maximum pulse width of 3 μs. Practical transmitter turn-on and turn-off times will cause the maximum pulse width to be somewhat less, perhaps 2.9 μs. If the maximum pulse width of 3 μs is used, the maximum duty cycle restriction results in a maximum PRF of 10 kHz. The difference between the maximum and minimum PRFs should be at least equal to the main-lobe clutter width, which in this example is 2.125 kHz, so that the set has a high P_c. Using a pulse width of 3 μs does not provide the desired degree of freedom of selection in the selection of the PRF. A constant pulse width and range gate separation of 2 μs will be used for the remainder of this example. The selection of the 2-μs pulse width is subject to verification to ensure that the required $P_d|C$ is achieved. The effect of duty cycle on PRF selection was described in Chapter 11. Detection calculations are discussed in detail in Chapter 14. Assume, for the purpose of this example, that the detection calculations indicate a duty cycle of 0.019 is satisfactory. The first major PRF, PRF_{a1}, and the corresponding number of range gates, N_{a1}, are established using

$$N_{a1} = \text{int}\left(\frac{1}{d_u}\right) = 52 \tag{12.2}$$

$$PRF_{a1} = \frac{1}{\tau N_{a1}} = 9.615 \text{ kHz} \qquad (12.3)$$

where

int( ) = integer part;
τ = range gate separation.

12.1.2 Selection of the First Set of Minor PRFs

Two criteria are used to select the two minor PRFs associated with the major PRF. First, the minor PRF should cause the apparent range of a target located in the region between R_u and $2R_u$ to change by at least two range bins. A shift of two range bins is recommended to improve range resolver accuracy, because the target may straddle range bins and be detected in both. Second, the interpulse period, as measured in number of range gates, should have factors that are relatively prime to ensure that the system unambiguous range after range resolving, R_{us}, is satisfactory. For this example, a design goal of $R_{us} \geq 150$ km (81 nmi) will be used.

Using the first criteria, the number of 2-μs range intervals, N_{a2} and N_{a3}, corresponding to minor PRFs PRF_{a2} and PRF_{a3} are established:

$$N_{a2} = 52 - 2 = 50$$
$$N_{a3} = 50 - 2 = 48$$

The coambiguous range is determined by calculating the LCM of each pair that can be formed from the three PRFs, which are

LCM of N_{a1}, N_{a2} = 1,300
LCM of N_{a2}, N_{a3} = 1,200
LCM of N_{a3}, N_{a1} = 624

The smallest LCM is 624, which translates to an R_{us} of 187 km (101 nmi.) The minor PRFs computed using (12.3) are

$$PRF_{a2} = 10.000 \text{ kHz}$$
$$PRF_{a3} = 10.417 \text{ kHz}$$

12.1.3 Selection of the Second Major PRF Set

The second major PRF, PRF_{b1}, is chosen to shift the clutter notch as shown in Figure 12.1. The width of the clutter notch was previously calculated as 2.125 kHz. The minimum value of PRF_{b1} is

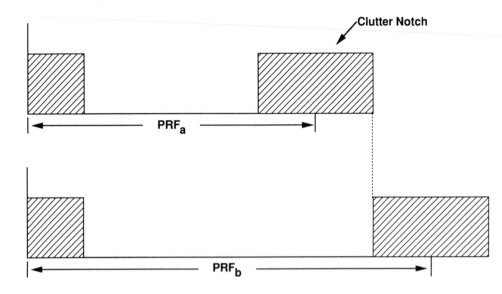

Figure 12.1 Shift of the clutter notch.

$$\text{PRF}_{b1} = (9.615 + 2.125)\,\text{kHz} = 11.740 \text{ kHz}$$

Again using the constant range gate interval of 2 μs, the integer $N_{b1} = 42$ yields

$$\text{PRF}_{b1} = 11.905 \text{ kHz}$$

The remainder of the PRFs in the set are

$$\text{PRF}_{b2} = 11.364 \text{ kHz} \qquad \text{N}_{b2} = 44$$
$$\text{PRF}_{b3} = 10.870 \text{ kHz} \qquad \text{N}_{b3} = 46$$

12.1.4 Selection of the Third Major PRF Set

If the third set is computed in the same way as the second (that is, adding 2.125 kHz to PRF_{b1}), the results are

$$\text{PRF}_{c1} = 13.889 \text{ kHz} \qquad N_{c1} = 36$$
$$\text{PRF}_{c2} = 13.158 \text{ kHz} \qquad N_{c2} = 38$$
$$\text{PRF}_{c3} = 12.500 \text{ kHz} \qquad N_{c3} = 40$$

Computation of the LCMs indicates that this set does not meet the requirements for R_{us}. PRF_{c1} can be increased slightly to yield

$$PRF_{c1} = 14.286 \text{ kHz} \quad N_{c1} = 35$$
$$PRF_{c2} = 13.514 \text{ kHz} \quad N_{c2} = 37$$
$$PRF_{c3} = 12.821 \text{ kHz} \quad N_{c3} = 39$$

The highest PRF of our nine-PRF ensemble is PRF_{c1}. Therefore, the maximum duty cycle d_u is

$$d_u = PRF_{c1}t = 1/N_{c1} = 0.0286$$

which is less than the maximum allowable value of 0.03.

12.2 *M:N* PRF SELECTION METHOD

The objectives and characteristics of the *M:N* method can be summarized as follows:

1. A set of *N* PRFs is selected to ensure that the target will be in a clear doppler region for at least *M*, and often *M* + 1, PRFs.
2. Detection on at least *M* PRFs is required for detection. Usually, *M* = 3.
3. Any three of the PRFs may be used for range resolving; that is, none is designated as major or minor.

Consider the following example, in which we design a system having eight PRFs instead of the nine required by the major-minor method. Assume the following definitions:

M:N	=	3:8;
τ	=	2 μs;
δf	=	2,125 Hz = clutter spread;
δPRF	=	$\delta f/4$ = 531 Hz = change between PRFs.

Figure 12.2 shows the relative frequency of the clutter notch at each of the eight PRFs in the region of the first doppler ambiguity. In the region of the second doppler ambiguity, the frequency separation of the clutter notch is 2 δPRF.

The maximum PRF permitted by the duty cycle limitation is 15 kHz. The first estimate of the set of PRFs is:

PRF_8	=	15.000 kHz
PRF_7	=	14.469 kHz
PRF_6	=	13.938 kHz
PRF_5	=	13.407 kHz
PRF_4	=	12.876 kHz
PRF_3	=	12.345 kHz
PRF_2	=	11.814 kHz

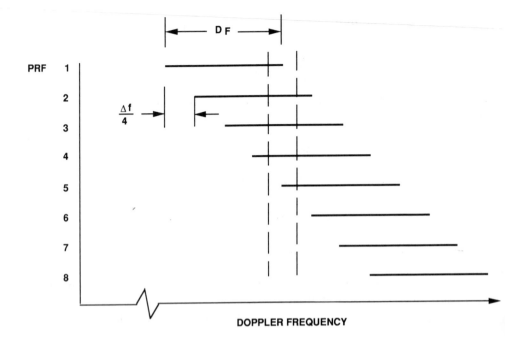

Figure 12.2 Clutter notch versus PRF.

$$PRF_1 = 11.283 \text{ kHz}$$

The next step is to adjust the PRFs to provide an integer number of range bins using

$$N_N = \text{int}\left(\frac{1}{PRF_N \tau}\right) \tag{12.4}$$

The PRFs become:

PRF_8	=	14.706 kHz	$N_8 = 34$
PRF_7	=	14.286 kHz	$N_7 = 35$
PRF_6	=	13.889 kHz	$N_6 = 36$
PRF_5	=	13.158 kHz	$N_5 = 38$
PRF_4	=	12.821 kHz	$N_4 = 39$
PRF_3	=	12.195 kHz	$N_3 = 41$
PRF_2	=	11.627 kHz	$N_2 = 43$
PRF_1	=	11.111 kHz	$N_1 = 45$

12.3 OTHER PRF SELECTION AND AMBIGUITY RESOLUTION METHODS

Thomas and Berg [1] have applied discrete combinatorics to the selection of the PRFs and to the evaluation of the blind zone performance. Mathematical relationships are presented that describe the location of potential ghosts and the number of clear PRFs at each of the unambiguous range gate positions. The blind zone charts presented in the following section were computed using a brute force approach.

Research is also under way to apply analog neural networks to resolve range ambiguities [2].

12.4 BLIND ZONE CHARTS

In both the major-minor example and the $M:N$ example, the differences in PRF were chosen to provide a minimum of three clear PRFs in the frequency region near the first ambiguity. One method for understanding the regions in which detections are possible is to make a range-doppler blind zone chart covering the entire range-doppler region of interest for each PRF similar to those presented in Figures 11.1 and 11.2. Then each of these maps could be overlaid to determine how many PRFs are clear at each range-doppler cell, what percentage of the total area meets the criterion of three clear PRFs, and where the blind zones of the PRF ensemble are located. Obviously, a detailed evaluation of the PRF set is best accomplished using a computer. A computer program, executed on a personal computer, was used to generate the blind zone illustrations. The computer program operation and a method for obtaining it is described in Appendix B.

Figure 12.3 shows the blind zone evaluation of the $M:N$ example. The probability of being clear on at least three of the eight PRFs is $P_c = 0.9708$.

Although the computer program was designed to be used with the $M:N$ strategy, an operating method is presented for using the program to evaluate PRF ensembles designed with the major-minor strategy. Each major-minor set can be evaluated individually by requiring detection on each of the three PRFs. P_c for each major set of the example is:

$$PRF_a \quad P_c = 0.6524$$
$$PRF_b \quad P_c = 0.5793$$
$$PRF_c \quad P_c = 0.6068$$

The P_c for the total of the three sets may be estimated using the computer program of Appendix B and the following procedure:

1. Require detection on one of three PRFs;
2. For the PRFs, use the center frequencies of each PRF set;
3. Increase the clutter notch width by the difference between the maximum and minimum PRFs in the set (for example, $PRF_{c1} - PRF_{c3}$).

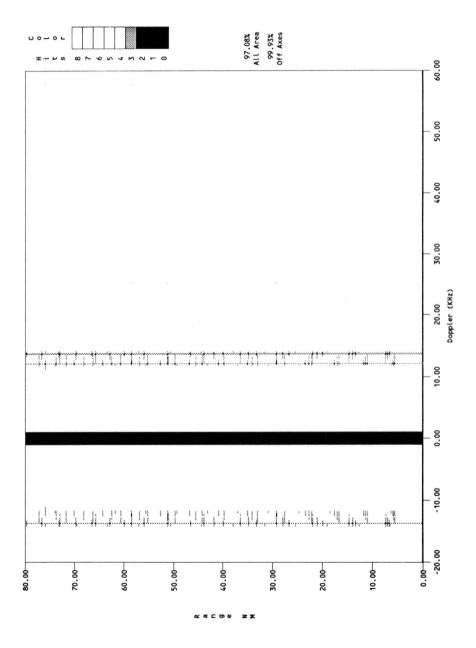

Figure 12.3 Blind zone chart showing regions clear on at least three of eight PRFs: M,N example.

This procedure yielded $P_c = 0.9632$ for the major-minor example. Figure 12.4 shows the blind zone chart.

12.5 COMPARISON OF *M:N* WITH MAJOR-MINOR

The major-minor method effectively widens the blind region of each major set, as shown in Figure 12.5, because detection is required on the major and both minor PRFs. Therefore, the effective clutter notch width of the example is not the 2,125-Hz region around each individual PRF. The notch is increased by an amount equal to the separation between PRF_{c1} and PRF_{c3} to a value of 3,116 Hz. Three major PRFs are usually required because of the widening of the notch.

The total antenna dwell time must be divided between the ensemble of PRFs. If the number of PRFs can be reduced, by an improved PRF selection strategy, without reducing P_c, then more time can be allotted to each PRF. This increase in dwell time increases the SNR and therefore increases P_d.

12.6 DWELL TIME ALLOCATION

Each time the PRF is changed, a "dead time" must be allowed equal to the round-trip propagation time to the maximum range of significant clutter. In our example, a propagation delay of 730 μs, which corresponds to approximately 110 km (60 nmi) will be used. Consider a radar with a mechanically scanned antenna, in which the time the target is within the 3-dB antenna beamwidth is a constant. All PRFs are used during the on-target time, regardless of the detection outcome. The average interpulse period for our *M:N* example is 77.25 μs. If 32 pulses are coherently integrated during each of the eight PRFs, then the minimum dwell time T_{dmin} required is given by

$$T_{dmin} = [(77.25)(32) + 730](8) = 25,616 \ \mu s$$

Some radars use the returns from a number of pulses at the beginning of each PRF to set the automatic gain control. (These pulses are not doppler-processed and are referred to as *burn* or *fill* pulses.) If a typical value of 16 fill pulses are used, the dwell time in our example, T_d, becomes

$$T_d = 35,504 \ \mu s$$

If the on-target time provided by the mechanically scanned antenna is longer that T_d, the alternatives for matching the time include increasing the antenna scan rate, performing two dwells at each PRF and noncoherently integrating the two, and increasing the number of pulses coherently integrated.

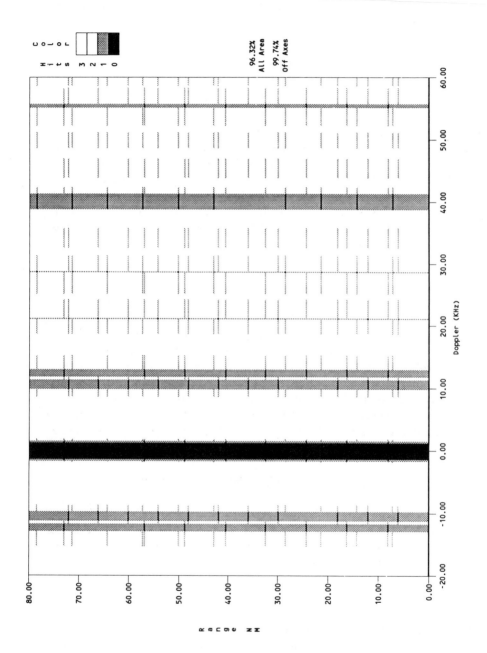

Figure 12.4 Blind zone chart: major-minor example.

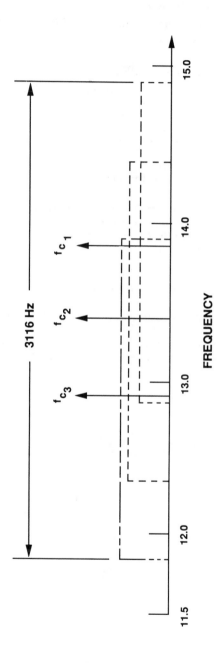

Figure 12.5 Blind zone broadening.

Now consider a radar with an electronically scanned antenna that permits adaptive control of the on-target time. One possible PRF control strategy based on the major-minor method is as follows:

1. Transmit using the three major PRFs;
2. Use the minor PRFs associated with a major PRF only if a target was detected with the major.

If no targets are detected on any major PRF, the dwell time is a minimum.

References

[1] Thomas, A. G., and M. C. Berg, "Medium PRF Set Selection: An Approach Through Combinatorics," *IEE Proc. Radar, Sonar, Navig.*, Vol. 141, No. 6, December 1994, p. 307.
[2] Wang, C.-J., and C.-H. Wu, "Analog Neural Networks Solve Ambiguity Problems in Medium PRF Radar Systems," *1993 IEEE International Conference on Neural Networks*, Vol. 1, p. 120.

Suggested Reading

Simpson, J., "PRF Set Selection for Pulse-Doppler Radars," *Proc. IEEE 1988 Region Five Conference*, March 1988.
Trunk, G. V., and M. W. Kim, "Ambiguity Resolution of Multiple Targets Using Pulse-Doppler Waveforms," *IEEE Trans. Aerospace and Electronics*, October 1994, Vol. 30, No. 4, p. 1130.

CHAPTER 13

Tracking Techniques

Guy V. Morris

The purpose of tracking is to improve the estimates of target position and velocity relative to those that can be provided in the search mode. The improvement in accuracy is achieved by a combination of the following techniques:

1. Maintaining narrow gates around the target to reject clutter and interference and improve the SNR;
2. Increasing the data rate;
3. Using many samples to derive smoothed estimates of present position and velocity and predictions of the position and velocity at the next sampling time.

Historically, the term *tracking* has been generally used to denote the single-target tracking (STT) mode, that is, controlling the doppler-tracking filter, range tracking gate, and antenna position to continuously illuminate the target. Many modern systems also provide multiple-target tracking (MTT) using a track-while-scan (TWS) mode. The tracking techniques that are discussed in this chapter are:

1. Angle tracking in STT;
2. Range tracking in STT;
3. Doppler tracking in STT;
4. TWS.

A basic principle of tracking (angle, range, or doppler) is that the target response must be measured at two or more positions (in angle, range, or doppler frequency). The two measurements are needed to determine in which direction and by what amount the tracking gates should be moved. The two measurements

---

(antenna beam positions, range gate positions, or doppler frequencies) can be achieved by using a single beam (range gate or doppler filter) time-shared at various positions or by using multiple simultaneous beams (range gates or doppler filters).

13.1 FUNDAMENTALS OF ANGLE TRACKING

13.1.1 Amplitude Comparison

The angle measurements can be made using either amplitude comparison or phase comparison. Amplitude comparison is illustrated in Figure 13.1. The antenna creates two similar beams that are separated in angle by a fraction of a beamwidth. The angle at which the amplitude of the two beams are equal is defined as the angle zero reference point. Let the voltages produced in the two beams by a point source be v_1 and v_2. A calibration curve can be constructed of voltage ratio (v_1/v_2) in decibels versus angle that is independent of target signal strength. When (v_1/v_2) of the target is measured, the angle away from zero may be determined using the calibration curve. The usual use of the angle-off measurement is to correct the antenna position to drive the angle-off to zero.

A single beam can be switched to the different positions. Several beam-switching methods are possible, but the most common method of time-sharing a single beam is a conical scan. The beam is moved continuously in a circular pattern, as shown in Figure 13.2. The return from a target is amplitude-modulated by the beam motion. The depth of the amplitude modulation is a function of the angle from the center of rotation θ. The phase of the modulation envelope relative to the angle reference is a function of ϕ.

The usual conical scan implementation moves both the transmit beam and the receive beam in space. The conical scan frequency is disclosed to the target, which, if unfriendly, can use this information to disrupt tracking. Therefore, most military radars no longer use conical scan. The measurements required for tracking can be achieved by moving only the receive beam. Some systems use schemes that produce a received signal that resembles that of a conical scan system. The methods are called *lobe on receive only* or *silent lobing*.

13.1.2 Phase Comparison

Phase comparison is illustrated in Figure 13.3. Two beams are used that have the same amplitude pattern and point in the same direction but have a displaced phase center. The amplitude of the signal received from the target is the same on each of the two beams. The direction of arrival is determined from the relative phase difference. Phase comparison systems of the type shown in Figure 13.3 use the antenna aperture inefficiently. Two separate beams are formed, each using half the aperture, instead of a single higher gain beam using the full aperture.

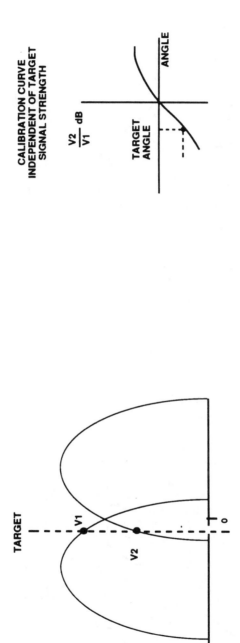

Figure 13.1 Amplitude comparison angle measurement.

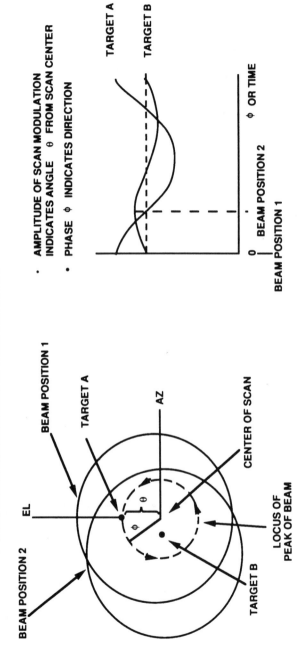

Figure 13.2 Conical scan.

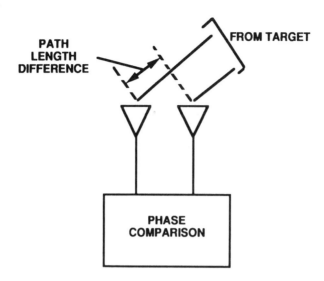

Figure 13.3 Phase comparison.

13.1.3 Monopulse

The most commonly used antenna in airborne systems is the monopulse planar array antenna. The name *monopulse* was chosen because it is theoretically possible to derive the angle error from a single pulse. As will be discussed later in this chapter, many pulses are processed in a doppler radar to derive the angle error. A planar array is one that is composed of a number of radiating slots cut in a plane surface. The planar array is subdivided into independent sections, usually quadrants, which are fed by a microwave sum and difference network. For simplicity, consider a planar array subdivided into two sections: A and B, as shown in Figure 13.4. The two sections constitute a phase comparison antenna. However, the microwave sum and difference network creates two beams, the sum (Σ) and difference (Δ). The gain and beamwidth of both the Δ and Σ patterns are determined by the total aperture (A + B). The null of the Δ pattern occurs at the peak of the Σ pattern.

A calibration curve can be constructed that is analogous to the amplitude comparison system. The ratio used is Δ/Σ, which is independent of target strength. The magnitude of the Δ pattern is symmetrical about zero angle error, but is in phase with the sum channel on one side of zero and 180° out of phase on the other. Since the sense of right versus left is contained in the phase information, the phase angle between the Δ and Σ channel must be maintained throughout doppler processing.

Figure 13.5 illustrates a four-quadrant monopulse antenna that is used to track in azimuth and elevation. The three beams, Δ_{el}, Δ_{az}, and Σ, each use the entire

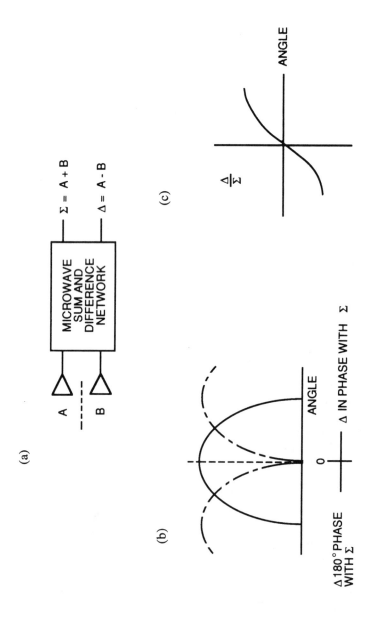

Figure 13.4 Monopulse: (a) generation of sum and difference; (b) gain; and (c) calibration curve.

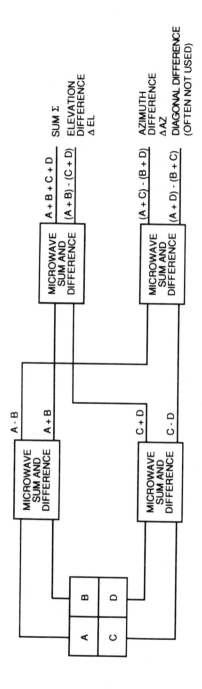

Figure 13.5 Simultaneous azimuth and elevation monopulse.

aperture. The book by Sherman [1] contains a detailed exposition of monopulse antennas.

13.1.4 Clutter and Multipath Interference

Airborne radars are required to angle-track low-flying targets in a look-down scenario. The return from the target will have to compete with clutter and multipath interference signals entering through the antenna main lobe. For smooth surfaces (e.g., a calm sea), the clutter, which is proportional to the backscatter coefficient, is low, but the multipath interference, which is a function of the forward scatter, is high. For rough surfaces, the converse is true.

The total clutter power can be reduced by using a narrow-beam antenna with low side lobes. The narrow main lobe attenuates the clutter in the range cell containing the target because the ground patch is not at the peak antenna gain. Narrow range gates also reduce the size of the ground clutter patch that competes with the target return. If the target has a component of velocity in the direction of the radar, then the doppler of the target differs from the ground patch. Therefore, doppler processing is very effective for eliminating clutter prior to the generation of the angle-tracking error signals.

Multipath interference is not so readily rejected in a doppler radar as clutter. The multipath interference creates a target image with an angular position that appears to be below the surface by the same distance as the true target ground clearance, as shown in Figure 13.6. The one-way difference between the direct path and the forward scatter path, ΔR, is approximately

$$\Delta R = \frac{2h_a h_t}{R} \tag{13.1}$$

where:

h_a = aircraft altitude;
h_t = target altitude;
R = range.

The strongest multipath interference results from a single ground reflection. Two single-bounce paths exist, one direct from the radar to the target and returning from the target via the ground, and the second from the radar to the target via the ground and returning directly to the radar. For a single-bounce path, the range difference is $\Delta R/2$. The double-bounce path (that is, from the radar to the target via the ground and returning to the radar via the ground) suffers a double reflection loss and has a range difference of ΔR. The range difference is often less than the range resolution of the radar, and therefore the multipath interference cannot be rejected by range gating. For example, for h_a = 3,000m (9,843 ft), h_t = 300m

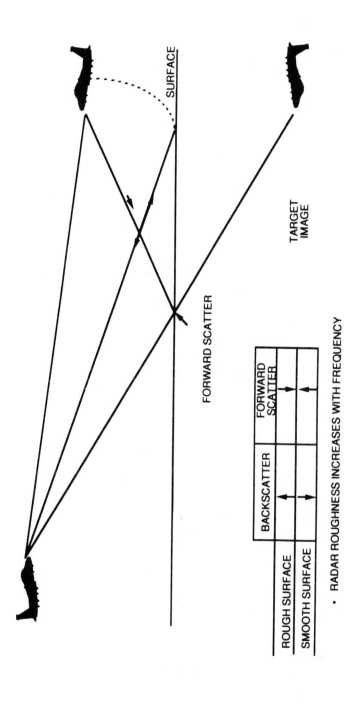

Figure 13.6 Clutter and interference geometry.

(984 ft), and $R = 75$ km (40.5 nmi), the single-bounce range difference $\Delta R/2 = 12$m (39 ft), which corresponds to a round-trip propagation time of 0.08 μs. Typical airborne radar pulse widths are 0.4 to 1.2 μs.

When ΔR is small, then the difference in doppler between the direct target return and the target image is small. Therefore, the image target cannot be rejected by doppler processing. The most effective way of reducing multipath interference is to use a narrow-beam antenna with low difference-pattern side lobes.

13.2 FUNDAMENTALS OF RANGE TRACKING

Range tracking is actually time delay tracking, of course. The early-late gate method of range tracking is widely used and will be described here. Consider the time waveforms shown in Figure 13.7. The range tracker is synchronized with the transmit pulse. A smoothed estimate of target range is available from the range tracker or from the acquisition circuits during the initial range lock-on sequence. Two sampling pulses are generated: one just prior to the time corresponding to the smoothed range estimate and one just after. The received video samples are used to form a time discriminator. The range-tracking loop adjusts the smoothed estimate of target range until the amplitudes of the early and late gate samples are equal.

A typical nondoppler range tracker is shown in Figure 13.8. The tracker is a double integrator; that is, a second-order control system. The integrators adjust the value of the smoothed range so that the gate is at the predicted position at the time of the next range sample. The system will track a target moving at constant velocity with zero range error. Note that the loop also provides a smoothed estimate of range rate. Other range gates, slaved to the center of the smoothed range, sample the Σ, Δ_{az}, and Δ_{el} video channels for use in generating the angle-tracking error signals.

The range tracker, which is shown functionally in Figure 13.8, is often implemented digitally. The range sweep function is a counter that is driven by a stable oscillator serving as the master clock. The comparators are AND gates. The integrator is an accumulator. The outputs of smoothed range are contained in registers and are often multiplexed with other data and sent to the system central computer.

13.3 DOPPLER TRACKING

Doppler frequency tracking provides clutter rejection that enables continuous range and angle tracking of a single moving target. The STT mode also provides the highest data rate and potentially the most accurate measurement of current target position. Although clutter rejection is the most common application, the tracking bandwidths can be narrowed considerably from those used for detection, providing an improvement in signal-to-noise or signal to interference ratio. Since velocity, or

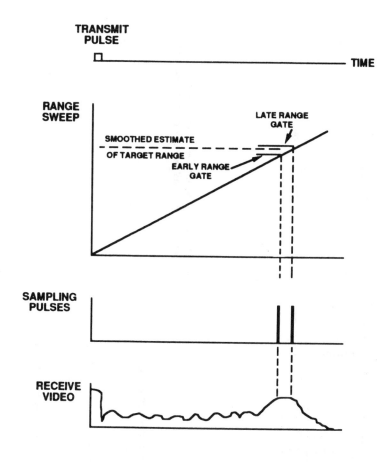

TRANSMIT PULSE

TIME

RANGE SWEEP

LATE RANGE GATE

SMOOTHED ESTIMATE OF TARGET RANGE

EARLY RANGE GATE

SAMPLING PULSES

RECEIVE VIDEO

- EARLY/LATE GATES FORM TIME DISCRIMINATOR
- USED TO ADJUST ESTIMATE OF TARGET RANGE

Figure 13.7 Range-tracking waveforms.

more precisely doppler frequency, is measured directly, predictions of future target position can be made more accurately.

Table 13.1 summarizes the types of radars, or radar modes, for which doppler tracking will be discussed. In the STT mode, the target accelerations between sample times may be small enough so that adequate tracking accuracy is provided by simple first-order tracking loops. However, at short ranges, target accelerations may necessitate higher order prediction methods. Kalman filters can greatly improve tracking accuracy. Kalman filters permit higher order modeling of the expected target motion and adaptively weight the current observation and track history.

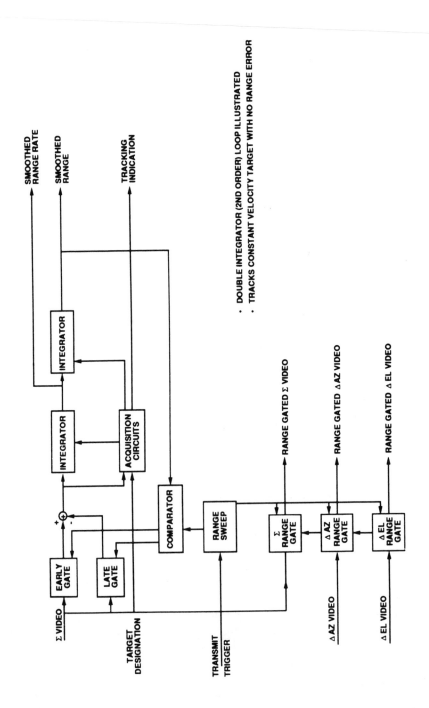

Figure 13.8 Range tracker block diagram.

Table 13.1
Typical Doppler-Tracking Methods

Tracking Mode	Radar Type	Sampling Rate		Doppler-Tracking Bandwidth	Typical Tracking Methods
Single target (Continuous Sampling)	CW	IF Bandwidth	(Highest)	Narrowest	Phase-lock loop
	High PRF	PRF			
	Low PRF	PRF			α-β; simplified Kalman filter
Multiple target (intermittent sampling)	Phased array	PRF burst: frequent revisit			
	Mechanical Scan	PRF Burst; less frequent revisit	(Lowest)	Broadest	Multistate Kalman filter

The α-β tracker is often used because of the relative simplicity of the calculations compared to the Kalman filter. The α-β tracker may be viewed as a special case of the Kalman filter, having nonadaptive weights.

Doppler frequency tracking will be explained by describing:

1. The fundamentals using a stationary CW radar example and then extending the concepts to high-PRF pulse-doppler radar;

2. The additional considerations applicable to low- and medium-PRF radars that also include range-tracking loops;

3. The adaptations of the tracking methods that are necessary for airborne radar.

Throughout this discussion, we will use the terms *measurement* (the current doppler frequency, as indicated by the frequency measurement method, such as a doppler filter bandwidth); *filtering*, also commonly called *smoothing* (combining a number of measurements to form an estimate of the current doppler that is presumably more accurate than a single measurement); and *prediction* (estimating the position, velocity, or acceleration at some time in the future).

13.3.1 CW Radar

A simplified block diagram of a CW tracking radar is shown in Figure 13.9. The doppler-tracking processor tracks the doppler frequency and performs other radar functions. For example, the signal-to-interference benefits of doppler tracking must be applied to the angle error channels as well. Also, when range must be measured, a secondary frequency modulation is applied to the transmit signal. Each of these functions must be accommodated by the doppler tracker.

A doppler-tracking loop is shown in Figure 13.10. For the moment, it will be described as if it were an analog mechanization. The primary tracking function is indicated by the bold signal flow path. A narrowband tracking filter is synthesized by down-converting the input signal with a voltage-controlled oscillator (VCO) so that the desired signal falls within a narrow fixed-frequency bandpass filter. This is the first in a series of successive steps of bandwidth reduction. The signal-to-receiver noise improvement is the ratio of the IF to the doppler-tracking filter bandwidth. Frequently, however, the most important benefit is the attenuation of large unwanted signals such as ground clutter, transmitter leakage, or other interference. The center frequency of the bandpass filter is a function of the technology used, but is chosen so that the ability to discriminate between opening and closing velocities (i.e., negative and positive doppler frequencies) is retained. A center frequency of 60 kHz might be enough for an X-band radar to preserve the doppler sense, but crystal filter technology can result in center frequencies of 10 MHz.

The elements from the phase-sensitive detector through the VCO represent a phase-locked loop. In the quiescent state, the VCO frequency is $f_D + f_2$ and represents a smoothed estimate of the target doppler plus a constant offset f_2. The measurement accuracy is established by the bandpass filter and phase-sensitive detector bandwidths. The low-pass filter and the integrator perform the filtering function. Figure 13.10 depicts a single-integrator loop, commonly called a *type 1 tracking loop*. This system tracks a constant-velocity target without bias but has a fixed bias error when tracking a target with constant acceleration. Double-integrator, or *type 2*, loops are often used when significant target accelerations are anticipated. The integrators also provide the prediction function. For example, if the signal is lost to a single integrator loop, the loop predicts that the target velocity is constant. Similarly, for a double integrator, the loop holds the previous acceleration. For most CW radars, these simple tracking loops are adequate because the high data rate ensures small changes in measured velocity between samples. More sophisticated filtering and prediction methods will be discussed later.

Several other functions are often included within the doppler frequency tracker. The acquisition circuits sweep the VCO about the designated frequency until a target is detected above the threshold. Matched bandpass filters, slaved to the doppler-tracking loop, are used in each of the monopulse angle-tracking channels.

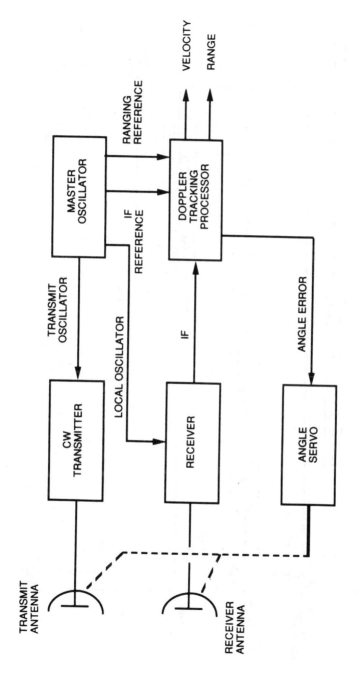

Figure 13.9 Block diagram of a CW radar.

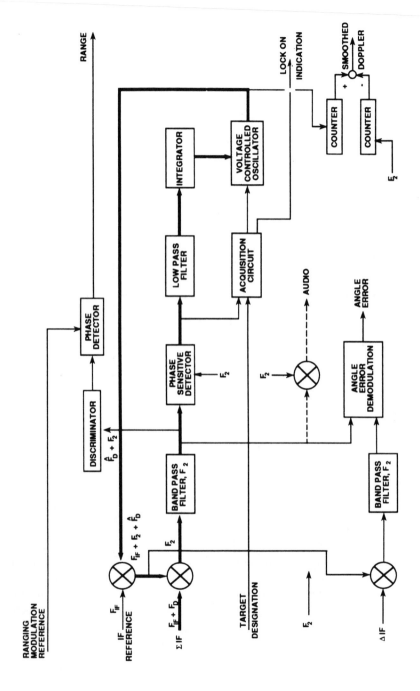

Figure 13.10 Doppler-tracking loop.

13.3.1.1 Ranging

After target tracking has been established, some systems add a ranging modulation to the transmitted signal. Sinusoidal FM is a common choice. The modulation frequency is chosen so that the range can be determined unambiguously by measuring the phase difference between the transmitted and received modulation waveforms. The target doppler has spectral sidebands at integer multiples of the modulating frequency. The target doppler may be translated to a convenient IF, and one of the sidebands can be detected for range measurements.

13.3.1.2 Audio Output

Complex targets with visible moving parts, such as jet engine compressor blades, or vibration modulate the amplitude and doppler frequency of the target return producing spectral sidebands. These same mechanical movements can produce sound waves in air. If the target doppler is translated to zero frequency, the resulting sidebands within the audio region can be input to the operator headset. Frequently, there is a close resemblance between the actual sound and that produced by the radar, and the operator is able to identify the type of target. Spectral analysis and pattern recognition processing can be used to perform the function automatically. Spectral sidebands are not always helpful and must be accounted for in the design. They can cause false lock-ons and other spurious effects.

13.3.1.3 Digital Implementations

The tracking implementation of Figure 13.10 can be implemented in digital form. Generally, the IF representing either the frequency of main-lobe clutter in low- and medium-PRF or the upper edge of side-lobe clutter in high-PRF radars is translated to zero frequency, as shown in Figure 8.1. The sequence of in-phase (I) and quadrature (Q) sample pairs preserves the ability to discern between opening and closing doppler frequencies. The A/D conversion rate must be greater than or equal to the Nyquist criterion of one sample pair every $1/B$ seconds, where B is the IF bandwidth. The digital equivalents of the functions of Figure 13.10 can be synthesized. For example, an integrator is an add and accumulate operation. Translating the spectrum of the time waveform by ω_s is achieved by rotating the phase of each of the complex sample pairs by sampled values of $(\cos \omega_s t + j \sin \omega_s t)$.

The filtering function may be performed using an FFT or with a specially designed multipole digital filter. The target response in two adjacent filters of an FFT can be used to provide the target velocity measurement or frequency discriminator function that is required for doppler tracking. One method is depicted in Figure 13.11. A calibration curve of A_r versus frequency (or relative velocity) can be constructed, where A_r is the ratio of the amplitude responses of two adjacent

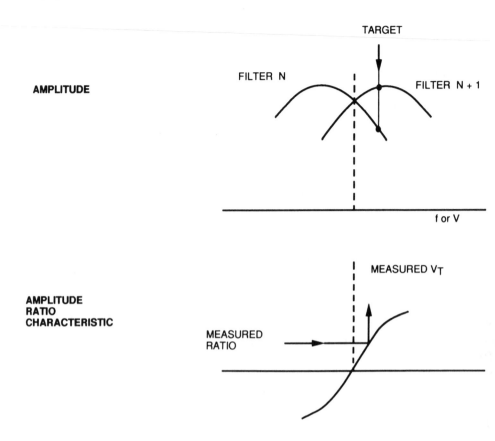

Figure 13.11 Target velocity measurement.

doppler filters. The measured value of A_r for a target is a measure of the velocity error relative to the filter crossover point. This error can be used to control the frequency of the oscillator used to make the down-conversion to video, and thus keep the target return at the crossover point. Alternatively, the error can be used to correct the velocity in a digital tracking loop such as the α-β tracker to be discussed later in this chapter.

The flexibility of digital processing often permits the same hardware to be used in several radar modes, such as surveillance and track. For some applications, an FFT spectral analysis, which synthesizes a contiguous bank of narrowband filters covering the entire doppler region, might be used for initial detection and for frequency measurement during tracking. The FFT can simultaneously provide information on modulation sidebands that result from ranging, conical scan, and target-induced effects. In practice, the optimum bandwidths necessary to accomplish tracking and sideband demodulation may be sufficiently different that separate FFTs with different sample sizes and time durations would be required.

13.3.1.4 Frequency Measurement Accuracy

The error in measuring a single isolated frequency in the presence of noise can be expressed as [2, pp. 7, 12]

$$\delta f = \frac{\sqrt{3}}{\pi \tau_0 \sqrt{2S/N}} = \frac{\sqrt{3}\Delta f}{\pi \sqrt{2S/N}} \tag{13.2}$$

where:

δf = rms frequency error;
Δf = spectral width;
S/N = SNR;
τ_0 = observation time.

This result is directly applicable to CW radar or to a single pulse if τ_0 is taken to be the pulse width. The book by Skolnik [2] contains a compendium of formulas derived for various cases, such as trapezoidal pulses. The usual situation in the frequency measurement section of pulse-doppler trackers is that some number of samples N_p of the return are analyzed by an FFT or other filter. Equation (13.2) is then interpreted by using $\tau_0 = N_p/\text{PRF}$ or $\Delta f = \text{PRF}/N_p$. If a number of independent frequency measurements N_0 are averaged and if the doppler frequency is constant, then the rms error will be further reduced by $\sqrt{N_0}$. An analysis of various sources of error is presented in [3].

13.3.2 High-PRF Pulse Doppler

Typical high-PRF systems operate at transmitter duty factors of 0.3 to 0.5 and with a PRF sufficiently high that the target doppler is unambiguous. A single antenna is duplexed between transmit and receive. A bandpass filter is placed in the receiver IF to eliminate all frequencies except those in a region around the central spectral line, as shown in Figure 13.12. The filter output is essentially CW, and tracking is accomplished as in Figure 13.10.

The most significant difference between CW and high-PRF tracking is that high-PRF radars are blind to targets at ranges such that the echo arrives when the receiver is gated off. This characteristic is called *eclipsing*. The time during which a target is eclipsed is inversely proportional to the closing rate. One solution is to continually switch sequentially among a set of slightly different PRFs chosen to ensure that the target is uneclipsed on at least one of the PRFs throughout the range of interest. A variation of this method is to switch between subsets, usually two, of the total PRFs available. The optimum PRF pairs for each range region are determined a priori and selected as a function of range. Another method is to monitor the signal strength in the doppler tracker and initiate a PRF change when approaching eclipse.

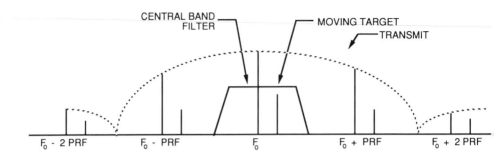

Figure 13.12 High-PRF spectra.

13.4 RANGE-GATED PULSE DOPPLER

In a range-gated pulse-doppler radar, the doppler-tracking loop and the range-tracking loop are interrelated. The doppler-tracking loop operates upon video sampled at the range indicated by the range tracker. Range tracking can be accomplished by several methods, but an early/late gate approach will be used for illustration. The primary differences between range tracking in a pulse-doppler radar and the nondoppler radar described in Section 13.2 are as follows:

1. Before the signals in the early and late gates are compared to derive the range-tracking error, each is passed through a doppler-tracking filter to reject returns from stationary objects at the same range.

2. The range-tracking loop of the nondoppler range tracker can be updated at the PRF. If an FFT that processes N_p pulses is used to perform the doppler analysis, only one output is available for N_p pulses, and the update frequency is reduced to PRF/N_p. Uncompensated lags in the tracking loop degrade accuracy and can cause loop instability.

Figure 13.13 is a simplified block diagram of range-gated doppler tracking. Initiation of tracking requires near-simultaneous initialization of all four tracking coordinates (range, two angles, and doppler frequency). The tracking acquisition sequence is usually aided or near-automatic. A typical acquisition sequence is as follows:

1. The operator, upon observing a target on which to initiate a track on the range versus azimuth display in the search mode, designates the target on the display using a cursor.

2. On the next antenna scan, if a target is detected within the designated area, the target parameters (range, azimuth angle elevation angle, and doppler frequency) are recorded.

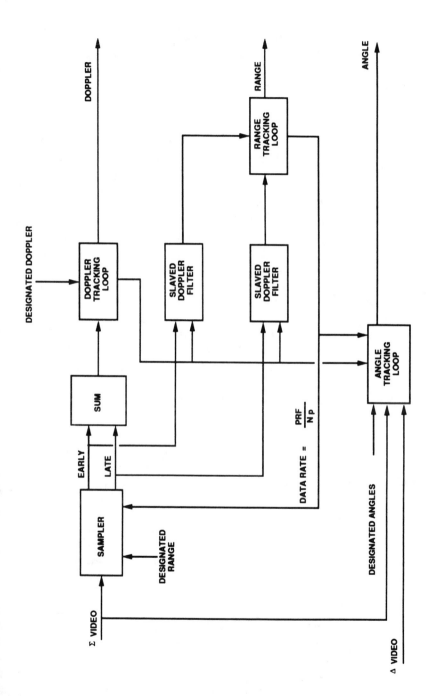

Figure 13.13 Range-gated doppler tracking.

3. The scan sequence is interrupted automatically by the acquisition processor, and the antenna is directed to return and illuminate the area where the target was detected.

4. The early-late range gate pair is moved to the estimated range of the target. A small sweep may be impressed upon the nominal position to compensate for the designation uncertainties.

5. The Σ-channel video received by the early and late gates is added together and processed by a doppler analyzer (e.g., an FFT). This video addition makes the doppler-tracking loop less sensitive to amplitude modulations caused by the range tracker. An alternative is to provide a third range gate, sometimes called the *target gate*, located between the early and late gates.

6. If a moving target is found at the designated range and is acceptably close to the doppler of the detection that triggered the acquisition sequence, then the doppler-tracking gate is placed over the target and the doppler-tracking loop is closed; that is, a frequency discriminator is used to maintain the target in the center of the tracking filter.

7. The video signals in both the early and late gates are analyzed by doppler filters that are slaved to the frequency of the doppler tracker. The range-tracking loop is closed (i.e., the difference in relative amplitudes of the early and late gates is used to control the range position of subsequent early-late sample pairs) after allowing a brief period for settling of the doppler-tracking loop.

8. The video signals in both the azimuth and elevation monopulse angle error channels are sampled by a range gate slaved to the target gate of the range tracker. The sampled video in each channel is filtered by a doppler filter slaved to the frequency of the doppler tracker. Thus, angle errors are derived only from the same range-doppler cell as the one containing the target. After a brief range tracker settling time, the angle-tracking loop is closed.

Frequently, a different set of bandwidths and gains is used during acquisition until a stable track is achieved. For many military applications, the transition from search to track is not a trivial problem. Pointing the antenna at an unfriendly target signals the intent to initiate tracking. The unfriendly target may attempt to prevent lock-on by maneuvers or electronic means.

13.4.1 Velocity Ambiguities

The choice of transmitter frequency and PRF often results in ambiguity of the measured doppler frequency. In a range-doppler-tracking radar, unambiguous radial velocity is also available from the range tracker as dR/dt and is usually sufficient to resolve the ambiguity. The ambiguity can also be resolved by varying the PRF and observing the shift in measured doppler.

13.5 DEMODULATION OF ANGLE ERRORS

13.5.1 Monopulse

Monopulse angle errors are derived by comparing the amplitudes of the output of the doppler-tracking filter in the sum Σ channel to filters at the same frequency in the difference Δ channels. In range-gated pulse-doppler systems, these doppler filters process only the data from the range being tracked by the range tracker. For the FFT implementation, a single update opportunity occurs each N_p pulses, where N_p is the number of pulses processed in a batch. The outputs of the S and D channels is shown in Figure 13.14(a). The angle error for each channel, δa, either azimuth or elevation, has the form:

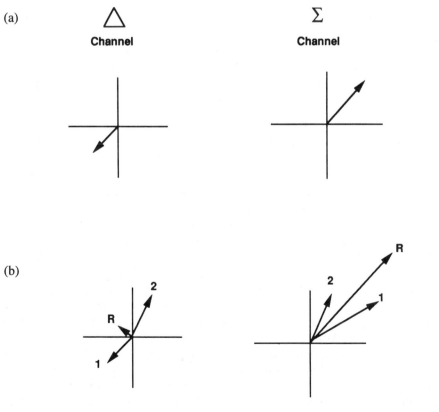

(a)

△

Channel

Σ

Channel

(b)

Figure 13.14 Monopulse angle error demodulation; (a) single signal and (b) two signals.

$$\delta a = k \frac{|\Delta|}{|\Sigma|} \cos \phi_d \qquad (13.3)$$

where:

Δ = difference channel filter output;
Σ = sum channel filter output;
ϕ_d = phase angle between sum and difference;
k = proportionality constant.

Equation 13.3 can be reformatted as

$$\delta a = k \frac{|\Delta||\Sigma|}{|\Sigma||\Sigma|} \cos \phi_d = \text{Re} \left| k \frac{\Delta\Sigma^*}{\Sigma\Sigma^*} \right| \qquad (13.4)$$

where Re denotes the real part and the asterisk superscript denotes the complex conjugate.

Ideally, ϕ_d is either 0 or 180 degrees and cos ϕ_d is either 1 or −1. Multipath interference or target jamming may result in multiple signals being present in the doppler filter. The error signal is incorrect if multiple signals are present as shown in Figure 13.14(b), and ϕ_d may be any value.

13.5.2 Conical Scan and Lobe-on-Receive

In the doppler-tracking loop of Figure 13.10, the signal being tracked is CW. A lobing frequency may be chosen so that the amplitude modulation sidebands are within the passband of the velocity tracker as shown in Figure 13.15(a). The amplitude modulation can be recovered by rectifying the signal and passing it through a filter tuned to the modulation frequency. The modulation signal must be normalized with respect to the target return so that the error signal is independent of target cross section. The normalization can be accomplished with a logarithmic element assuming the clutter has been filtered out prior to the introduction of the nonlinearity. The demodulated signal contains both the azimuth and elevation error components which then must be resolved. Figure 13.16 is a block diagram of the demodulation process.

The modulation frequency may be chosen high with respect to the track filter bandwidth, as shown in Figure 13.15(b). Separate filters are necessary to capture the sidebands. An FFT can be used to form a filter bank which provides both the tracking and angle error demodulation filters. Figure 13.15(c) depicts the target signal vector from the tracking filter overlayed with the modulation signal vector. The angle error signal must be resolved into components that are in-phase and quadrature, and normalized with respect to the target signal.

(a)

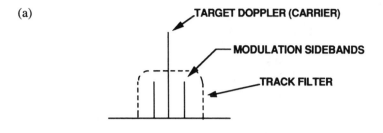

(b)

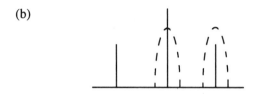

(c)

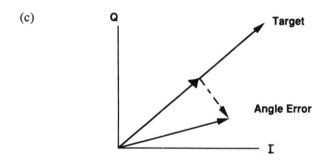

Figure 13.15 Lobe-on-receive angle error demodulation: (a) low modulation frequency; (b) high modulation frequency; and (c) filter outputs.

13.6 CLUTTER CONSIDERATIONS

13.6.1 High-PRF Radar

Figure 13.17(a) is a simplified representation of the clutter spreading that results from the radar being in motion. Side-lobe clutter in the high-PRF radar is confined

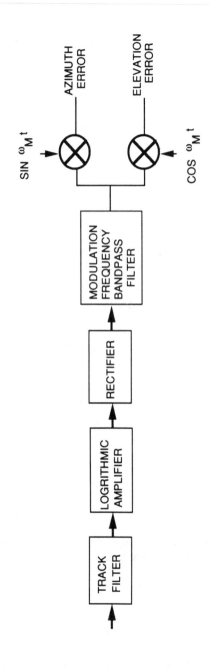

Figure 13.16 Lobe-on-receive angle error demodulation block diagram.

(a)

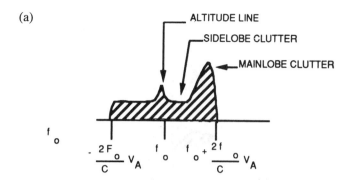

Figure 13.17 Received clutter spectra: (a) CW or central line of high-PRF radar and (b) range-gated pulse-doppler clutter.

to the frequency region that corresponds to $\pm V_a$, the velocity of the aircraft carrying the radar. The primary use of the high-PRF mode is in nose-on encounters, where initial detection and tracking is accomplished in the clutter-free region. However, the target may maneuver and cause its return to fall in the side-lobe clutter region. Therefore, it is desirable to provide tracking in the side-lobe clutter region. The strength of the side-lobe clutter is a function of the range to the ground, the ground backscatter coefficient, and the side lobes of the antenna. The ability to track in the side-lobe clutter region is obviously enhanced by providing a low-side-lobe antenna. The tracking capability is also enhanced by using a narrowband doppler-tracking filter to optimize the target-to-side-lobe clutter ratio. Equation 2.8 can be used to estimate this ratio. Generally, target tracking in the side-lobe clutter region is possible, but the sensitivity is reduced by at least the clutter-to-receiver noise ratio as measured in a doppler filter bandwidth.

The altitude line and the main-lobe clutter signals are generally of such great magnitude that true tracking of the target return is not possible while it is in the vicinity of these interfering signals. In fact, the spectrum of the signal presented to the tracking loop has often been prefiltered to remove the altitude line and main-lobe clutter signals via notch filters. The target signal is removed by the notch filter also, of course, if it is within the reject band. The integrators that are a part of the tracking loops provide a rate memory that extrapolates the rate of change of doppler frequency and positions the tracking filter at the expected target doppler frequency. If the target return reappears at the expected frequency, then normal tracking is resumed, and the extrapolation process may be invisible to the operator.

13.6.2 Medium- and Low-PRF Radar

Figure 13.18 shows an idealized representation of the main-lobe clutter spectrum received by a low- or medium-PRF radar. Side-lobe clutter can be present throughout

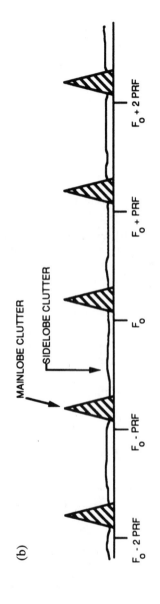

Figure 13.17 (continued)

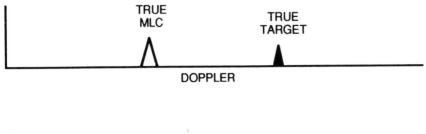

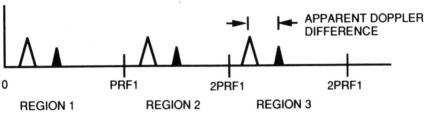

Figure 13.18 PRF control: main-lobe clutter avoidance.

the frequency region between the ambiguous replications of main-lobe clutter. The strength of the side-lobe clutter depends primarily upon aircraft altitude, terrain characteristics, and the side-lobe gain of the antenna. Generally, there is little difficulty acquiring and tracking a target that has been detected in the search mode as long as the target doppler does not coincide with the one of the main-lobe clutter returns. The frequency region where main-lobe clutter is expected is usually notched out before the signal is presented to the tracking loop.

If the radial component of the target velocity approaches zero, then the target doppler will approach the frequency of main-lobe clutter. The integrators in the tracking loop can be used to provide extrapolation through the region of main-lobe clutter in a manner similar to that described for the high-PRF mode. Because of the doppler ambiguity, the apparent frequencies of the target doppler and main-lobe clutter are equal at times other than when the target radial velocity is zero, as illustrated in Figure 13.18. The conditions for main-lobe clutter obscuration can be expressed as

$$(f_{mlc} - \delta f_{mlc}/2)_{\text{mod}} > (f_d)_{\text{mod}} > (f_{mlc} + \delta f_{mlc}/2)_{\text{mod}} \tag{13.5}$$

where:

f_{mlc} = center frequency of main-lobe clutter from (2.6);
δf_{mlc} = main-lobe clutter spectral width from (2.7);
f_d = target doppler frequency.

and the operator mod means that the number is modulo PRF.

13.6.3 Control of Doppler Blind Zones

The apparent target doppler and the main-lobe clutter frequencies can be separated by changing the PRF if either the target doppler or the main-lobe clutter is ambiguous. In contrast to the search mode in which the target doppler is unknown and the system switches sequentially through the set of PRFs, PRF switching in the track mode can be done in a deterministic manner. The true velocity of main-lobe clutter, V_{mlc}, can be calculated using the antenna angles and speed and attitude from the aircraft inertial reference system. The true closing rate, V_C, is available from the range-tracking loop. The apparent doppler difference and the number of doppler ambiguities for both the target doppler and main-lobe clutter can be calculated for each PRF. The PRF can be switched to increase the doppler difference. The frequency position of the doppler-tracking filters must be repositioned to the expected value of the apparent target doppler at the new PRF.

Defining the following terms,

$$N_{mlc} = \text{int}\left(\frac{2V_{mlc}}{\lambda\text{PRF}}\right)$$

$$N_c = \text{int}\left(\frac{2V_c}{\lambda\text{PRF}}\right)$$

the effect of a relative change in PRF (i.e., an increase or decrease) on the apparent doppler difference is shown in Table 13.2. The entry for $N_c = N_{mlc}$ requires explanation. Changing the PRF does not change the difference between the two ambiguous dopplers. However, one property of the FFT is that the filter spacings decrease as the PRF is reduced. Therefore, the apparent separation as measured in number of doppler filters increases.

13.6.4 Control of Range Blind Zones in Medium-PRF

Changing the PRF may also be used to move the range blind zones and facilitate tracking in the medium-PRF mode. The relationship between the true time delay

Table 13.2
PRF Control: Main-Lobe Clutter Avoidance

Ambiguity Relationship	Encounter	PRF Change	Apparent Doppler Difference
$N_c > N_{mlc}$	Nose	Decrease	Increase
$N_c < N_{mlc}$	Tail closing	Increase	Increase
$N_c = N_{mlc}$	Near abeam	Decrease	Increase

and the apparent time delay is shown in Figure 13.19. The range-resolving process provides the true range. The range bin number corresponding to the target can be calculated for each of the PRFs. The PRF may be switched when the apparent time delay approaches zero. The early and late range gates used for range tracking must be repositioned any time the PRF is changed whether it is motivated by range blind zone or doppler blind zone control.

The transmitter duty factor is on the order of 0.01 for systems that use an uncompressed pulse. PRF changes caused by range blind zones are infrequent, since the radar is blind during only 1% of the range interval. If a 13-bit pulse compression is used, resulting in a duty factor of 0.13, then PRF changes will be more frequent.

13.7 KALMAN AND α-β TRACKERS

Kalman filters and α-β trackers are widely used in both TWS and STT modes. Both are readily implemented using recursive digital computer algorithms. The benefits of Kalman filtering have been known for some time [4]. Implementation of the Kalman filter can impose heavy computational demands. The α-β tracker [5], which is now recognized as a simplified subset of the Kalman filter, has seen wide application because of its simplicity. Also, a body of theory for suboptimal filters, each with simplifications appropriate for a certain class of applications, has been developed to reduce the computational load [6]. Two technological factors have created a growth in Kalman filter applications during the 1990s. First, low-cost, high-speed digital computing capability has made Kalman filters practical for more applications. Second, an increase in electronically scanned antennas has resulted in more MTT systems.

Velocity tracking will be described by first considering an α-β tracker applied to a system that makes only velocity measurements, such as a high-PRF radar. Then an application will be illustrated in which range and velocity are measured independently. Finally, the relationship between the α-β tracker and a two-state Kalman filter will be shown.

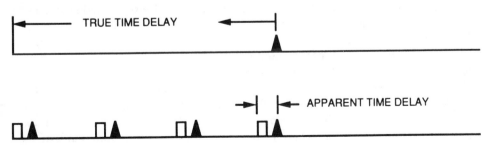

Figure 13.19 PRF control: range blind zones.

13.7.1 α-β Tracker

The α-β equations when applied to velocity become:

Smoothing:

$$\hat{v}_N = \hat{v}_{PN} + a(v_N - \hat{v}_{PN})\tag{13.6}$$

$$\hat{a}_N = \hat{a}_{PN} + \beta\frac{(v_N - \hat{v}_{PN})}{T}\tag{13.7}$$

Prediction:

$$\hat{v}_{P(N+1)} = \hat{v}_N + \hat{a}_N T\tag{13.8}$$

$$\hat{a}_{P(N+1)} = \hat{a}_N\tag{13.9}$$

where:

$\hat{v}_N$	=	smoothed estimate of current velocity;
$\hat{a}_N$	=	smoothed estimate of current acceleration;
v_N	=	measured velocity;
T	=	time between samples;
$\hat{v}_{P(N+1)}$	=	predicted velocity T seconds later;
α, β	=	predetermined smoothing constants;
$\hat{v}_{PN}$	=	predicted velocity at the time of measurement.

Equations (13.6) and (13.7) are used to compute the smoothed estimates of the current velocity and acceleration. These values are computed immediately after the current measurement of velocity vN and are frequently part of the radar output data stream. The physical interpretation of (13.6) is straightforward. The quantity $v_N - v_{PN}$ is the difference between the measured and predicted velocities. A portion of that difference determined by α is added to the prediction to form the smoothed estimate. This same difference when divided by T represents an acceleration difference that is weighted by β in an analogous manner. The performance of the tracker depends on the choice of α and β, but the choices are not independent. For $\alpha = \beta = 0$, the current measurement is ignored. For $\alpha = \beta = 1$, the current estimate is simply the current measurement, and no smoothing is provided.

The filter reduces the variance of measurement noise at steady state (constant acceleration). The variance reduction ratio (VRR) is given by

$$\text{VRR} = \frac{2\alpha^2 + 2\beta + \alpha\beta}{\alpha(4 - 2\alpha - \beta)}$$

Clearly $(4 - 2\alpha - \beta) > 0$ is required for stability. The selection of α and β is application-dependent and usually represents a compromise between smoothing (variance reduction) and transient response. Benedict and Bordner's criterion, when applied to velocity tracking, minimizes the total velocity output error (steady-state variance plus transient error due to a step change in acceleration) through the choice of $\beta = \alpha^2/(2 - \alpha)$.

The computational sequence for the α-β tracker is to (1) use the current measurement to update smoothed velocity and acceleration ((13.6) and (13.7)) and (2) predict the velocity and acceleration at a time T seconds later ((13.8) and (13.9)). The value of T may range from a few milliseconds for a continuous tracking or agile beam radar to several seconds for the TWS mode of a radar with a mechanically scanned antenna.

The α-β tracker described above is capable of tracking a target having a constant acceleration with a mean velocity error of zero, and thus can be regarded as a type 2 tracking loop. Equation (13.9), although it appears trivial, clearly states that the system model predicts constant acceleration. The constants α and β are usually predetermined, although multiple sets and some criteria, such as range and operator selection, to select the proper set could be used.

13.7.2 Extension to Kalman Filter Notation

A heuristic simplified view of the Kalman filter is one in which (1) the optimum weighting coefficients (somewhat analogous to α and β) are dynamically computed at each update cycle and (2) a more precise model of the target dynamics can be used. The benefits of this additional complexity include an improvement in tracking accuracy, a running measure of the accuracy with which target coordinates are being estimated, and a method for handling measurements of variable accuracy, nonuniform sample rate, or missing samples.

Kalman filtering is usually described by matrix notation, which is used below to provide a transition between the α-β tracker and more sophisticated models. Prediction equations (13.8) and (13.9) may be rewritten compactly as

$$\hat{v}_{P(N+1)} = \Phi \hat{v}_N \tag{13.10}$$

where:

$$\hat{v}_N = \begin{vmatrix} \hat{v}_N \\ \hat{a}_N \end{vmatrix} = \text{current (smoothed) state vector;}$$

$$\hat{v}_{P(N+1)} = \begin{vmatrix} \hat{v}_{P(N+1)} \\ \hat{a}_{P(N+1)} \end{vmatrix} = \text{predicted state vector;}$$

$$\Phi = \begin{vmatrix} 1 & T \\ 0 & 1 \end{vmatrix} = \text{transition matrix.}$$

The transition matrix contains the assumed system model, that is, calculations used to predict the state at the next sampling time. Selection of a system model is a compromise between computational complexity and tracking accuracy. Approximations in the model and target maneuvers not in accordance with the model introduce tracking errors. In a more general formulation, an additional term, a modeling noise vector, is added to the right side of (13.10).

Smoothing equations (13.6) and (13.7) may be rewritten as

$$\hat{v}_N = \hat{v}_{PN} + B_N(Mv_N - M\hat{v}_{PN}) \tag{13.11}$$

where:

$$B_N = \begin{vmatrix} \alpha & 0 \\ \dfrac{\beta}{T} & 0 \end{vmatrix} = \text{weighting coefficients matrix;}$$

$$v_N = \begin{vmatrix} v_N \\ a_N \end{vmatrix} = \text{measurement vector;}$$

$$M = \begin{vmatrix} 1 & 0 \\ 0 & 0 \end{vmatrix} = \text{measurement matrix.}$$

and the other terms are as previously defined.

For notational completeness, the measurement vector contains all of the system states. The product $Y_N = Mv_N$ specifies which quantities are actually measured at each sample time (only v in this example). The Kalman formulation adds an additional term, the observation error vector N_N, to the formulation of Y_N.

The power (and the computational complexity) of the Kalman filter results from the dynamic computation of the weighting coefficient matrix B_N. The filter then computes the weights such that if the error in the measurement, Y, increases relative to the error of the smoothed estimate, then the coefficients will be reduced to weight the measurement less heavily. The weighting coefficient matrix is computed using the matrix equation:

$$B_N = C_{PN}M^T(E_N + MC_{PN}M^T)^{-1} \qquad (13.12)$$

where:

$$
C_{PN} = \begin{vmatrix}
\begin{array}{ll}
\text{mean square} & \text{cross-correlation} \\
\text{velocity error} & \text{between velocity and} \\
 & \text{acceleration errors} \\[1em]
\text{cross-correlation} & \text{mean square} \\
\text{between velocity and} & \text{acceleration error} \\
\text{acceleration errors} &
\end{array}
\end{vmatrix}
$$

$\qquad\qquad\qquad$ = predicted covariance matrix;
superscript T $\quad$ = transposed matrix;
superscript -1 $\quad$ = inverse matrix;
E_N $\qquad\qquad$ = covariance of measurement error.

The Kalman filter has an additional prediction and an additional correction or smoothing equation. These provide a running measure of tracking accuracy.

Correction:

$$C_N^+ = [I - B_N M]\, C_{PN} \qquad (13.13)$$

Prediction:

$$C_{P(N+1)} = \Phi C_{N+1} \Phi^T \qquad (13.14)$$

The computational sequence is summarized in Figure 13.20.

Kalman filters permit, and are most often applied to, higher-order system models. The previous discussion used a simple model. Only one quantity, velocity, was measured directly, and the model predicts constant acceleration. One example of a higher-order model is a range-gated pulse-doppler radar which measures both range and doppler [7]. Another example is an airborne radar in which measurement of own-ship velocity is used to improve the prediction and tracking accuracy. The reader is referred to [8] for additional information on prediction.

13.8 COMPENSATION FOR OWN-VEHICLE MOTION

Consider the geometry of Figure 13.21. Even when the radar and the target being tracked are moving with constant velocity vectors, the relative geometry between the two can produce a nonconstant velocity. In general, the acceleration, rate of change of acceleration, and so on are also not constant. In short-range encounters, the accelerations can be several g's (g is the acceleration due to gravity). The

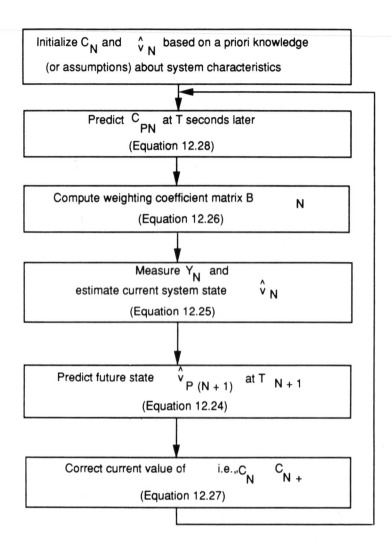

Figure 13.20 Kalman filter computation sequence.

simplest of the velocity tracker prediction methods discussed earlier, the single integrator, predicts zero acceleration between updates. It depends on frequent updates and a doppler measurement bandwidth wide enough to accommodate the shift in doppler between updates to maintain the track. The more sophisticated tracking methods such as Kalman filtering can model the geometry and make a prediction that includes geometric effects. The models generally chosen predict that both the target and the radar continue to follow the trajectory established by the measurement history.

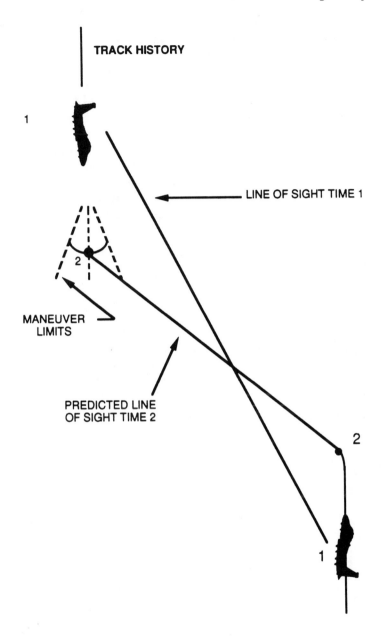

TRACK HISTORY

1

LINE OF SIGHT TIME 1

2

MANEUVER
LIMITS

PREDICTED LINE
OF SIGHT TIME 2

2

1

Figure 13.21 Airborne tracking geometry.

If either the radar or the target initiates a maneuver, this constitutes a transient input to the tracking loop. If the radar and target are at position 1 at one update time, then at the next update time they could be anywhere on arc 2 representing the maneuver limits of the aircraft. The target maneuvers must be measured and tracked by the radar. However, it is not necessary to widen the doppler measurement bandwidth to accommodate the part of the doppler frequency shift that is due to the radar motion. An inertial measurement of own-vehicle acceleration in the direction of the target can be used as an additional input to the integrator that controls the VCO in the tracking loops of Figure 13.10 or as an additional direct measurement in a Kalman filter tracking mechanization. An accelerometer may be mounted on the antenna to provide the input directly in the proper coordinate system, or input may be supplied by the aircraft navigation system.

References

[1] Sherman, S. M., *Monopulse Principles and Techniques*, Norwood, MA: Artech House, 1986.

[2] Skolnik, M. I., Chap. 4 in *Radar Handbook*, New York: McGraw-Hill, 1970.

[3] Barton, D. K., Chap. 12 in *Radar Systems Analysis*, Englewood Cliffs, NJ: Prentice-Hall, 1964.

[4] Kalman, R. E., "New Results in Linear Filtering and Prediction Theory," *Trans. ASME*, Vol. 83D, March 1961, pp. 95–105.

[5] Benedict, R. T., and G. W. Bordner, "Synthesis of an Optimal Set of Track-While-Scan Smoothing Equations," *IRE Trans. Automated Control*, Vol. AC-7, July 1962, pp. 27–32.

[6] Faruqi, F. A., and R. C. Davis, "Kalman Filtering Design for Target Tracking," *IEEE Trans. Aerospace and Electronic Systems*, Vol. AES-16, No. 4, July 1980, pp. 500–508, including "Corrections to . . . ," Vol. AES-16, p. 740.

[7] Fitzgerald, R. J., "Simple Tracking Filters: Position and Velocity Measurement," *IEEE Trans. Aerospace and Electronic Systems*, Vol. AES-18, September 1982, pp. 531–537.

[8] Blackman, R. B., *Linear Data-Smoothing and Prediction in Theory and Practice*, Reading, MA: Addison-Wesley, 1965.

CHAPTER 14

Target Detection by Airborne Radars

Melvin L. Belcher, Jr.

The basic principles of detection theory are reviewed in this chapter with emphasis on pulse-doppler radar systems. This chapter summarizes estimation of noise-limited detection performance using the radar range equation. A number of topics will be introduced for more thorough examination in later chapters. Chapter 15 describes the effects of target signature fluctuation phenomena on detection performance. Clutter-limited detection performance is examined in Chapter 16. Automatic detection processing for pulse-doppler radar systems is surveyed in Chapter 17.

14.1 MECHANIZATION OF TARGET DETECTION

14.1.1 Operational Considerations

Fundamentally, a sensor system only performs two functions: detection and estimation. Detection is emphasized in this section, since it is often the limiting factor in airborne radar performance and it must precede any attempt to estimate target parameters via tracking or signature characterization. Upon initiating track, measurement quality generally improves significantly due in part to suppression of beamshape loss and signal-processing losses, resulting in an increase in signal-to-noise ratio (SNR). Airborne radar applications generally do not demand stressing angle and range measurement accuracy, but high probability of maintaining firm track and illumination on selected targets is important. Gun fire accuracy is largely determined by factors other than radar tracking performance such as gun scatter and recoil effects. Pulse-doppler radars are employed for airborne target detection by virtue of their capability to detect small targets at long ranges over a wide field of view, insensitivity to environmental conditions including weather and surface clutter, and operational flexibility. These capabilities have increased with developments in digital processing and RF power-generation technology.

Airborne radar target detection is generally associated with either dedicated search dwells or TWS returns. In both cases, the radar repetitively scans a raster over the search volume. The primary difference between search and TWS is the associated signal and data processing. Search processing generally only reports gross target positions. In contrast, airborne early warning (AEW) and multiple-target airborne intercept (AI) radars generally have the capability to employ TWS techniques, where detection processing is followed by additional signal-data processing to refine estimates of target range, doppler, azimuth angle, and elevation angle. Returns are correlated within a scan and scan to scan to generate track files. Phased-array radars may also employ track-during-scan (TDS), where dedicated agile-beam track dwells are interleaved with search dwells so as to provide high-data-rate track on selected targets while maintaining search or TWS functions.

These detection operations on a given target by the radar system may be very brief in comparison to the total period the target may be tracked by the radar. However, under a typical operational scenario, most of the radar resources are devoted to search. Effective employment of radar resources for detection requires efficient mission planning as well as a well-balanced radar system design.

Effective mission planning is necessary to ensure adequate target detection coverage of a specified air volume over some period of time dictated by scenario requirements. For example, a patrol pattern of three AEW aircraft acting in concert may be devised to provide near total coverage of a boundary region paralleling a national border or a defended area. Figure 14.1 depicts such a patrol pattern layout that would be suitable for AEW aircraft such as the E-3A Sentry. The small spatial coverage blind zones depicted in this diagram could not be exploited in practice by an adversary aircraft attempting undetected penetration due to his lack of a priori knowledge of the AEW patrol pattern and location of aircraft at specific times. Note that these coverage blind zones are independent of the range-doppler blind zones associated with specific radar system parameters.

Interceptor aircraft generally receive a hand-off vector from an AEW radar or long-range surface-based search radar to reduce the attendant search requirements on the AI radar as much as possible. In addition, combat air patrol missions may be flown along anticipated adversary approach corridors chosen to provide an acceptable probability of intercept by the AI radar without benefit of hand-off.

Scan coverage effects can be included in the system engineering process by estimating the probability of a given target being within the field of view (raster coverage) of the radar. Calculation of this probability is highly dependent on the specific scenario but generally includes consideration of (1) accuracy of any hand-off data used to initialize the search and (2) the anticipated defender and aggressor velocities and flight paths in conjunction with radar scan period to estimate the number of looks the radar will possess against a potential target. The composite probability of target acquisition is the product of this coverage probability and the detection probability derived from the target, environment, and radar system parameters.

(a)

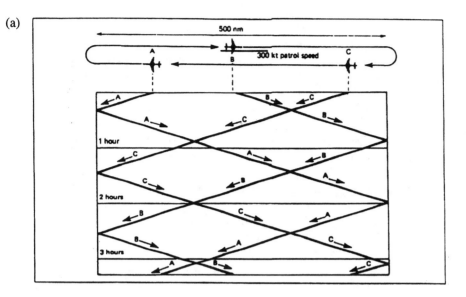

(b)

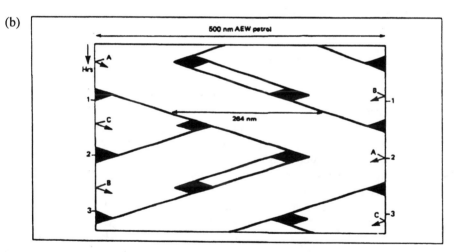

Figure 14.1 AEW operational requirements: (a) Using several AEW aircraft in one patrol pattern—Three AEW aircraft, regularly spaced in a 500-nm-long control pattern and each traveling at 300 knots, will complete one orbit each during a 3 hr, 20 min period. To use each aircraft with equivalent effect they should be spaced regularly in the pattern. (b) Predicting AEW coverage gaps—In the example illustrated previously, if each AEW aircraft could detect low-flying aircraft along 264 nm at the border, its protective capability could be visualized as a ribbon of the same width, as shown above, criss-crossing along the patrol pattern. The dark areas show where coverage gaps occur and their total duration. These areas can be reduced or eliminated by varying the patrol pattern length, AEW aircraft altitude, patrol speed, or number of AEW aircraft.

System level performance is often specified in terms of detection range R_n against a target of specified RCS. The subscript n denotes the cumulative probability of detection as a percentage. Obviously, this quantity is scenario-dependent in that it assumes a specific closing velocity between the radar platform and the target. The closing velocity in conjunction with the radar raster scanning period and the maximum line-of-sight range determines the number of looks the radar will obtain against the target.

Detection range can degrade markedly under conditions of clutter-limited or jamming-limited detection. Clutter-limited detection means that the total clutter return power in the range-doppler-angle resolution cell occupied by the target return is significantly greater than any other interference source at the receiver output. Under clutter-limited detection, detection range will vary with the relative geometry among the target, radar platform, and clutter sources as well as the RCS distribution of the clutter return sources. Similarly, jamming-limited detection occurs when the interference power received from a jammer is significantly greater than either the clutter returns or the thermal noise component in the radar receiver output. Effective jamming of a pulse-doppler radar usually requires that the jammer signal be spread across a bandwidth sufficient to cover all the doppler filters that could potentially contain a target return.

14.1.2 Implementation of Detection Processing

The relationships between the detection algorithms and the corresponding radar subsystem requirements are particularly critical in an airborne pulse-doppler radar, due to the presence of a complex time and spatially varying interference process composed of thermal noise, surface and volume clutter, and potentially jamming. In particular, automatic detection can be stressed by the necessity of filtering target returns from clutter returns. Detection implementation and performance characteristics vary significantly among low-PRF, medium-PRF, and high-PRF radar systems.

Detection can generally be considered a three-stage process in modern pulse-doppler radar systems, as illustrated in Figure 14.2. Range gating and doppler filtering are depicted as occurring in the signal processor after analog-to-digital conversion, since digital processing is commonly used for those functions in modern radar systems. The operator interface is depicted with the radar system through the data processor. Manual target detection using a raw video display, such as a plan position indicator or amplitude-versus-doppler display, may be effective in a single-target, low signal-to-clutter ratio (SCR) environment but degrades rapidly otherwise. The details of the detection process will vary with specific system design and the PRF selection as noted below.

The receiver and signal processor range-doppler processing that constitutes the first stage is designed to maximize the signal-to-interference ratio (SIR) subject to hardware limitations and uncertainty in the characteristics of the interference.

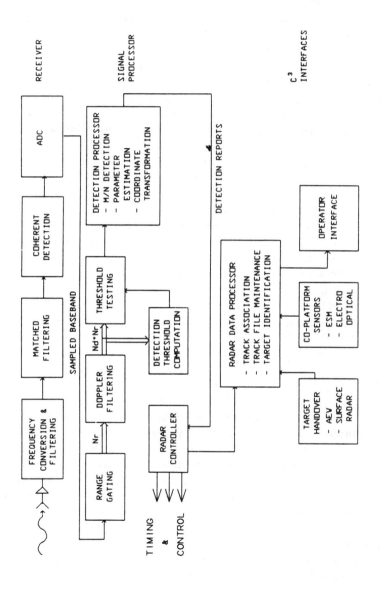

Figure 14.2 Detection processing.

Interference is the composite signal resulting from thermal noise, clutter returns, and jamming.

Interference produces false alarms with resulting desensitization of the radar system. False alarms occur when the noise or clutter return magnitude in a given range-doppler cell exceeds the threshold defined for declaration of a detection (alarm). False target reports are generated from discrete clutter returns that persist from one integration period to the next subject to geometry constraints. Such persistent false alarms are encountered in medium-PRF radar operation due to such conditions as the presence of human-made structures having large RCS in the surface area illuminated by the radar. In addition, spectral impurities in the radar system may also produce false targets at specific doppler frequencies.

This initial processing should fully preserve the information content of a given return. The received signal parameters should not be corrupted prior to decision processing. Deviations from linear processing, such as limiting, generally degrade detection performance.

The second stage of detection processing consists of (1) detection threshold computation and (2) the detection decision. As described in Chapter 17, a threshold voltage is computed by sampling the receiver signal at the output of the range gates and doppler filters. This value is computed from examination of the range and doppler cells surrounding the cell under test or observing the time history of the cell under test. Much effort to develop clutter-limited detection techniques can be characterized as seeking more robust ways to compute this threshold. The radar data processor generally possesses the capability to command the threshold computation method and parameters. A detection is declared if the test cell voltage exceeds the threshold. The signal processor may incorporate a postprocessor unit to provide estimates of target parameters, unambiguous range doppler, and azimuth-elevation bearings. The postprocessor could also be designed to detect the presence of jamming signals and alert the radar data processor.

Finally, the postprocessor may generate an appropriately formatted detection report containing coordinate-corrected target position for transmission to the radar data processor via the radar controller. Coordinate correction may include both correction of aircraft rotational motion—pitch, yaw, and roll—and conversion of the estimated target position from radar coordinates—range, azimuth, elevation, and doppler—to the coordinate system employed in the radar data processor. Delegation of coordinate conversions to a dedicated postprocessor can significantly reduce the computational demands on the general-purpose computer employed as the radar data processor.

The detection processing just described could potentially suffice for a single-target tracking radar, but multiple-target airborne radars require a third stage of detection processing within the data processor to incorporate consideration of known target tracks and other information into the detection process. Moreover, the combination of decreasing target RCS and lower-altitude flight imposing larger clutter returns may mandate that the detection threshold be sufficiently low so as to pass numerous detection reports not associated with true targets. If this is the case, then numerous detection reports must be consolidated into likely target tracks while simultaneously eliminating clutter returns and false alarms. This process of associating measurements with potential tracks tend to be very computationally intensive and generally demands specialized computational capability.

In addition, redundant tracks could be (and in practice sometimes are) established on the same targets, since a given target can be detected by multiple dwells. A TWS or TDS radar system data processor may reject detections that occur within a predefined range-range rate-azimuth-elevation measurement space distance relative to the estimated position of a target under track. Multiple unresolved targets are not detected until their measurement space separation is sufficient to satisfy this criterion. AI radar systems may employ high-resolution doppler processing for raid assessment against targets so as to detect the presence of multiple aircraft and provide some measure of their number prior to attaining adequate resolution to implement multiple track files. As addressed in Chapter 16, engine modulation of the doppler return may also induce redundant detection reports and tracks.

Note that Figure 14.2 also depicts command, control, and communication (C^3) data processor interfaces for targets handed over from another radar system, such as an AEW radar system, or collocated sensors such as radar warning receivers or electro-optical sensors. Target registration among dissimilar sensors may be extremely difficult in a multiple-target or clutter-dominated environment. In addition, airborne radars often contain an integrated IFF system to interrogate the identity of targets from onboard transponders.

14.2 NOISE-LIMITED TARGET DETECTION

The detection performance of any radar system is ultimately limited by the thermal noise that is inherently generated in the antenna and receiver. As indicated in Table 14.1, pulse-doppler radar systems can provide noise-limited performance under specific scenarios.

14.2.1 Single-Pulse Detection

The output signal of a radar receiver consists primarily of the sum of target and clutter returns, thermal noise, and potential jamming. Description of clutter effects on detection processing will be deferred until the next chapter. The effects of jamming on signal detection may be treated in the same way as thermal noise, assuming barrage (noise) jamming over a bandwidth completely overlapping that of the victim radar system. Radar detection is depicted in Figure 14.3 for the above conditions. Following conventional terminology, the term *receiver* includes functions such as pulse compression and doppler filtering that are typically performed in a digital signal processor in modern systems such as those described in Chapter 8.

Table 14.1
Scenarios for Noise-Limited Detection

Low-PRF Operation
  Target range less than platform altitude
  Look-up scenario (targets at higher altitudes than radar)
  Long range, low grazing angle (low backscatter)
  Surface mapping
Medium-PRF Operation
  Low side-lobe clutter
High-PRF Operation
  Closing velocity greater than radar platform velocity

In the following discussion, the output of the receiver is defined as a coherently demodulated baseband sample which represents the output of a specific range gate and doppler filter. Equivalent results can be derived using the sinusoid signal model extensively utilized in classical radar detection theory. The sinusoid signal receiver model physically corresponds to an IF processor implemented with analog components.

The noise depicted as injected at the front of the receiver is the sum of external and internal noise sources referred to that point by multiplication of the gains and losses encountered along the receiver chain. Thermal noise is characterized by a white power spectral density and a Gaussian amplitude probability density. More accurately, radar detection theory deals with band-limited thermal noise due to the spectral filtering of the receiver front end. However, the resulting noise process is "nearly white" in that it is independent sample to sample if the receiver output sampling interval is greater than or equal to the inverse of the receiver bandwidth. Each sample of the receiver output can be represented as a random variable generated from the random noise signal model.

As indicated in Figure 14.3, I_{ij} and Q_{ij} are the in-phase and quadrature voltage components of the receiver output at the ith range gate and jth doppler filter, given a received signal $s(t)$ and the noise process $n(t)$. High-PRF radar systems typically employ a single range gate, while low-PRF systems often employ a single doppler filter implemented with MTI processing. The sum of the return and thermal noise in the in-phase and quadrature channels is

$$I_{ij} = A \cos(\theta) + x_{ij}$$
$$Q_{ij} = A \sin(\theta) + y_{ij} \tag{14.1}$$

for a return of magnitude A and relative phase θ and at the output of the receiver. The noise components x_{ij} and y_{ij} are independent zero mean random variables with variances σ_x^2 and σ_y^2 each equal to one-half the noise power P_n such that

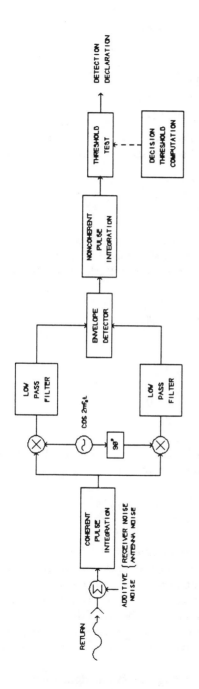

Figure 14.3 Noise-limited detection model.

$$2\sigma_x^2 = 2\sigma_y^2 = P_n = B_n k T_s = \sigma^2 \tag{14.2}$$

where:

B_n = the receiver noise bandwidth;
k = Boltzmann constant (1.38×10^{-23} J/K);
T_s = the equivalent system temperature (K).

As previously noted, the probability density function of the magnitude m in both the in-phase and quadrature channels due to Gaussian noise alone is given by

$$f_{pn}(m) = \frac{1}{\sqrt{2\pi\sigma^2}} = e^{-m2/2\sigma^2} \tag{14.3}$$

Decision thresholding is performed on the magnitude of the sample pair from a given range-doppler cell, as depicted in Figure 14.3. The resulting magnitude of a sample pair with no signal present possesses a Rayleigh probability density function as given by

$$f_n(m) = \frac{m}{\sigma^2} e^{-m2/2\sigma^2} U(m) \tag{14.4}$$

where $U(x)$ is the unit step function defined as

$$U(x) = 1; \ x > 0$$
$$= 0; \ x < 0.$$

The probability density function of a sample pair from a cell containing the target return denoted by (14.1) is Rician, as given by

$$f_{sn}(m) = \frac{m}{\sigma^2} I_0\left(\frac{Am}{\sigma^2}\right) e^{-m2/2\sigma^2} U(m) \tag{14.5}$$

The term I_0 is the zero-order modified Bessel function. In the absence of a return, $A = 0$, the Rician density is equal to the Rayleigh density. For high SNR, the Rician density approaches a Gaussian density of variance σ^2 and mean A.

As described previously, the magnitude of the receiver output at a specific range-doppler cell (the test cell) is compared to a decision threshold to determine if a return is present. A representative pair of probability density functions representing the noise magnitude and the signal-plus-noise magnitude, respectively, are depicted in Figure 14.4.

The term P_d is calculated by integrating the Rician signal-plus-noise density function such that

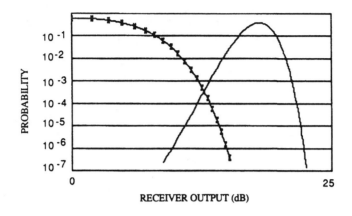

S/N = 15 dB

----- PDF of Signal-plus-Noise

✕-✕-✕ PDF of Noise

Figure 14.4 Noise and noise-plus-signal density functions.

$$P_d = \int_\tau^\infty f_{sn}(m) \; dm \qquad (14.6)$$

where τ is the threshold setting.

P_{fa} is calculated by integrating the Rayleigh noise density function such that

$$P_{fa} = \int_\tau^\infty f_n(m) \; dm \qquad (14.7)$$

P_d can be increased while maintaining a constant P_{fa} only by decreasing the overlap between the density functions via increasing the SNR. In principle, the decision threshold is computed so as to provide the maximum false alarm rate that can be accommodated by the radar system signal-data processor or operator. In practice, the threshold is computed as a function of a lower design value than that required to avoid system saturation so as to provide a margin against sporadic increases in false alarm rate induced by heterogeneous clutter sources. Increasing the detection threshold results in desensitization to target returns.

SNR requirements as a function of P_d and P_{fa} are depicted in Figure 14.5. Note that for high probability of detection, a decrease of 1 dB in SNR results in the probability of false alarm increasing approximately an order of magnitude.

From (14.7), it can be readily discerned that the probability of false alarm is the probability that a single sample of noise will exceed the decision threshold.

Assuming each doppler filter of each range gate is being sampled at the end of each integration period bandwidth, the resulting system level false alarm rate can be defined as

$$\text{FAR} = \frac{P_{fa} N_{rg} N_{df}}{T_i} < \frac{P_{fa}(\text{PRI} - T_p)}{\text{PRI} T_p} \tag{14.8}$$

where:

N_{rg} = the number of range gates;
N_{df} = is the number of doppler filters;
T_p = the transmit pulse width;
T_i = the coherent integration time.

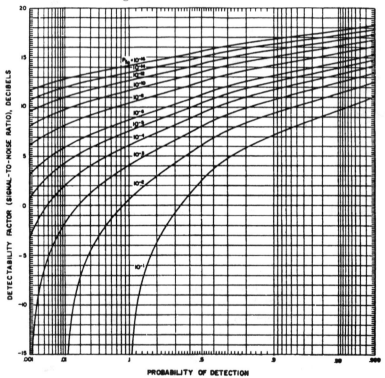

Figure 14.5 Required SNR for single-pulse detection [1].

The inequality is due to the practice of blanking specific range gate or doppler filter outputs known to contain main-lobe or altitude line clutter. Equivalently, the average false alarm time is defined as equal to the inverse of the false alarm rate specified in (14.8). Unfortunately, the false alarm time T_{fa} is defined in a number of classic detection works as set forth by Marcum [2] as the period of time over which the probability of a false alarm occurring is 0.5. This definition employs Marcum's false alarm number, n, such that $(1 - P_{fa})\,n = 0.5$, so

$$n = \ln(0.5)/\ln(1 - P_{fa}) \approx 0.69/P_{fa} \tag{14.9}$$

The false alarm number for a pulse-doppler radar in terms of Marcum's definition can be given approximately by

$$n_{pd} \approx (0.69\ N_{rg}N_{df})/(\text{FAR}\ T_i) \tag{14.10}$$

The Marcum false alarm number is not used, and this description is provided only to provide a link to the earlier works.

14.2.2 Matched Filtering

Under the condition of noise-limited detection, the receiver and signal processor should be designed to maximize SNR. Specifically, a composite transfer function $h(t)$ must be devised to process the input sum of the return and noise so as to maximize SNR. This transfer function is the well-known matched filter as derived below [3]. The peak power of the receiver output in response to $s(t)$ is

$$A^2 = \left(\int S(f)H(f)\ df\right)^2 \tag{14.11}$$

with $S(f)$ and $H(f)$ being the Fourier transforms of $s(t)$ and $h(t)$, respectively. Similarly, the output of the receiver in response to the equivalent thermal noise input is

$$P_n = kT_s\int_{-\infty}^{\infty} H(f)^2\ df \tag{14.12}$$

SNR is equal to A^2/P_n so that

$$\frac{S}{N} = \frac{[\int S(f)H(f)]^2}{kT_s\int_{-\infty}^{\infty} H(f)^2\ df} \tag{14.13}$$

Schwartz's inequality for integrals of complex functions reveals that

$$\left[\int_{-\infty}^{\infty} x(t)z(t)\ dt\right]^2 \leq \int_{-\infty}^{\infty} x(t)^2\ dt \int_{-\infty}^{\infty} z(t)^2\ dt \tag{14.14}$$

The equality is satisfied for $z(t) = K_x^*(t)$, where K is an arbitrary constant. It follows that SNR is maximized by devising a system response function such that

$$H(f) = S(f)^*$$

so that

$$\left(\frac{S}{N}\right)_{peak} = \frac{\int S(f)^2 \, df}{kT_s} \tag{14.15}$$

where the energy of the received signal E is given by

$$E = \int S(f)^2 \, df = \int s(t)^2 \, dt \tag{14.16}$$

It follows that a matched-filter receiver provides the peak SNR possible so that

$$S/N = E/(kT_s)$$

The factor kT_s is the power spectral density (PSD) of the noise. Note that no constraint is placed on the probability density function of the noise amplitude. Hence, the matched filter is the optimal linear filter for any white PSD noise process in the sense that it produces the maximum SNR. Under the constraint of Gaussian noise, the matched filter is optimal across all nonlinear filters. The matched-filter derivation can be generalized to include colored noise as well.

It follows that the frequency response of the radar receiver should emulate the conjugate of the transmitted pulse spectrum to the extent possible with reasonable hardware implementation requirements. Deviations from this ideal result in SNR losses at the receiver output. The mismatched filtering loss L_m is greater than or equal to unity and can be defined as the inverse of the ratio of mismatch loss in signal power to mismatch loss in noise power such that

$$L_m = \frac{\int S(f) H(f) \, df}{\int |H(f)^2| \, df}$$

This mismatch loss is included in the system loss factor, which will be discussed later, to account for deviation from matched-filter performance.

14.2.3 Extension to Pulse-Doppler Radar Detection

Pulse integration may be performed coherently or noncoherently. Coherent integration consists of summing the magnitude and phase of a train of pulses from a given range gate. Noncoherent integration consists of summing the magnitudes of a train of pulses. In practice, the range-gate-doppler-filter outputs of a coherent pulse integrator may be noncoherently integrated across multiple coherent integration periods.

Coherent integration is inherent in pulse-doppler radar operation. A pulse-doppler radar pulse train is generated by a coherent transmitter and mixed down to IF and baseband by a succession of phase-locked coherent oscillator outputs. A scanning pulse-doppler radar typically transmits a continuous pulse train. The associated integration period of the coherent dwell is determined by the inverse of the doppler filter bandwidth.

A doppler offset f_d produces a pulse-to-pulse phase change of $2\pi f_d T$ rad, where T is equal to the PRI. This modulation is effectively removed by doppler filtering. A doppler filter with a bandwidth equal to approximately the inverse of the integration period and a center frequency equal to f_d approximates a matched filter. From the definition of matched filtering in the previous section, a coherent pulse integrator can be conceived as a tapped delay line with a phase weight on each tap corresponding to the conjugate of the interpulse phase shift imposed for a specific doppler filter. As illustrated in Figure 14.6, the combination of pulse matched filtering and coherent pulse integration constitutes a matched filter to a coherent pulse train.

The returns from each range gate must be processed through a bank of doppler filters spanning the range of possible ambiguous doppler frequencies. The filters should be spaced in frequency no farther than the approximate doppler resolution in order to provide acceptable straddling losses. The waveform energy associated with returns is the sum of the single-pulse returns' energy across the integration period.

Noncoherent integration consists of simply summing the magnitudes of the pulse returns for a given range gate. The absence of interpulse doppler compensation results in simplified implementation requirements in comparison to coherent integration. Moreover, a coherent integration period must be limited to less than the decorrelation time of the target in order to prevent the return from spreading among several doppler filters. The duration of noncoherent integration is restricted only by range gate fly-through. Range gate fly-through occurs when the composite return from a target is spread among multiple range gates due to the product of the closing velocity and the integration time being substantially greater than the range gate extent. Integration efficiency α is related to the SNR integration gain of N pulses, G_I, by

$$G_I = N^\alpha$$

Coherent integration of a nonfluctuating target with no signal-processing losses provides α equal to unity. Equivalently, it is commonplace to define the integration SNR loss L_I given by

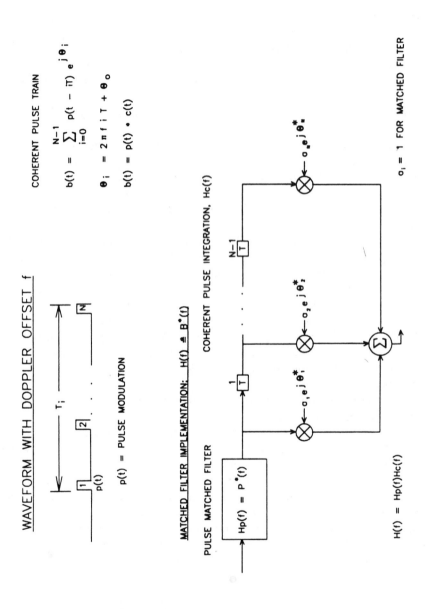

Figure 14.6 Matched filtering for pulse-doppler radar.

$$L_I = -10 \log (G_I/N) \tag{14.17}$$

Coherent integration of N pulses results in a signal gain of N^2 and a noise gain of N for a net SNR integration gain of N corresponding to an ideal integration efficiency of unity. In contrast, noncoherent integration provides an SNR gain between N and $N^{1/2}$. Noncoherent integration effectively smoothes the interference as well as the signal so that integration gain should be computed relative to composite effect on target detectability. Barton [4] has reviewed a number of noncoherent integration results and cited sources indicating noncoherent integration gain efficiency approaches 1 for integration of a small number of pulses and can be approximated as 0.8 for integration of N between 10 and 1,000 with reasonable accuracy for nonfluctuating targets.

High-PRF and medium-PRF systems may be designed to noncoherently integrate the range-gate-doppler-filter outputs of several coherently integrated pulse trains in order to increase SNR. The resulting doppler resolution is inversely proportional to the duration of the coherent integration period. This technique is attractive to prevent doppler filter spreading of target returns as well as to reduce signal-processing requirements. Single-pulse detection performance can readily be generalized to include integrated pulse detection by multiplying the single-pulse SNR by the integration gain. Thus, the receiver output magnitude A of (14.1) can be interpreted as the output of a pulse integrator in the event the detection thresholding is performed on an integrated pulse train. Accordingly, the associated false alarm rate is determined by the number of independent output samples at the pulse integrator output, as reflected in (14.8).

14.3 THE PULSE-DOPPLER RADAR RANGE EQUATION

The modern definition of the single-pulse radar range equation is generally given as

$$\left(\frac{S}{N}\right)_p = \frac{P_p G A \sigma}{(4\pi)^2 R^4 k T_s B_n L_s} \tag{14.18}$$

where:

$(S/N)_p =$ single-pulse SNR;
$P_p \quad =$ peak rms transmitter power;
$G \quad =$ transmitting antenna directive gain;
$A \quad =$ receiving antenna area;
$\sigma \quad =$ target RCS;
$R \quad =$ range to the target;
$L_s \quad =$ system loss factor.

The system temperature T_s and noise bandwidth B_n have been previously defined in the context of noise power at the receiver output.

The radar range equation can be written in general form for pulse-doppler radar systems as

$$\left(\frac{S}{N}\right)_{pd} = \frac{P_p GA\sigma N_p}{(4\pi)^2 R^4 k T_s B_n L_s} \tag{14.19}$$

with N_p being the number of pulses coherently integrated. N_p can be defined simply as the product of the PRF and the integration period T_i. The system loss factor has been implicitly expanded to incorporate additional losses associated with pulse-doppler operation. The noise bandwidth of a radar receiver may be generally approximated as the inverse of the pulse duration T_p. These substitutions result in

$$\left(\frac{S}{N}\right)_{pd} = \frac{P_p GA\sigma \mathrm{PRF}\, T_i T_p}{(4\pi)^2 R^4 k T_s L_s} \tag{14.20}$$

Peak power P_p is related to average power P_{av} by the transmit duty cycle, PRF $\cdot\ T_p$, so that

$$\left(\frac{S}{N}\right)_{pd} = \frac{P_{\mathrm{av}} GA\sigma T_i}{(4\pi)^2 R^4 k T_s L_s} \tag{14.21}$$

The previous expressions may all be characterized as time domain computations of SNR. An equivalent frequency domain expression can also be written for high-PRF and medium-PRF radar systems. Time domain and frequency domain variants of the radar range equation provide the same system performance estimates but differ in utility depending on the application. For example, an analog doppler filter bank implementation of a high-PRF radar processes the central line component of the pulse train. Typically, the array of filters will cover a portion of the bandwidth between the central line and the first doppler ambiguity that occurs at the PRF. Hence, it is useful to relate SNR to the central line power and doppler bandwidth.

Manipulation of the Fourier transform of a coherent pulse train reveals that the central line voltage is equal to the peak voltage of the time domain pulse train multiplied by the transmit duty cycle d_t. It follows that the central line power P_{cl} is equal to the time domain power multiplied by the transmit duty cycle squared. The doppler bandwidth of the system, B_d, can be approximately defined as the inverse of the coherent integration period. The frequency domain radar range equation can be given as

$$\left(\frac{S}{N}\right)_{pd} = \frac{P_{cl} GA\sigma}{(4\pi)^2 R^4 k T_s L_s B_d d_r} \tag{14.22}$$

where d_r is the receive duty cycle, which is greater than or equal to the transmit duty cycle. Eclipsing loss increases as a function of receive duty cycle for a fixed transmit duty cycle.

In practice, estimation of single-scan detection range is more complex than simply using the radar range equation. High-PRF and medium-PRF radars may noncoherently integrate several coherent dwells. Medium-PRF radars employ M out of N detection processing. Moreover, medium-PRF detection performance is constrained by side-lobe clutter under most operational scenarios. However, the radar range equation can be employed to compute the SNR available in relatively clear range doppler areas that afford noise-limited detection performance.

The general form of the radar range equation is sufficiently commonplace as to not require a detailed description. However, an understanding of the physical significance of the variables is essential to accurately estimate the detection performance of a given radar design.

14.3.1 Target Parameters

The two target-dependent parameters in the above expression are the RCS and the range. Monostatic RCS is generally defined as 4π times the power per unit solid angle scattered along the observation vector divided by the power per unit area in a plane wave impinging on the target along the observation vector. In the absence of propagation losses, this quantity may be written as

$$\sigma = \lim_{R \to \infty} \left(4\pi R^2 \frac{|E_r|^2}{|E_t|^2} \right)$$

where:

E_r = target-scattered electric field intensity observed at the radar;
E_t = incident field intensity at the target.

Target RCS is generally a complex function of the target shape, size, aspect angle, and electrical characteristics and is dependent on the radar frequency. Moreover, aircraft returns can possess a relatively complex doppler spectra due to engine doppler modulation of the return. Petts [5] has published tables for estimating jet engine aircraft RCS as a function of physical dimensions.

More generally, radar system detection performance is assessed by modeling target RCS as a random variable. Appropriate PDFs and the resulting effects on detection performance are described in Chapter 15.

14.3.2 Power Gain

Transmitter power is specified as the rms measure of power over a pulse at the transmitter output. Losses that occur in the transmission line between the transmitter output and the antenna terminals due to the microwave impedances are included in the system loss factor, L_s.

Note that G represents antenna directive gain rather than power gain. The directive gain is the ratio of the maximum radiation intensity over that of an isotopic source radiating the same power. In contrast, the power gain of an antenna is the ratio of the maximum radiation intensity over that of an isotopic source with the same RF input power as the antenna. Power gain incorporates consideration of both ohmic losses and active amplification within the antenna.

The usage of directive gain rather than power gain significantly simplifies estimation of system noise temperature and system loss factor in radar systems employing active aperture solid-state transmitters such as those described in [6]. Antenna losses are incorporated into the system loss factor (transmit) and system noise temperature (receive).

For the case of a uniformly illuminated antenna aperture, the directive gain is related to the antenna aperture area by

$$G = \frac{4\pi A}{\lambda^2} \tag{14.23}$$

14.3.3 System Loss Factor

The system loss factor is typically calculated as the sum (in decibels) of a loss budget that lists the individual losses encountered at each stage of the transmission-reception path. Estimation of an accurate loss budget is difficult and requires detailed assessment of the specific design. The magnitude of the entries varies with selection of PRF and technology. In general, the estimate of the system loss factor represents the expected average loss; individual loss components may be significantly greater under worst-case conditions.

For purposes of example, a loss budget representative of losses encountered in an AI radar is presented in Table 14.2. Aronoff and Greenblatt [7] have published representative loss budgets for high-, medium-, and low-PRF airborne radar systems. Petts [5] has produced a table illustrating the range of loss values encountered in modern radar systems.

The system loss factor includes two-way (transmit and receive) losses due to the radome losses and atmospheric absorption and transmit losses due to the microwave hardware. Receiver ohmic losses are included in the calculation of system noise temperature. Receiver losses include mismatch loss due to not satisfying the matched-filter condition for the transmitted pulse train. The system loss factor also contains nonohmic receiver losses due to pulse integration efficiency, eclipsing, doppler-filter-range-gate straddling, collapsing, beamshape, doppler side-lobe suppression weighting, and detection processing, as described below. Antenna taper losses are included in the transmit and receiver RF losses.

Table 14.2
Airborne Pulse-Doppler Loss Budget (Single Coherent Dwell)

		Type of Radar System	
Losses (dB)	*Low-PRF*	*Medium-PRF*	*High-PRF*
RF-IF hardware losses:	2.00	2.00	2.00
Transmit RF			
Nonohmic receive RF	1.50	2.00	1.50
Radome (two-way)	1.50	1.50	1.50
Frequency instability	0.00	0.00	0.00
Signal-processing losses:			
Range gate straddling	0.50	0.50	0.50
Doppler filter straddling	N/A	0.30	.30
Range collapsing	0.00	0.00	0.40
Doppler side-lobe weighting	0.80	0.80	0.80
Detection thresholding	1.50	2.50	1.50
External losses:	0.50	0.50	0.50
Atmospheric absorption			
Beamshape loss	3.20	3.50	3.50
Eclipsing	0.00	0.50	2.50
Subtotal losses	11.50	14.10	14.60
Receiver ohmic losses	2.00	2.00	2.00
Preamp noise figure (db)	4.00	4.00	4.00
System temperature (K)	965	965	965

Eclipsing loss is due to the arrival of a return during the period when the receiver is turned off while the radar is transmitting. Pulse-doppler radar systems generally use the same antenna for transmit and receive so that it is necessary to shut off the receiver during transmission. Eclipsing loss is particularly severe for high-PRF radar systems, since the associated receiver duty cycle is typically on the order of 50. The worst-case scenario is that the times of arrival of returns closely coincides with pulse transmissions so that the target returns are completely eclipsed.

The PRF may be adjusted during target tracking so that the ambiguous range of the target is clear from eclipsing. For search processing, eclipsing loss can be estimated for a given receiver duty cycle by modeling the time of arrival of the return as uniformly distributed over a PRI. The receiver duty cycle d_r may be defined as

$$d_r = 1 - \frac{T_p}{T} - \frac{\tau_s}{T} \qquad (14.24)$$

with transmit:receive switching time τ_s. Neglecting transmit-receive switching time, the resulting average eclipsing loss is plotted in Figure 14.7.

(NEGLIGIBLE SWITCHING TIME)

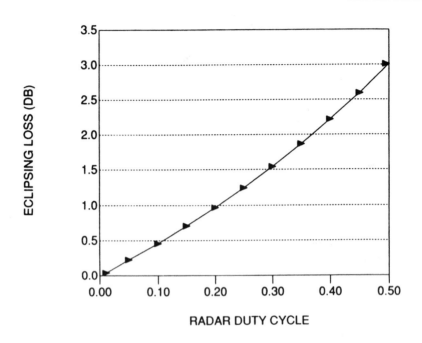

Figure 14.7 Eclipsing loss.

Straddling and beamshape losses arise from the same circumstance: a return does not occur at the peak response of a range gate, doppler filter, and beam. This loss is estimated by computing the average loss associated with a return uniformly distributed between the response peak and the crossover point with the adjoining gate, filter, or beam position. Straddling loss is generally only associated with search and TWS radar operation. Radars employing dedicated track dwells will typically maintain the returns near peak response via range, doppler, and angle tracking.

Peak and average straddling loss as a function of normalized peak response separation x for two representative response functions are plotted in Figures 14.8 and 14.9. The sinc$(x)^2$ function $[\sin(\pi x)/(\pi x)]^2$ is representative of the power response of unweighted doppler filters or the one-way beamshape of antennas with no taper. The Hamming power response function is the square of the Fourier transform of $[0.54 + 0.46\cos(2\pi x)$ over the interval of ±0.5. Both response functions are plotted in Figure 14.10. Note the 13.5-dB peak side lobes of the sinc$(x)^2$ function in comparison to the 42.8-dB peak side lobes of the Hamming response. However, the 3-dB main-lobe resolution of the Hamming function is approximately 50% larger than that of sinc(x). In practice, the side-lobe and resolution characteristics of most response functions will lie between those of these two response functions.

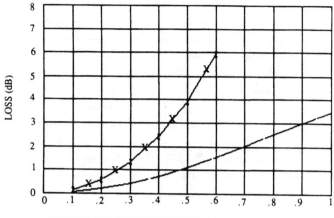

NORMALIZED RESPONSE-FUNCTION CROSSOVER ORDINATE

```
-----   Average Loss
X-X-X   Peak Loss (at Crossover)
```

Figure 14.8 Straddling loss of unweighted response.

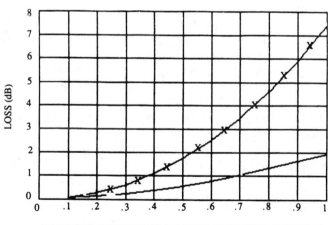

NORMALIZED RESPONSE-FUNCTION CROSSOVER ORDINATE

```
-----   Average Loss
X-X-X   Peak Loss (at Crossover)
```

Figure 14.9 Straddling loss of Hamming-weighted power response.

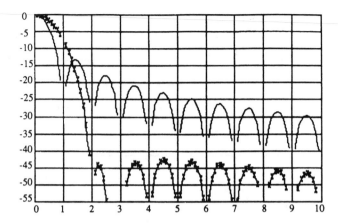

Figure 14.10 Frequency response of representative aperture functions.

Similarly, beamshape losses occur when the target is not at the center of the transmit-receive beam. Estimation of the average beamshape loss requires the mean loss to be computed as a function of the scanning main-lobe antenna pattern for a target uniformly distributed within the radar raster.

Weighting losses are encountered when the radar system is deliberately mismatched to the received signal in order to suppress side lobes. Antennas are weighted with an illumination taper designed to suppress angle side lobes so as to reduce vulnerability to clutter returns and jamming. Both analog and digital filters may employ weighting to suppress doppler side lobes. Pulse compression of LFM pulses may incorporate weighting to suppress range side lobes. For example, a Taylor weighting designed to achieve 30-dB doppler side-lobe suppression imposes an SNR loss of approximately 0.7 dB, while a 40-dB Taylor weight imposes approximately a 1.2-dB loss. Detection losses are typically due to threshold estimation errors, as discussed in Chapter 17.

Digital quantization during sampling and signal processing typically results in small losses on the order of 0.1 to 0.3 dB.

14.3.4 System Noise Temperature

Following the definition of Blake [1], receiver ohmic losses L_r are incorporated into the equivalent system noise temperature. Equivalent system temperature can be estimated as

$$T_S = T_a + T_t(L_r - 1) + L_r T_0 (F - 1) \qquad (14.25)$$

where:

T_a = antenna noise temperature;
T_t = thermodynamic (actual) temperature of the antenna-receiver transmission network;
T_0 = standard reference temperature (290K);
F = receiver noise figure.

If the transmission network is at ambient temperature, then T_t is approximately T_0, so that T_s can be approximated as

$$T_S = T_a + T_0(L_r F - 1) \tag{14.26}$$

The antenna noise temperature is a measure of the noise power received from external radiating sources. Unless the radar system main lobe intercepts the sun or specular solar reflections, the antenna temperature is generally low at microwave frequencies relative to the noise temperature of the receiver front end.

14.3.5 Application of the Radar Range Equation

14.3.5.1 Noise Jamming

The pulse-doppler radar range equation may be readily extended to incorporate consideration of noise jamming. (Deceptive electronic countermeasures that induce targetlike returns using coherent repeater or other techniques can be analyzed as generating false target returns. The associated mismatch loss in jammer power at the output of the signal processor can be calculated using the previously given approach for computing mismatch loss.) The noise jammer should possess a power spectral density that is almost white over the radar bandwidth. Application of the standard detection curves presupposes that the amplitude PDF is Gaussian. The received jammer power at the output of the radar receiver is given by

$$J = \left(\frac{P_j G_j g_{rj} A}{4\pi R_j^2 L_j}\right)\left(\frac{B_d}{B_j}\right) \tag{14.27}$$

where:

P_j = jammer transmitter power;
G_j = jammer antenna gain;
B_j = jammer bandwidth;
g_{rj} = relative gain of the radar antenna in the jammer direction;
R_j^2 = range of the jammer;
L_j = jammer loss factor;
B_d = victim radar doppler bandwidth;
B_j = jammer bandwidth ($B_j \geq B_d$).

The relative gain g_{rj} denotes the suppression of jammers located within the radar antenna side lobes by the side-lobe gain relative to the peak antenna gain. The jammer loss factor includes internal jammer transmit losses, one-way propagation and radar radome losses, and polarization mismatch between the jammer antenna and the radar antenna.

In the case of a stand-off jammer at a different range or angle from the radar than the target, the resulting signal-to-jammer power ratio, S/J, is given by

$$\left(\frac{S}{J}\right)_{so} = \left(\frac{P_{av}G\sigma}{4\pi R^4 L_s g_{rj}}\right)\left(\frac{R_j^2 L_j}{P_j G_j}\right)\left(\frac{B_j}{B_d}\right) \tag{14.28}$$

In the event of a self-screening jammer, R_j equals R and g_{rj} is approximately unity so that

$$\left(\frac{S}{J}\right)_{ss} = \left(\frac{P_{av}G\sigma}{4\pi R^2 L_s}\right)\left(\frac{L_j}{P_j G_j}\right)\left(\frac{B_j}{B_d}\right) \tag{14.29}$$

The composite interference due to jamming and noise can be calculated by defining the net radar system noise temperature T_s' such that

$$T_s' = T_s + \frac{J}{kB_j} \tag{14.30}$$

In many ECM scenarios of interest, the received jammer power will be much greater than the radar system noise so that the composite signal-to-interference ratio can be approximated by S/J.

Jamming-limited detection range can be estimated by substituting S/J for SNR and employing Figure 14.5. The accuracy of this estimate is heavily dependent on the fidelity of the jamming signal characteristics relative to thermal noise. Deviation of the jammer modulation from thermal noise characteristics can significantly elevate false alarm rates against the victim radar.

14.3.5.2 Radar Range Equation for Search

The radar range equation can readily be rewritten to incorporate variables encountered in implementing search requirements. Search requirements are generally specified by stating that a given angular extent must be searched within a specified search frame time T_F. A minimum detection range R_{min} is specified on a target of a certain RCS σ_t with a specified closing velocity v_c.

The solid angle search requirement, Ω, maybe defined as

$$\Omega = \delta Az[\sin(El_{max}) - \sin(El_{min})] \tag{14.31}$$

where δAz is the azimuth extent of the required search raster, El_{max} is the upper raster elevation bound of the raster, and El_{min} is the lower raster elevation bound. Assuming N_c coherent dwells and N_n noncoherent integrations per beam position, the approximate number of coherent dwells Q_d to cover this solid angle extent with a radar of beamwidth $\delta\phi$ is given by

$$Q_d \approx \frac{4\Omega}{\delta\phi^2\pi} \approx \frac{4\Omega}{\pi}\left(\frac{D}{1.2\lambda}\right)^2 \qquad (14.32)$$

for a circular antenna of diameter D and moderate amplitude taper. Exploiting the definition of directive gain in (14.23), the number of dwells may be written as

$$Q_d \approx \frac{4\Omega G}{1.2^2\pi^3} \qquad (14.33)$$

The search requirements imply a single-scan detection range capability of

$$R_{max} = R_{min} + v_c T_F \qquad (14.34)$$

in order to ensure that the target is detected prior to penetrating the minimum range during the frame time T_F.

The required signal-to-noise ratio, $(S/N)_{req}$, is estimated as a function of the desired probability of detection and probability of false alarm.

The required dwell rate per second, F_d, is Q_d/T_f. In order to approximate matched-filter processing, the doppler filter bandwidth should be set equal to F_d. This condition corresponds to choosing an integration period equal to the duration of a coherent dwell. The resulting number of pulses integrated per dwell, N_p, is determined by the PRF selection.

Substituting these variables into (14.21) provides

$$\frac{P_{av}A}{T_s L_s} \geq \frac{14\Omega(S/N)_{req}(R_{min} + v_c T_f)^4 k}{T_f \sigma} \qquad (14.35)$$

This expression can be modified by observing $P_{av} = P_p N_p T_p B_d$. This relation reaffirms that PRF and doppler filter bandwidth may be selected so as to maximize S/C without impacting noise-limited detection performance as long as matched filtering is preserved. It follows that

$$\frac{P_p A N_p T_p B_d}{T_s L_s} \geq \frac{14\Omega(S/N)_{req}(R_{min} + v_c T_f)^4 k}{T_f \sigma_t} \qquad (14.36)$$

A similar expression can be generated for jammer-limited pulse-doppler radar detection such that

$$\frac{P_{av}A}{g_{rj}L_s} \geq (S/J)_{req}\left(\frac{\Omega}{T_f\sigma_t}\right)\left(\frac{P_jG_j}{B_jL_j}\right)\left(\frac{(R_{min} + v_cT_f)^4}{R_j^2}\right)$$ (14.37)

Note that all the radar parameters are grouped on the left-hand side of the equation, while the right-hand side contains target and operational requirements. Antenna illumination tapering for side-lobe reduction decreases SNR as it increases the system loss, but increases S/J against side-lobe jammers as well as increasing side-lobe clutter rejection. Increasing average power enhances detection performance in both noise-limited and jammer-limited environments. Application of (14.35) through (14.37) may be limited to short-range applications due to the necessity of enforcing a minimum time on target to support clutter filtering.

14.3.5.3 Cumulative Probability of Detection

In practice, search requirements for airborne radars are often defined in terms of cumulative detection rather than single-scan detection. As subsequently demonstrated, cumulative detection significantly reduces radar system requirements to attain a specified probability of detection by a given range.

The preceding sections have presupposed single-scan detection; that is, the detection process described has been based entirely on examination of a single receiver output sample containing a target return. In reality, target detection is generally a cumulative process in that the radar will receive returns from an approaching target on multiple scans. In particular, airborne radars typically command large line-of-sight ranges against aircraft at medium to high altitudes as indicated in Figure 14.11. Hence, multiple detection opportunities are available even at high closing velocities.

As noted previously, airborne radar detection performance is generally specified in terms of a cumulative detection range against a target of stated RCS and closing velocity. The cumulative probability of detection may be defined as

$$P_{cd} = 1 - \prod_{I=1}^{M}\left[1 - P_i\left(\frac{S}{N}\right)\right]$$ (14.38)

for the Mth observation opportunity (scan) with the probability of single-pulse detection P_i as a function of the SNR at that range. Application of (14.38) to evaluating radar performance implicitly presupposes that track can be initiated and maintained on the basis of a single detection. Phased-array radars can provide track initiation range on par with detection range, but this performance can be difficult to achieve in the case of mechanically scanning radars operating in severe clutter or ECM environments.

For example, using the AI radar system parameters of Table 14.3, we can estimate detection range performance against an incoming aircraft with an RCS of

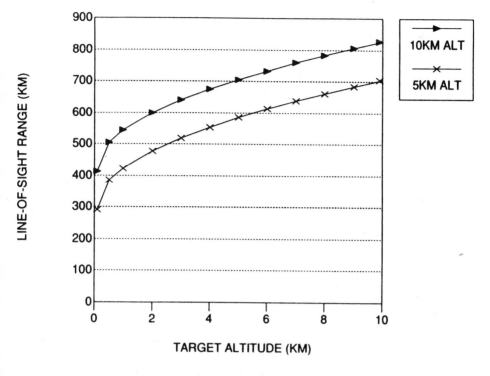

Figure 14.11 Maximum line-of-sight range.

Table 14.3
Example of High-PRF Pulse-Doppler Parameters

Parameter	Unit	Value
Transmitter average power	kW	4
Antenna diameter	m	0.8
Carrier frequency	GHz	9.4
Search raster frame time	sec	10
System losses	dB	14.6
System temperature	K	965

5 m^2 and a closing velocity of 1,000 m/s. What is the approximate range at which 90% cumulative probability of detection can be achieved against this target?

For the purposes of this application, the probability of detection of a nonfluctuating target can be approximated by modifying Albersheim's detection equation [8] so that

$$P_i(\text{SNRdB}_i) = \frac{e^{b\,(\text{SNRdB})}}{1 + e^{b\,(\text{SNRdB})}} \qquad (14.39)$$

where SNRdB is the signal-to-noise ratio in decibels and

$$b(\text{SNRdB}) = \frac{10^{\text{SNRdB}/10} - a}{0.12a + 1.7} \qquad (14.40)$$

In turn, a is defined as

$$a = \ln(0.62/P_{fa})$$

For convenience, assume that the detection processes begins at a range corresponding approximately with a 10% probability of detection. Then (14.37) can be iteratively applied to compute the cumulative probability of detection using (14.38) to compute the single-scan detection probability on each scan. The resulting single-pulse and cumulative probability of detection is depicted in Figure 14.12, which reveals that R_{95} is approximately 131 km. In this example, cumulative detection provides a range advantage of almost 10% over single-pulse detection. Alternatively, a given detection range requirement could be met with less average power antenna aperture product than that predicted by the single-scan radar range equation.

Cumulative detection probability effectively reduces the radar search requirements from the R^4 dependency implied by (14.35) to a range power between 3

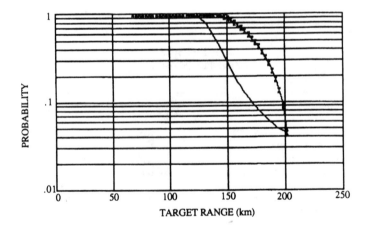

-----   Single-scan probability-of-detection
X-X-X   Cumulative probability-of-detection

Figure 14.12 Comparison of single-scan and cumulative detection.

and 4. As described in detail in the next chapter, cumulative detection is even more effective against fluctuating RCS model targets than against the less realistic constant RCS model target used in this example.

14.4 IMPLEMENTATION OF SYSTEM DETECTION REQUIREMENTS

The preceding sections of this chapter have described the search requirements and performance estimation for airborne pulse-doppler radar systems. These elements constitute a unified process for defining the subsystem performance required to achieve system level target detection requirements as summarized in Figure 14.13. At the top level, the radar designer is presented with an engagement scenario that includes target RCS and anticipated encounter altitudes and target aspects. In

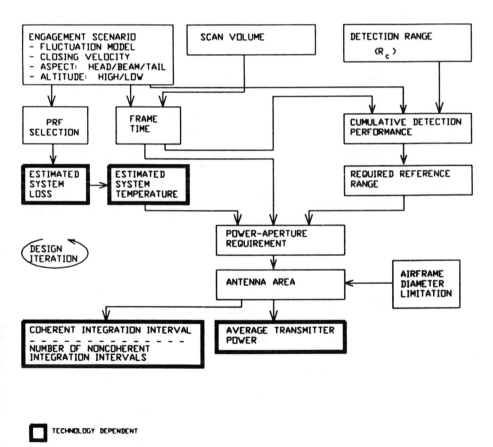

Figure 14.13 Detection requirement estimation.

addition, a search coverage volume may be specified in terms of the required detection range and scan angle extent that must be covered.

PRF selection is generally driven by clutter-limited detection considerations. The combination of the maximum anticipated closing velocity and the scan volume extent determine the frame time duration over which the antenna scans a single raster. Cumulative detection range requirements are determined from target signature characteristics and closing velocity in conjunction with the frame time. A reference range at which a specified SNR is achieved against the target is derived from the cumulative detection analysis.

The specifications of the waveform parameters and the reference range are used to establish the subsystem designs. In practice, the design process iterates among various trade-offs to attain the specified detection performance.

References

[1] Blake, L. V., *Radar Range-Performance Analysis*, Norwood, MA: Artech House, 1986.
[2] Marcum, J. I., "A Statistical Theory of Target Detection by Pulsed Radar," *Trans. IRE Prof. Group on Information Theory*, Vol. IT-6, No. 2, April 1960, pp. 59–144.
[3] Schwartz, M., *Information, Transmission, Modulation, and Noise*, New York: McGraw-Hill, 3rd edition, 1980.
[4] Barton, D. K., *Radar System Analysis*, Dedham, MA: Artech House, 1976.
[5] Petts, G. E., III, "Radar Systems," in *Handbook of Electronic Systems Engineering*, C. A. Harper, ed., New York: McGraw-Hill, 1980.
[6] Logan, R., "Airborne Solid State Phased Arrays: A System Engineering Perspective," *Radar 87*, 19–21 October 1987, London.
[7] Aronoff, E., and N. M. Greenblatt, "Medium PRF Radar Design and Performance," *Radars*, Vol. 7, D. K. Barton, ed., Dedham, MA: Artech House, 1978.
[8] Tufts, D. W., and A. J. Cann, "On Albersheim's Detection Equation,". *IEEE Trans. Aerospace and Electronic Systems*, Vol. AES-19, No. 4, July 1983, pp. 643–646.

CHAPTER 15

Effects of Clutter on Detection Performance

Melvin L. Belcher, Jr.

15.1 INTRODUCTION

The primary motivation for employing pulse-doppler radar is to detect targets that would otherwise be masked by clutter returns. The emphasis of this chapter is estimation of pulse-doppler radar detection performance under conditions of severe clutter such as those encountered in an AI look-down, shoot-down engagement. A thorough description of clutter return characteristics is not attempted, since the characteristics of clutter vary as widely as its sources and there is a large body of literature dedicated to this subject. However, some general properties and commonly accepted models are presented.

A recurrent theme in this description is that the radar characteristics constitute an integral component of the clutter modeling process. The clutter RCS characteristics are a function of the radar frequency and polarization. Clutter returns are mapped into ambiguous range-doppler outputs determined by the radar PRF. The short-term frequency stability of the radar transmitter-receiver chain may determine the degree of clutter spectral broadening present at the receiver output. Radar system parameters determine the necessary fidelity of the clutter modeling process.

This chapter introduces these concepts. Chapter 16 provides a synthesis of target, clutter, and radar system characteristics in the context of detection performance assessment.

15.2 CLUTTER DOPPLER SPECTRUM

The doppler offset of a stationary scatterer relative to an airborne radar can be defined as

$$f_d(\theta_a, \theta_e) = (2V_a/\lambda) \cos(\theta_a) \cos(\theta_e) \qquad (15.1)$$

where:

V_a = aircraft velocity;
θ_a = azimuth angle;
θ_e = elevation angle of the scatterer relative to the radar velocity vector.

The doppler bandwidth B_c associated with a given angular extent, $\pm\beta/2$, can be defined as

$$B_c = (2V_a/\lambda)|[\cos(\alpha - \beta/2) - \cos(\alpha + \beta/2)]|$$
$$B_c = (2V_a/\lambda) \sin(\alpha) \sin(\beta/2) \qquad (15.2)$$

for $\beta < 2\alpha$ and as

$$B_c = (2V_a/\lambda)[1 - \cos(\alpha + \beta/2)] \qquad (15.3)$$

for $\beta > 2\alpha$. An approximation for the null-to-null beamwidth, β, given in Chapter 2 is $\beta = 2.5 \cdot$ (3-dB beamwidth).

Hence, the clutter doppler center frequency decreases and the bandwidth increases with increasing scan angle. This relationship is illustrated in Figure 15.1 for typical AI and AEW radar system parameters. A common beamwidth of 4 degrees is assumed for purposes of comparison. The maximum scan angle for an AI radar is generally between 60 degrees and 70 degrees, while AEW radars typically rotate 360 degrees in azimuth. The AI radar is assumed to transmit at a wavelength of 0.03m, while the AEW radar transmits at 0.3m.

Low-to-moderate altitude AI radar performance can be modeled using the flat-Earth approximation as illustrated in Figure 15.2. The clutter RCS measured at the output of the single doppler filter of the receiver consists of the incident power density on the surface multiplied by the product of the clutter patch area A_c and the surface backscattering coefficient σ degrees. Using the radar range equation variables defined in Chapter 14, the total clutter power C can be defined as the summation of an ensemble of clutter patches such that

$$C = \frac{PGA}{(4\pi)^2 L_{cs}} \sum_i \frac{A_{ci}\sigma_{ci}^0(g_{sli})^2}{(R_{ci})^4} \qquad (15.4)$$

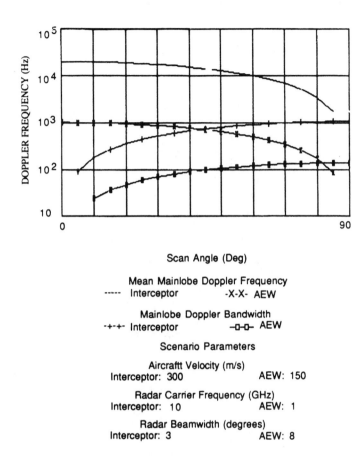

Figure 15.1 Clutter doppler variation with scan angle.

where R_c is the range to the clutter patch. The side-lobe gain, g_{sli}, is the radar antenna gain relative to the boresight gain and is normally given in spherical coordinates. In practice, the fidelity of the antenna representation ranges from a simple two-dimensional ensemble of step functions to the measured pattern of the radar antenna. The clutter system loss, L_{cs}, is computed similarly to the target system loss term described in Chapter 14 with the exception that straddling losses are generally neglected, since the clutter is assumed locally contiguous in range and doppler.

If the cell area is sufficiently small, then each clutter contribution can be modeled as uniquely associated with a given range and doppler frequency. The resulting two-dimensional clutter distribution, $C(R, f_d)$, is mapped in terms of range R and doppler frequency f_d. Hence, single-pulse clutter calculation consists of summing the clutter power across doppler over a given range extent. The clutter

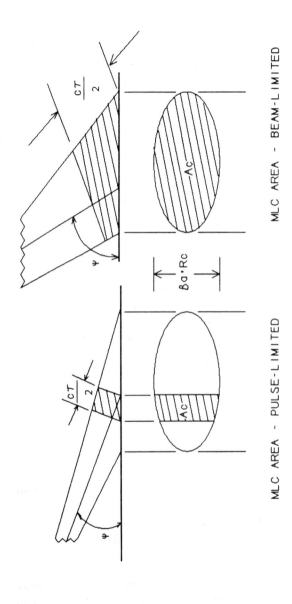

Figure 15.2 Main-lobe clutter area.

summation is nominally conducted over the entire field of view of the radar antenna. As described in Section 15.3, pulse-doppler radar performance modeling requires efficient calculation of the composite clutter return in the ambiguous range-doppler output space defined by the pulse-doppler radar parameters.

The clutter returns associated with airborne pulse-doppler radar operation are generally divided into three categories: main-lobe, side-lobe, and altitude. The main-lobe clutter is the clutter return component due to the surface illuminated by the radar antenna main lobe. A related clutter return source is the reflection of the main lobe that is produced by transmission through the radome. This effect is particularly significant in AI radars due to the sharply pointed radome shape and the associated free-space mismatch difficulties. The reflection lobe generally points on the opposite side of the aircraft from the main lobe and is somewhat broader in extent. The range-doppler coordinates of the clutter returns due to the main beam differ from those due to main-lobe reflection and each varies with scan angle.

Strictly speaking, main-lobe surface returns are clutter only when detection of airborne or discrete surface targets is being attempted. The main-lobe surface return is effectively the intended target in ground-mapping and doppler-navigation applications.

Side-lobe clutter is due to clutter return occurring through the two-way antenna pattern, as indicated by the side-lobe weighting term in (15.4). Since side lobes exist throughout the antenna field of view, side-lobe clutter returns are extensively distributed in range and doppler. Side-lobe clutter power generally falls off with increasing range.

Altitude return clutter is actually a subset of the side-lobe clutter, and is the clutter return due to the region almost directly below the aircraft, so that it occurs at a slant range equal to the aircraft altitude and a doppler frequency of approximately zero. The altitude return is commonly termed the *altitude line*, since it possesses a relatively narrow spectral extent. The altitude clutter return is specified as a separate clutter class due to its strong magnitude and fixed range-doppler coordinates. The near 90-degree grazing angle associated with the altitude return results in a large-clutter backscatter RCS. The combination of the relatively short range with the large-clutter RCS results in the altitude return being significantly greater than the nominal side-lobe clutter level.

The range interval from zero to the aircraft altitude is free of clutter. This clutter-free extent permits single-pulse radar modes to be employed to support short-range functions such as air combat maneuvering. Prior to the development of airborne MTI and pulse-doppler radar techniques, the altitude line severely interfered with detection and tracking of targets at ranges close to the altitude.

As an example of clutter return magnitude, consider the clutter return due solely to the main lobe, since it dominates any side-lobe clutter contributions at the coinciding slant range, assuming the target range is beyond the altitude return region. The illuminated surface area can be modeled as elliptical in shape, with the cross-range extent limited by the main-lobe azimuth extent. The along-range extent will be determined by the lesser of the main-lobe elevation range extent or the pulse range extent projected along the surface. Hence, the surface area A_c may be defined such that

$$A_c = (\pi/4)(R^2\beta_e\beta_a)/\sin(\theta_e) \text{ for } |\tan(\theta_e)| > (\beta_e R)/(c\tau/2) \qquad (15.5)$$

and

$$A_c = (c\tau/2)(R\beta_a)/\sin(\theta_e) \text{ for } |\tan(\theta_e)| < (\beta_e R)/(c\tau/2) \qquad (15.6)$$

The radar azimuth and elevation beamwidths are given by β_a and β_e respectively, while the azimuth and elevation coordinates for the antenna scan position are given by θ_a and θ_e, respectively. Equation (15.5) corresponds to the case of the illuminated ground patch being constrained in the along-track dimension by the elevation beamwidth extent, while (15.6) denotes that the along-track patch extent is limited by the pulse width.

The backscatter coefficient, σ^0, is defined as the radar reflectivity per unit area (square meters/square meters). This parameter can vary markedly with terrain, polarization, frequency, and grazing angle. A number of empirical studies have indicated that the reflection coefficient can be approximated as $\sigma^0 = \lambda \sin(\varphi)$ over a wide angular extent. The basis for this expression is that it compensates for the foreshortening of surface reference area that occurs with increasing grazing angle. However, the constant-γ approximation is inaccurate as the grazing angle approaches either 0 degrees or 90 degrees. Table 15.1 lists typical γ values for various types of surfaces. More detailed clutter models are presented in the next section.

Similar relations apply to volumetric clutter such as rain and chaff. The composite clutter return power for contiguously distributed volumetric clutter is the product of the volumetric backscatter coefficient and the revolution volume of the radar, V_c defined by

$$V_c = (\pi/4)(\beta_a\beta_e R_c^2)(c\tau/2) \qquad (15.7)$$

Table 15.1
Typical X-Band γ Values [1]

Surface	γ (dB)
Smooth water	−45.4
Desert	−12.4
Wooded area	−7.4
Urban	0.6

The product of V_{ci} and the appropriate volumetric backscatter coefficient would simply replace $A_{ci}\sigma^\circ$ in (15.4). Unlike surface clutter, the range-doppler extent of volumetric clutter cannot be computed on the basis of a priori data, since it is derived from dynamic phenomena such as weather. Hence, volumetric clutter can contribute significantly to the false alarm rate.

Computation of the clutter power obtained in the main lobe from a single pulse provides understanding of the necessity for pulse-doppler radar to detect airborne targets in a look-down geometry. Consider the example of an AI radar with a 3-degree beamwidth and 50-μs pulse duration onboard an aircraft flying at an altitude of 8 km, attempting to acquire a low-flying target at −30-degree elevation and 16-km range. This combination of parameters results in the main-lobe clutter patch being limited by the elevation beamwidth in the along-track dimension. Application of (15.5) indicates a ground area of approximately 10^6 m^2. Assuming a wooded terrain, γ equal to −7.4 dB can be used, resulting in a clutter RCS of approximately 54 dBsm. (RCS is often reported in units of dBsm, which is simply the ratio of the RCS referenced to 1 m^2 in decibels.) Hence, the single-pulse S/C for a target RCS of 0 dBsm is approximately −54 dB. Single-pulse S/C of −50 to −70 dB is common in airborne radar applications. In order to achieve reliable target detection, the S/C must be improved via doppler filtering in excess of 75 dB.

Decreasing the beamwidth through a larger antenna aperture would decrease the clutter return magnitude. Decreasing the pulse width increases the S/C if the radar range-resolves the clutter sources from the target return. However, SNR degrades linearly with decreasing pulse width (increasing bandwidth). Doppler filtering via coherent integration effectively improves SNR and S/C, as described below.

Assuming a radar frequency of 9.4 GHz, the main-lobe clutter return would possess a center frequency of 12.7 kHz and a bandwidth of 513 Hz. Hence, target detection would require that the target's velocity relative to the radar platform be greater than 220 m/s. S/C is being improved to an adequate level for detection by spectrally resolving the target return from the main-lobe clutter return. The target return would be competing with side-lobe clutter returns for any relative velocity less than the interceptor's velocity projected along the ground.

In practice, a significantly larger spectral separation between the target doppler and main-lobe clutter is required than that predicted by the inverse of the dwell duration. Instabilities in the radar transmitter-receiver chain broaden the main-lobe clutter spectral extent, as explained in Section 15.5. As described in Chapter 8, the doppler filters may be designed with a spectral side-lobe weighting that results in resolution degradation. The detection-threshold computation mechanism may unintentionally suppress target detections near strong range-doppler clutter regions, as described in Chapter 17.

In addition to the above considerations, airborne pulse-doppler radar systems may be designed to "blank out" the main-lobe clutter return with spectral notch filtering. The doppler center frequency of the main-lobe clutter return can be estimated in real time from (15.1) and the receiver down-conversion adjusted to heterodyne the main-lobe clutter component into the notch filter.

This main-lobe clutter rejection filter must be of sufficient spectral width to suppress the clutter doppler spread corresponding to the maximum scan angle of the antenna and to suppress moving surface target returns. Unfortunately, design of the filter to accommodate the maximum scan angle results in unnecessarily suppressed signals at smaller scan angles. The spectral spread of the main-lobe clutter is known from (15.2) and (15.3) so that the notch filter bandwidth can be changed with scan angle to minimize unnecessary signal blanking. Adaptive filtering is attractive for AEW systems that employ 360-degree azimuth scans and possess relatively complex signal processors. AI radar systems operate over smaller scan angles and often employ less complex signal processor implementations due to volume and weight constraints. An additional consideration in choosing the main-lobe clutter notch bandwidth is rejection of returns from moving surface objects such as automobiles. Airborne pulse-doppler radar operation over areas with extensive vehicle traffic may require an effective main-lobe clutter notch width on the order of ± 30 m/s corresponding to approximately ± 1.8 kHz at AI radar frequencies.

15.3 CALCULATION OF CLUTTER POWER

The unambiguous range-doppler geometry encountered by an airborne radar is depicted in Figure 15.3. The concentric circles representing a constant range from the radar are termed *isorange contours*. Isorange contours are simply formed by the intersection of a sphere of specified radius relative to the radar with the surface of the Earth.

Isodoppler (constant doppler) contours are more complex to map, since they consist of conic intersections with the Earth's surface. The vertex of the cone is at the radar, while the conic axis is aligned with the aircraft velocity vector.

The isodoppler contours are all hyperbolas for a level-flying aircraft ($\epsilon = 0$). However, the velocity vector of a diving aircraft ($\epsilon < 0$) intersects the surface so that (1) elliptical conic sections are formed for $0 < \theta < \epsilon$; (2) a parabolic isodoppler contour is generated at $\theta = \epsilon$; and (3) hyperbolic conic sections are formed for $\epsilon < \theta < (\pi - \epsilon)$. Similarly, a climbing aircraft ($\epsilon < 0$) results in (1) elliptical conic sections for $(\pi + \epsilon) < \theta < \pi$; (2) a parabola at $\theta = (\pi + \epsilon)$; and (3) hyperbolic conic sections for $-\epsilon < \theta < (\pi + \epsilon)$ [2]. The isodoppler contour shapes are distorted from these planar conic intersection expressions if a spherical Earth model is employed. If the aircraft altitude is sufficient to require use of a spherical Earth model, the isodoppler contours are conic intersections with a sphere.

Pulse-doppler radar performance modeling consists largely of computing the range-doppler distribution of the clutter power present at the receiver output. Isorange-isodoppler contour mapping results in surface clutter mapped into a range-doppler $(R - f_d)$ coordinate system. The clutter return associated with an intersection of a given isorange contour and isodoppler contour would contribute to the clutter power present in the corresponding range-gate-doppler-filter outputs. More generally, the isorange contours and isodoppler contours are computed with sufficient resolution to ensure that the clutter range-doppler extent of the patches formed by the intersection of adjacent contours is less then the radar range-doppler resolution.

Each elemental patch of ground clutter A_{ci} used to calculate the total clutter power according to (15.4) is in general at a different doppler frequency, which can be calculated using (15.1). To be useful in predicting the performance, ambiguous range-doppler returns must be mapped into the appropriate range-gate-doppler-filter by the relation

$$C(R_a, f_{da}) = \sum_i \sum_j C\left(R_a \pm \frac{ci}{2\text{PRF}}, f_{da} \pm j\text{PRF} \right) \qquad (15.8)$$

where R_a and f_{da} are the ambiguous range and doppler coordinates of the surface patch.

In the general case, estimation of the incremental surface area illuminated by a an airborne radar with arbitrary dive angle and altitude requires numerical integration techniques. (Arbitrary altitude includes the high-altitude radar case which necessitates the spherical earth surface approximation.) Hence, the composite estimation of the incremental-area clutter returns is computationally intensive due to the large number of patch area integrations required.

Computational requirements can be reduced by exploiting symmetry and significance properties, efficient coordinate system selection, and integration approximations. The number of area cells can be halved by simply noting the symmetry of the isorange and isodoppler contours about the intersection of the velocity vector and the Earth's surface. In the context of Figure 15.3, it is only necessary to estimate the cell areas in the positive y half of the $x - y$ plane.

Only a limited portion of the area cells contribute significantly to the composite clutter return. Side-lobe area cells from beyond the range at which side-lobe returns fall below the radar thermal noise floor can be neglected. Moreover, cell areas associated with the main-lobe and altitude line clutter returns can be neglected if the radar system blanks out the associated range-gate-doppler-cell outputs as is common in practice. Appendix A contains a Fortran program listing that calculates the ambiguous range-doppler clutter map and that can be executed on a personal computer.

The selection of the integration-summation coordinate system can have a heavy impact on computational requirements. Ringel [3] has noted that the coordinate system should satisfy three requirements: (1) provide an area expression readily amenable to numerical integration, (2) readily support selection of step size across the integration area, and (3) provide a relatively simple definition of area-cell

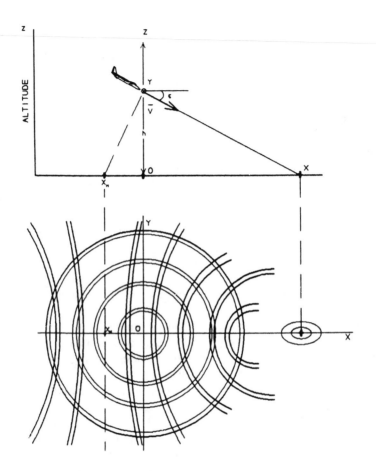

Figure 15.3 Range-doppler geometry.

boundary. On the basis of these criteria, Ringel selected a polar coordinate system. He notes difficulties with singularities and step size selection associated with the nominal (R,F) coordinate system as described by Friedlander and Greenstein [4].

Computational requirements can be significantly reduced by efficient cell-area estimation routines. Ringel [3] has described an effort to characterize the myriad different range-doppler cell shapes possible in the general case. He notes that this variation in shape can significantly complicate the cell boundary definition and area estimation process. As noted in the preceding paragraph, the coordinate system should be chosen to minimize these problems. At moderate to low aircraft altitudes, the flat Earth approximation is adequate. This simplifies the estimation process somewhat by reducing the number of cell shapes.

Closed-form expressions have also been suggested to eliminate the need for numerical integration. Jao and Goggins [2] have devised closed-form expressions for estimating cell areas using a combination of Cartesian and spherical coordinate system expressions under a flat Earth model. Sandhu [5] has exploited various numerical approximations and nonuniform sampling in the range-doppler coordinate system to produce an efficient clutter spectrum model suitable for real-time hardware-in-the-loop simulations.

15.4 CLUTTER MODELS

15.4.1 Overview of Clutter Characteristics

This section provides a brief overview of airborne clutter characteristics. Sea and land backscatter coefficient models suitable for estimating clutter return power are presented in this chapter. More detailed descriptions of clutter characteristics are provided in [6,7].

Conceptually, the clutter backscatter coefficient may be interpreted as the area-normalized product of the specified surface's reflectivity and retrodirectivity. Reflectivity characterizes the portion of incident illumination power that is neither absorbed by nor penetrates through the surface. The retrodirectivity is that portion of the reflected power that is radiated along the incident illumination's angle of arrival. Some surfaces, notably the sea, possess a high reflectivity but a low retrodirectivity at grazing angles not approaching 90 degrees. Hence, the sea possesses a relatively small (monostatic) backscattering coefficient but a relatively large forward scattering coefficient.

The backscattering coefficient varies with frequency, grazing angle, and polarization. In addition, backscatter coefficient tables represent quantities that are both temporally and spatially averaged. Temporal averaging is the consequence of the fact that a clutter cell return is the coherent sum of a large number of unresolved discrete scatterers. The resulting composite RCS fluctuates as the relative range of these scatterers vary with wind and other environmental effects. Similarly, the clutter return characteristics will vary with terrain and local surface scatterer properties.

To a first approximation, sea clutter can be characterized as spatially homogeneous and temporally varying while land clutter is spatially heterogeneous and temporally stationary. For a given frequency, polarization, and grazing angle, sea clutter can be reasonably well related to sea state. Sea state is a measure of the local sea roughness; increasing sea state number is correlated with increasing wind speed. Unfortunately, there is no corresponding unifying parameter for land clutter that varies markedly with heterogeneous land classifications.

Urban areas are particularly noteworthy due to the associated high average backscatter coefficient and a relatively large number of specular reflectors such as metallic planar surfaces. Specular returns can produce false detections in any radar design since they provide a strong return of limited range and doppler extent. Side-lobe specular returns are particularly stressing to medium-PRF radars since they produce "ghost" targets at the output of the multiple-PRF range-doppler ambiguity resolution processing. Ghost targets can be detected by a guard channel that is used to effectively blank out returns that are stronger in the side lobes than the main lobe. Specular return sources are not limited to urban areas but are generally associated with man-made structures.

The validity of any clutter model should be questioned at grazing angles of less than a few degrees or greater than about 85 degrees. Generally, the backscatter coefficient markedly decreases as the grazing angle approaches 0 degrees and markedly increases as the grazing angle approaches 90 degrees.

A clutter model should not be any more complex than what is demanded by the fidelity of the data. Hence, it seems worthwhile to implement a relatively complex sea clutter model since it can be modeled with acceptable accuracy. Conversely, most land clutter models are less detailed due to the larger variance encountered in measured data.

System evaluation simulation should incorporate some measure of the expected clutter return variation so as to provide a more robust measure of detection performance. A detailed appraisal of pulse-doppler radar clutter return characteristics must incorporate the associated resolution of clutter spectral components, which is not generally performed in clutter measurements.

15.4.2 Sea Clutter Model

Petts [8] has devised a relatively complete model for sea clutter. His compilation is reproduced in Appendix C. Each tabular entry corresponds to $-\sigma^0$ (dB) or, equivalently, the inverse of the magnitude of σ degrees converted to decibels. Note the decrease in backscatter associated with vertical polarization rather than horizontal polarization. The backscatter coefficient can be estimated using I and Δ listed in the table as

$$\sigma^0 = I + (3.376\Delta \log(f_0/0.3132)) \tag{15.9}$$

where σ^0 is in decibels and f_0 is the radar carrier frequency in gigahertz.

Sea clutter backscatter is due to the presence of waves, which disrupt the surface smoothness. Hence, the sea clutter backscatter coefficient increases with frequency, grazing angle, and sea state, since all three quantities increase the apparent roughness of the sea surface.

15.4.3 Land Clutter Model

A land clutter model compiled by researchers at Georgia Institute of Technology [7] is presented in Table 15.2. The backscatter coefficient is evaluated as

Table 15.2
Land Clutter Model Constants

Constant	Frequency	Soil/Sand	Grass	Tall Grass Crops	Trees	Urban	Wet Snow	Dry Snow
	3	0.0045	0.0071	0.0071	0.00054	0.362	—	—
	5	0.0096	0.015	0.015	0.0012	0.779	—	—
	10	0.25	0.23	0.006	0.002	2.0	0.0246	0.195
A	15	0.05	0.079	0.079	0.019	2.0	—	—
	35	—	0.125	0.301	0.036	—	0.195	2.45
	95	—	—	—	3.6	—	1.138	3.6
	3	0.83	1.5	1.5	0.64	1.8	—	—
	5	0.83	1.5	1.5	0.64	1.8	—	—
	10	0.83	1.5	1.5	0.64	1.8	1.7	1.7
B	15	0.83	1.5	1.5	0.64	—	—	—
	35	—	1.5	1.5	0.64	—	1.7	1.7
	95	—	1.5	1.5	0.64	—	0.83	0.83
	3	0.0013	0.012	0.012	0.002	0.015	—	—
	5	0.0013	0.012	0.012	0.002	0.015	—	—
	10	0.0013	0.012	0.012	0.002	0.015	0.0016	0.0016
C	15	0.0013	0.012	0.012	0.002	0.015	—	—
	35	—	0.012	0.012	0.012	—	0.008	0.0016
	95	—	0.012	0.012	0.012	—	0.008	0.0016
	3	2.3	0.0	0.0	0.0	0.0	—	—
	5	2.3	0.0	0.0	0.0	0.0	—	—
	10	2.3	0.0	0.0	0.0	0.0	0.0	0.0
D	15	2.3	0.0	0.0	0.0	0.0	—	—
	35	—	0.0	0.0	0.0	—	0.0	0.0
	95	—	0.0	0.0	0.0	—	0.0	0.0

$$\sigma^\circ = A(\psi + C)^B \exp\left[-D \middle/ \left(1 + \frac{0.1\sigma_h}{\lambda}\right)\right] \tag{15.10}$$

where σ_h is the standard deviation of the surface in centimeters. Nathanson [6] provides less detailed surface backscatter models for radar bands from UHF to Ka-band.

15.5 PERFORMANCE DEGRADATION DUE TO TRANSMITTER STABILITY

15.5.1 Clutter Sideband Masking

As noted previously, the composite clutter return can be represented as the sum of range-delayed and doppler-offset sinusoids with an associated amplitude weighting corresponding to the clutter patch RCS. The frequency domain output of the radar receiver may be modeled as the convolution of the system spectral response with this ensemble of discrete returns. The spectral response of the system can be characterized as the spectrum of a single-scatterer return generated over an arbitrarily long coherent integration period.

As the coherent integration period is increased, the spectral width of the reference return will decrease until it reaches the resolution limit imposed by the short-term frequency instabilities of the radar system. More importantly, the system spectral response will possess sidebands composed of discrete tonal components and contiguous noise modulation. Discrete spurious spectral components of sufficient magnitude produce false alarms. The sidebands associated with strong clutter returns can mask the returns from targets located in nominal clear regions of the ambiguity surface. In particular, the sideband masking from the main-lobe clutter return and the altitude return may define the effective dynamic range of the receiver so that the associated detection sensitivity may be significantly poorer than that indicated by the thermal noise floor. These concerns are illustrated in Figure 15.4. Note that notch filtering to suppress the nominal main-lobe clutter and altitude return does not suppress the associated sideband components.

Spurious spectral signal suppression is a driving concern in pulse-doppler radar design. In the case of one operational airborne pulse-doppler radar, it has been estimated that approximately 50% of the complaints of inadequate performance were due to spurious signal effects [9].

15.5.2 Sources of Spurious Components

Spurious sideband components can potentially be generated at any point in the radar transmit-receive chain, as indicated in Figure 15.5. Note that spurious sideband modulation may be independent in that its source is unique to either the transmit or receive chains or common to both.

In the context of this description, the exciter subsystem includes the pulse modulator and the local oscillator (LO) frequency synthesis. Frequency synthesis via frequency-multiplied crystal oscillator output is common in pulse-doppler radar design. Since low-noise crystal oscillators generally operate at a frequency between 1 and 100 MHz, it is necessary to derive the LO signals by feeding this reference signal through a frequency multiplier. A frequency multiplier is a nonlinear device which provides an output frequency that is M times signal input frequency. Frequency multipliers generally employ a highly nonlinear amplifier to produce the Mth-order harmonic of the input; other harmonic components are suppressed by filtering. The output signal bandwidth is also increased by a factor of M relative to the input.

The magnitude and frequency of tonal interference is generally deterministic in that it can be anticipated during the radar design. Discrete spurious signals are

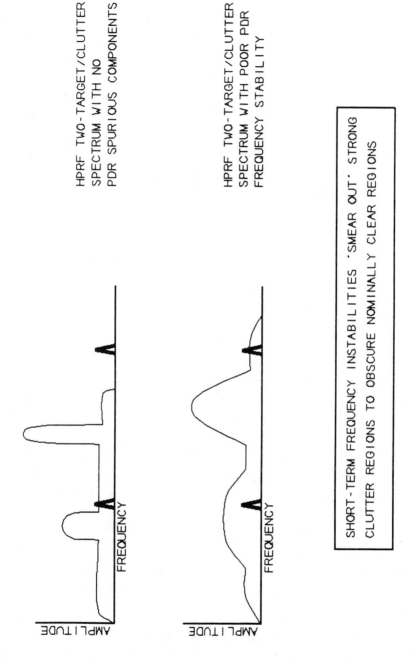

Figure 15.4 Effect of frequency instability.

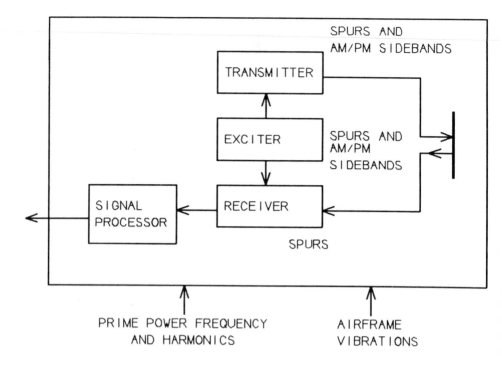

Figure 15.5 Primary sources of spurious signal components.

due to such sources as power line coupling, receiver intermodulation products, and microphonic effects. The prime-power frequency and its harmonics are generally difficult to suppress, since it can enter the radar system through transmitter power supply ripple and electromagnetic coupling from transformers and other radiating structures. Stringent power supply buffering and electromagnetic shielding can generally mitigate discrete power line leakage. Receiver intermodulation products, or spurs, can be mitigated by appropriate specification of mixer and filter characteristics. The frequency synthesis design community is well versed in techniques to suppress discrete spectral signals [10].

Microphonic signals are due to vibrations coupling into the pulse-doppler radar receiver. This source of interference is particularly stressing to airborne pulse-doppler radar systems, due to the airframe vibration environment. Development of effective isolation mounts for components sensitive to microphonic coupling was a significant milestone in achieving adequate airborne radar performance [11,12].

In the general case, the error modulation is characterized as a sum of amplitude modulation (AM) and narrowband phase modulation (PM) components. FM expressions can simply be formulated as the time derivative of the corresponding PM expression divided by 2.

Sideband suppression performance is typically limited by spurious phase modulation. Most radar transmitter chains possess a limiter nonlinearity at some point which suppresses AM while increasing the magnitude and spreading the spectral extent of PM. A limiter provides a constant output for all inputs above some specified level. Operation of the final amplifier stage of the transmitter in saturation for maximum efficiency results in limiting action. Exciter frequency multiplication imposes limiting on reference frequency synthesis in both the transmit and receive chains. The AM error component is typically reported to be insignificant in comparison to the PM component. Hence, the following development will neglect AM error as well as deterministic tonal components.

The general form of the transmitted signal referenced to the final IF stage with associated transmit and receive PM error modulation, $\eta_t(t)$ and $h_r(t)$, may be written as

$$s_t(t) = \cos[2\pi f_{if} t + \eta_t(t) - \eta_r(t - T_R)]$$
$$= \cos[2\pi f_{if} t + \Delta_\eta(t)] \tag{15.11a}$$

where f_{if} is the radar IF and T_R is the range delay of the return. This expression presupposes that the received coherent pulse train has been converted to CW by bandpass filtering of the center spectral line.

In most cases of interest, the error PM within a narrow bandwidth can be approximated as a sinusoidal phase modulation such that

$$\eta(t) = m \cos(2\pi f_m t) \tag{15.11b}$$

where m is the modulation index and f_m is the modulation frequency. For the narrowband PM constraint, $m \ll 1$ rad, this signal component produces sidebands offset from the modulation carrier frequency f_{if} by ^+f_m with an SSB-to-carrier power ratio of $(m/2)^2$. The modulation frequency corresponds to a specific doppler frequency offset from a given return.

The SSB-to-carrier ratio, $S_{ssb}(f)/C_a$, resulting from common sinusoidal PM can then be expressed as

$$S_{ssb}(f_m)/C_a = [2 \sin(\pi f_m T_R)]^2 (m/2)^2 \tag{15.12}$$

As noted by Raven [13], the cancellation term $(2 \sin(\pi f_m T_R))^2$ is due to the correlation inherent in common spurious PM. This cancellation is particularly significant in ground-based pulse-doppler radar operation due to the magnitude of short-range, low-doppler-frequency surface clutter, where the $f_m T_R$ product is relatively small.

15.5.3 Estimation of Subclutter Visibility

Subclutter visibility can be defined as the effective dynamic range from the maximum return power to the associated spurious sideband "floor" in the receiver. As indicated above, this floor is often determined by the phase noise performance of the radar exciter. Following the development of Robbins [14], contiguous phase noise can be modeled as an ensemble of narrowband sinusoidal PM components associated with arbitrarily narrow bandwidth. The corresponding modulation index is equal to the standard deviation of the phase within that bandwidth, σ_η, under the narrowband PM approximation.

The performance of specific exciter sources is generally specified as a power spectral density, as depicted in Figure 15.6. The SSB-to-carrier power ratio, S_{ssb}/C_a, is computed for a reference bandwidth (nominally 1 Hz) as a function of carrier offset frequency. The S_{ssb}/C_a is typically calculated in terms of the SSB power within the reference bandwidth in decibels relative to the carrier (dBc/Hz). The C/N_{op} of Figure 15.6 is equivalent to this measure. Application of these curves to estimating radar performance requires that the nominal power spectral density (PSD) of the exciter be convolved with the doppler-filter response of the radar receiver. The composite spurious SSB power at a given doppler frequency is calculated by (1) summation across all clutter returns in range and doppler with appropriate weighting by the product of S_{ssb}/C_a and the cancellation factor, followed by (2) ambiguous range-doppler mapping in the receiver output.

Equation (15.12) provides insight into the corresponding short-term phase stability requirements, except for offset frequencies near the carrier where the narrowband PM approximation is inaccurate. Specifically, the modulation index m can be approximated as $\sqrt{2\sigma_\eta^2}$, where σ_η^2 is the variance of the phase jitter in units of rad^2/Hz.

The airborne intercept example of Section 15.2 can be extended to provide insight into phase noise performance requirements while illustrating a back-of-the-envelope approach to estimating SSB suppression requirements. Consider the same intercept geometry, but assume high-PRF pulse-doppler radar operation with a 200-kHz PRF, 2-μs pulse duration, and 200-Hz doppler resolution. On the basis of high-PRF operation, spectral aliasing of the spurious SSB of the main-lobe clutter shall be presupposed insignificant.

Assume that the target doppler frequency relative to the main-lobe clutter is approximately 5.9 kHz and the target return is outside the side-lobe clutter region. The associated correlation cancellation of the main-lobe clutter is approximately 5 dB. The single-pulse signal-to-clutter ratio, $S/C|sp$, is approximately -51 dB. The convolution of the nominal phase noise PSD with the doppler filter response of the subject radar can be approximated by assuming a rectangular bandpass of 200 Hz. It follows that the nominal C_a/S_{ssb} can simply be multiplied by the ratio of the radar doppler bandwidth to the reference bandwidth (200/1) to estimate

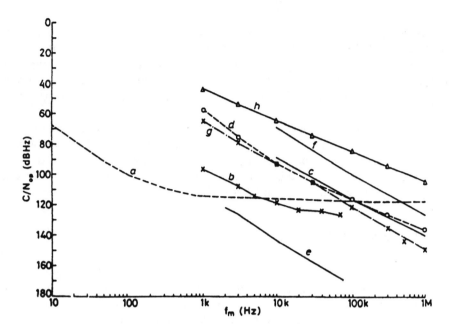

a) 5 MHz crystal oscillator followed by 1600x frequency multiplier (8 GHz)

b) 120 MHz crystal oscillator followed by 72x frequency multiplier (8.64 MHz)

c) 1.2 GHz bipolar transistor cavity oscillator followed by 6x frequency multiplier (7.2 GHz)

d) Reflex klystron (0.4 GHz)

e) Two-cavity klystron oscillator

f) GaAs FET oscillator (10.14 GHz)

g) Gunn oscillator (11.1 GHz)

h) Impatt oscillator

Figure 15.6 X-band phase noise performance comparison. (*Source:* [14].)

the relative SSB power present in a single doppler filter. For adequate detection performance, assume that the required signal-to-clutter-SSB-ratio, S/C_{ssb}, is 15 dB. The required maximum C_a/S_{ssb} value to achieve the specified S/C_{ssb} in decibels is given by

$$S_{ssb}/C_a = -S/C_{ssb} - 10 \log(Bd) + 10 \log \frac{S}{\sum_m \sum_n C_{mn}(2 \sin \pi f_m T_{Rn})^2} \quad (15.13)$$

Evaluation of this expression results in a required S_{ssb}/C_a of approximately 92 dBc/Hz at 5.9-kHz offset from the carrier. Comparison of this requirement with Figure 15.6 indicates that this exciter performance is achievable. Assuming 5 dB of correlation cancellation, this SSB-to-carrier ratio corresponds to an allowable phase variance of approximately 2×10^{-9} rad^2/Hz over the period corresponding to the range delay.

In practice, detection performance must be assessed by summing the product of the correlation cancellation coefficient and appropriate SSB-to-carrier weight across each range-doppler patch composing the main-lobe clutter, altitude, and side-lobe returns. In addition, this summation must be mapped into the ambiguous range or frequency coordinates required by medium-PRF radar modeling [15].

15.6 CLUTTER-LIMITED DETECTION

Ideally, the interference in the pulse-doppler radar range-gate-doppler-filter containing a target return will be dominated by thermal noise. Noise-limited detection is associated with certain radar-target-surface scenarios, as noted in the previous chapter. However, clutter returns will constitute the dominant interference source in many instances. In particular, high-PRF doppler filter outputs in the side-lobe region and most medium-PRF range-gate-doppler-filters contain significant amounts of clutter power. This subsection briefly notes some related detection consideration.

The central limit theorem (CLT) implies that if a range-doppler resolution cell contains a large number of discrete scatterers and the scatterers are of comparable RCS, then the composite return fluctuations will follow a Gaussian probability density function (PDF). Hence, the magnitude of the composite return at the output of the radar receiver can be modeled as possessing a Rayleigh PDF. Unfortunately, this approximation is inaccurate in describing the fluctuation characteristics of specific clutter return sources such as the sea and urban terrain.

At the fundamental level, the failure of the Rayleigh magnitude PDF to adequately model clutter statistics implies that one or both of the CLT conditions are violated. A pulse-doppler radar can violate the first condition by spectrally resolving clutter into a small effective number of individual scatterers. Scatterer shadowing as determined by aspect geometry is also an important consideration in determining the effective number of scatterers. In the second case, specular returns from sources such as man-made structures can increase the nominal false alarm rate considerably over some terrain. Clutter with non-Rayleigh PDFs can be characterized as "heavy tailed," denoting a higher ratio of mean to median. This condition results in a significant increase in false alarm rate corresponding to a greater occurrence of high-amplitude "spikes" for a given average clutter power level.

Accurate modeling of clutter RCS statistics is important in controlling the false alarm rate to an acceptable level without undue sacrifice of detection performance. This concern is discussed further in the context of automatic detection in Chapter 17.

References

[1] Stimson, G. W., *Introduction to Airborne Radar*, Hughes Aircraft Co., El Segundo, CA, 1983.

[2] Jao, J. K., and W. B. Goggins, "Efficient, Closed-Form Computation of Airborne Pulse-Doppler Radar Clutter," *Record of the IEEE 1985 Radar Conference*, Arlington, VA, pp. 17–22.

[3] Ringel, M. B., "An Advanced Computer Calculation of Ground Clutter in an Airborne Pulse Doppler Radar," *NAECON 1977 Record*, pp. 921–928.

[4] Friedlander, A. L., and L. J. Greenstein, "A Generalized Clutter Computation Procedure for Airborne Pulse Doppler Radars," *IEEE Trans.*, Vol. AES-6, No. 1, January 1970, pp. 51–61.

[5] Sandhu, G. S., "A Real-Time Clutter Model for an Airborne Pulse Doppler Radar," Southeastcon 1982, pp. 316–321.

[6] Nathanson, F. E., *Radar Design Principles: Signal Processing and the Environment*, New York: McGraill Book Co., 1969.

[7] Eaves, J. L., and E. K. Reedy, eds., *Principles of Modern Radar*, New York: Van Nostrand Reinhold, 1987.

[8] Petts, G. E., III, "Radar Systems," in *Handbook of Electronic System Design*, C. A. Harper, ed., New York: McGraw-Hill, 1980.

[9] Gray, M., et al., "Stability Measurement Problems and Techniques for Operational Airborne Pulse Doppler Radar," *IEEE Trans.*, Vol. AES-5, No. 4, July, 1969, pp. 632–637.

[10] Manassewitsch, V., *Frequency Synthesizers Theory and Design*, 2nd edition, New York: John Wiley & Sons, 1980.

[11] Leeson, D. B., and G. F. Johnson, "Short-Term Stability for a Doppler Radar: Requirements, Measurements, and Techniques," *Proc. IEEE*, Vol. 54, No. 2, February 1966, pp. 244–248.

[12] Rossman, H., and J. J. Chino, "Crystal Oscillator Vibration Isolation for Airborne Radar Applications," *IEEE Mechanical Engineering in Radar*, pp. 72–74.

[13] Raven, R. S., "Requirements for Master Oscillators for Coherent Radar," in *Radars Volume 7, CW and Doppler Radar*, D. K. Barton, ed., Norwood, MA: Artech House, 1978.

[14] Robbins, W. P., *Phase Noise in Signal Sources (Theory and Applications)*, London: Peter Peregrinus, 1982.

[15] Belcher, M. L., and G. V. Morris, "Pulsed Doppler Radar," in *Coherent Radar Performance Estimation*, J. A. Scheer and J. L. Kurtz, eds., Norwood: MA: Artech House, 1993.

CHAPTER 16

Target Fluctuation Effects

Melvin L. Belcher, Jr.

16.1 COMPLEX TARGET RCS

The preceding two chapters have generally presupposed a nonfluctuating target in estimating detection performance. This condition effectively assumes that a target consists of a single scatterer with aspect-insensitive RCS. Although unrealistic, this condition provides a measure of performance for single-pulse detection as noted in Chapter 14. In reality, target RCS can fluctuate markedly and significantly impacts detection performance.

Most targets of interest lie within the optical scattering region; that is, the target dimensions are much greater than the radar wavelength. In addition, the range resolution cells of pulse-doppler waveforms are generally larger than the target dimensions. Hence, the composite RCS of a complex target is actually the sum of returns from the individual scattering centers making up the target. This summation may be written as

$$
\sigma_t = \left| \sum_i \sqrt{\sigma_{tri}} \, \exp(\phi_{tri}) \, \exp[\, j\theta_i(t)\,] \right|^2
$$

$$
= \left| \sum_i \sqrt{\sigma_{tri}} \, \exp(\phi_{tri}) \left\{ j\left[\frac{4\pi R_i(t)}{\lambda} \right]_{\mathrm{mod}(2\pi)} \right\} \right|^2 \tag{16.1}
$$

where σ_{tr} and ϕ_{tr} are the RCS and relative phase of the scatterer for radar transmit polarization t and receive polarization r, respectively. The time-varying relative phase of the scatterer, $\theta(t)$, is defined as a function of the time-varying radar range, $R(t)$, and the carrier frequency. The mod(*) operation denotes taking the modulus of the preceding quantity relative to the argument. Variation among the scatterers in the relative line-of-sight range results in coherent constructive and destructive addition. As a result, the RCS of a complex target varies markedly with aspect angle, since line-of-sight rotation alters the phasing relationship among the scatterer returns. The sensitivity of RCS to aspect is illustrated in Figure 16.1, where a fraction of a degree variation in aspect can result in over 10 dB of variation in RCS. In operational scenarios, aspect angle variation is created by target maneuvers as well as relative translational motion between the radar platform and the target. Aircraft maneuvers correspond to rotation around the pitch, yaw, or roll axes.

Airframe vibrations also induce relative line-of-sight range variation among individual scatterer returns. Airframe vibration sources include flexing, flutter, and engine vibration. In addition, doppler modulation imposed by propeller blades and jet engine fans imposes distinctive sidebands on a pulse-doppler target return.

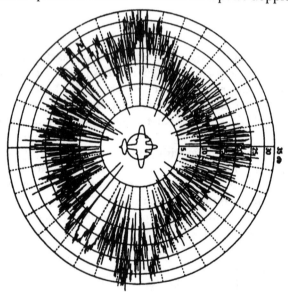

Figure 16.1 RCS versus azimuth for B-26 Aircraft at S-band [1].

As indicated by the polarization subscript on RCS, polarization changes can also contribute to variation in RCS with target aspect. The RCS and relative phase of individual scatterers can be a function of the transmit and receive polarization of the radar and the orientation of the individual scatterer relative to the radar line of sight. Differing characteristics of target and clutter returns can potentially be exploited to enhance target detection in a clutter-limited environment [2].

Specular returns existing over limited angular extents also contribute to fluctuations of RCS with aspect angle. Specular RCS may be defined as

$$\sigma_s = 4\pi [\, \gamma(\lambda)\,]^2 A_s^2 / \lambda^2 \qquad (16.2)$$

where A_s is the effective area of the reflecting surface and $\gamma(\lambda)$ is the voltage reflection coefficient of the surface at the radar carrier wavelength. The reflection coefficient is less than or equal to unity. Specular returns can occur near beam and other aspects that present large reflective surfaces orthogonal to the radar line of sight.

Aircraft antennas can provide very strong specular returns, providing that a significant portion of the incident power is reflected [3]. The reflection coefficient is often near unity (high voltage standing wave ratio (VSWR)) if the radar is operating at a frequency outside the operating bandwidth of the antenna. The retroreflectivity of an antenna is also determined by the polarization response to the incident radar illumination. Aperture antennas are often highly reflective at incident polarization orthogonal to that of antenna operation.

The angular extent of a retroreflective return from an aperture surface of length L_s is approximately $60\lambda / L_s$ degrees, where L_s and λ are in consistent units. For example, an X-band return from a circular antenna of 0.5m diameter and unity reflection coefficient would result in a specular RCS of 500 m^2 across an angular extent of approximately 4 degrees. Antenna RCS reduction is a significant concern in modern aircraft design.

For the reasons just outlined above, the RCS variation associated with a complex scatterer target such as an aircraft under arbitrary motion is not readily modeled in a deterministic fashion. It is customary to employ statistical RCS fluctuation models that are indicative of the RCS variation anticipated with operational targets to assess radar system performance. Specifically, target RCS is typically modeled as a random variable with a specified PDF.

Characterization of RCS as a random variable is generally insufficient for assessing radar detection performance. The associated PDF corresponds to the statistical properties of the target RCS associated with a given observation. It is also necessary to define the statistical variation of RCS over the observation interval. In the general case, RCS can be represented as a time-varying random variable. The resulting stochastic process representation is defined as a function of the aforementioned PDF and a temporal autocorrelation function [4].

16.2 FLUCTUATION MODELS

16.2.1 Fluctuation PDFs

Swerling's classic work [5] suggested two PDFs and two temporal autocorrelation conditions, resulting in four combinations of target fluctuation models as described below. Both PDFs are drawn from the chi-square family. Swerling classified the temporal autocorrelation of a fluctuating target in terms of its decorrelation time relative to the PRI and scan interval of the subject radar.

A chi-square random variable of $2n$ degrees of freedom is formed from the sum of $2n$ mutually independent Gaussian random variables with zero mean and common variance σ_x. The variable k, the duo-degrees of freedom, will be used to denote the quantity $2n$. As defined by Swerling, the general chi-square PDF, $f_c(x)$, may be given as

$$f_c(x) = \left(\frac{1}{\Gamma(k)}\right)\left(\frac{k}{\bar{x}}\right)^k x^{k-1} \exp\left(\frac{-kx}{\bar{x}}\right) U(x) \tag{16.3}$$

where:

Γ = gamma function;
$\bar{x}$ = mean RCS;
$U(x)$ = unit step = 1, $x \geq 0$
= 0, $x < 0$.

The random variable x corresponds to the target RCS, or, equivalently, the return power received at the radar.

Swerling cases 1 and 2 both employ a chi-square PDF with $k = 1$, which is the Rayleigh PDF. The resulting PDF, $f_{sw1/2}$, can be written as

$$f_{sw1/2}(x) = (1/\bar{x}) \exp[-x/\bar{x}] U(x) \tag{16.4}$$

The unity duo-degrees PDF represents the fluctuation associated with a target composed of a large number of randomly distributed scatterers of comparable magnitude RCS; that is, no single scatterer is dominant. This PDF is generally considered to be reasonably representative of aircraft return magnitude fluctuations. In addition, it represents the RCS fluctuations of clutter returns from rain cells and most terrain.

The Swerling 1 and 2 cases are distinguished by the associated temporal autocorrelation constraint. Under case 1, the RCS of a given target is assumed completely correlated over the time interval of observation in a single scan, but independent from scan to scan. Under case 2, the RCS is assumed uncorrelated from sample to sample. Hence, the RCS of a Swerling 1 target return would be constant over a single dwell, whereas a Swerling 2 target return would fluctuate pulse to pulse within a dwell. The Swerling 1 and 2 cases are identical for single-pulse detection.

The PDF for Swerling cases 3 and 4 is the two duo-degrees chi-square PDF and is given by

$$f_{sw3/4}(x) = (4x/\bar{x}^2) \exp[-2x/\bar{x}] U(x) \tag{16.5}$$

Case 3 assumes interpulse correlation and interscan decorrelation; Case 4 assumes interpulse decorrelation. The two duo-degrees PDF approximates the RCS fluctuations of complex targets composed of one dominant constant reflector and multiple smaller reflectors. Missiles and satellites are often cited as exhibiting fluctuations corresponding to this case.

Increasing the degrees of freedom of the chi-square PDF corresponds to increasing the coherent component of the target fluctuation model. Hence, the relative RCS fluctuation decreases with increasing k. A nonfluctuating target corresponds to a chi-square density with infinite degrees of freedom.

Weinstock fluctuation models consist of chi-square PDFs with duo-degrees of freedom ranging from 0.6 to 4. The specific value of k is computed as a function of geometric shape and the range of aspect angles.

Experimental data have indicated a number of targets exhibit a significantly higher percentage of large returns than predicted with chi-square PDFs given the median RCS of the target. These types of target fluctuation histories can be characterized as possessing a high mean-to-median ratio. It has been suggested that this phenomenon arises from the occurrence of specular returns at specific angular positions. Returns from large aircraft over limited angular extents at certain aspects and from ships reportedly display these characteristics. The log-normal density has been suggested for modeling targets with these fluctuation characteristics. The log-normal PDF, f_{ln}, is given as

$$f_{ln} = \frac{1}{\sigma_s x \sqrt{2\pi}} \exp\left\{\frac{[\ln(x) - \ln(x_m)]^2}{2\sigma_s^2}\right\} U(x) \tag{16.6}$$

where σ_s is the standard deviation of $\ln(x)$ and x_m is the median of x.

Selection of a fluctuation model for detection performance analysis should reflect the following considerations. The autocorrelation time constraints of the Swerling cases are seldom applicable to actual radar targets. More often, a given series of pulses will be partially correlated. The Swerling 1-2 and 3-4 cases provide limiting performance bounds.

A sequence of returns may be modeled more accurately as block-correlated. Given an N-pulse return sequence, the correlation interval is estimated so that the N_b pulses observed over a single correlation interval are assumed perfectly correlated and each block is modeled as independent. The effective number of independent returns is approximated as the number of blocks within the total observation interval.

Use of the unity duo-degree chi-square PDF (Swerling 1) provides a conservative performance estimate. This function imposes relatively high fluctuation losses and does not assume the relatively large PDF tails inherent in the log-normal density. Estimation of the target RCS should be in terms of the median rather than the mean. Use of the median RCS estimation has been shown to result in less performance estimation error [6] than use of the mean when the data set used to estimate RCS is sparse.

16.2.2 Extension to Pulse-Doppler Radar

Pulse-doppler radars impose additional considerations in assessing target detection, since a complex target is often resolved into separate spectral components. Aircraft doppler spectra may generally be considered to consist of two major signal components. The first component is the airframe return, which constitutes the nominal "true" return as it corresponds to the actual target range rate. The second component is the ensemble of sidebands due to engine modulation, which surround the airframe return. This division of the composite return reduces SNR, since detection thresholding is performed on the output of a given doppler filter.

Radar detection performance should normally be assessed in terms of Swerling 1 or Swerling 3 target models. In the context of the previous subsection, the output of a given doppler filter at the end of a coherent processing interval is treated as a single pulse. The target is assumed to possess a constant RCS over a coherent integration interval. This supposition can be empirically justified for the airframe return by consideration of aircraft-return spectra.

If we assume that RCS fluctuation can be represented as ergodic, the associated power spectrum density can be estimated from the averaged doppler spectrum. The autocorrelation of a variate is equal to the Fourier transform of the power spectrum density. Hence, the decorrelation interval can be estimated as the inverse of the fluctuation spectrum bandwidth. Nathanson [6] has estimated decorrelation intervals ranging from 30 to 300 ms from examination of noncoherent aircraft return spectra. The target return will be distributed across multiple doppler filters if its spectrum is significantly wider than the radar doppler resolution, resulting in SNR loss. However, the airframe return is generally considered as narrowband relative to the radar doppler resolution and referred to as a *single doppler line*.

16.2.3 Engine Modulation

Engine modulation produces discrete spectral sidebands as a function of the number of blades and rotation rate. Discrete sidebands can potentially be detected at

$$f_{mn} = r \cdot m \cdot n \tag{16.7}$$

where r is the engine-shaft rotation rate in revolutions per second, m is the number of blades, and n signifies the nth signed harmonic. The sign of the harmonic denotes whether the spectral component is above or below the airframe doppler line. The fundamental turbine frequency ($n = \pm 1$) has typically been reported to be on the order of 1 kHz. Note that the sideband offset frequency is independent of radar wavelength. However, the size of the blades relative to the radar wavelength can be a significant factor in determining the jet engine modulation (JEM) line magnitude.

Engine modulation is associated with both propeller-driven and jet-engine aircraft, but JEM is typically of greater interest in evaluating airborne radar performance. Gardner [7,8] reported significant JEM lines at up to 20 kHz from the airframe doppler line. The air intake ducts of jet engines are large in comparison to most airborne radar wavelengths, so radar pulses are ducted to the compressor fans and back in the case of forward quarter illumination of the target aircraft. Illumination of the rear quarter can result in similar internal propagation to the turbine blades and back. In the context of (16.7), JEM lines can be produced at frequencies corresponding to multiple compressor stages as well as the turbine blades. An example of JEM sidebands is reproduced in Figure 16.2.

JEM is commonly encountered in operational scenarios. Significant modulation components have been detected at up to 60 degrees off the nose of the target aircraft. Analysis of pulse-doppler detection data has revealed JEM detection in 80% of the head-on target contacts [9].

Gardner's measurements of specific single-engine turbojet aircraft indicated that the lower sideband data were 5 dB below the airframe return on the average.

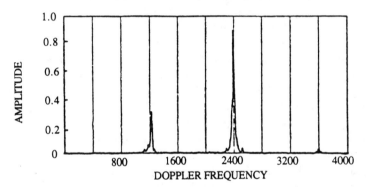

Doppler spectrum of an approaching single-engine turbojet aircraft

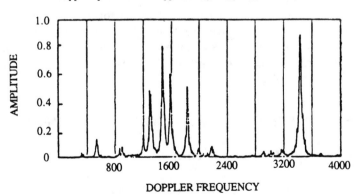

Figure 16.2 Skin return and JEM sidebands of approaching single-engine turbojet aircraft [7].

The upper-sideband return was generally much lower. Only a single pair of significant JEM sidebands were noted in this data set. Other data have reportedly indicated up to approximately 10 detectable JEM lines on either side of the airframe return with the strongest one or two dominating the remainder. In some instances, JEM sideband magnitude has been significantly greater than the airframe return.

JEM line magnitude fluctuates over time. Empirical data have suggested the decorrelation interval corresponds to a variation in aspect angle on the order of 1 degree. As would be expected, JEM is most prominent near head-on and tail aspects. JEM sidebands must be accounted for in the design of automatic detection processors. Effectively, the presence of JEM causes a single target to appear as a complex of targets collocated in range but with a large range rate spread. The postdetection processing outlined in Chapter 14 should be capable of reducing this apparent multiple-target return into a single detection report or track initiation request. Comparison of apparent doppler with calculated range rate is one potential discriminant for eliminating JEM detection reports.

CFAR operation may also be degraded by the presence of JEM lines. As noted in Chapter 17, multiple target lines can effectively act as interference in a doppler-averaging CFAR threshold computation so as to degrade target detectability by raising the decision threshold. Tests of the AN/APG-66 airborne intercept radar revealed an interesting false detection source termed *skinless sidebands* [10]. These returns were determined to originate from idling aircraft sitting on airport runways. The aircraft was stationary so that the airframe was rejected by the main-lobe clutter notch. However, the spectral extent of the JEM sidebands extended outside the notch and appeared as valid target returns.

Doppler tracking may also be degraded by the presence of JEM lines [9]. Target maneuvering can potentially result in the tracking filter locking onto a JEM sideband, resulting in significant degradation of smoothing and prediction performance. Modification of the tracking filter to detect target maneuvers and impose appropriate track constraints has been proposed to mitigate this problem.

Although JEM lines may degrade detection performance, they can potentially provide a means of noncooperative target identification. The spectral composition of the JEM sidebands constitutes a signature of a specific aircraft and engine design. Moreover, the target aircraft presents near-head-on aspects to the radar in many operational scenarios of interest.

16.3 DETECTION PERFORMANCE AGAINST FLUCTUATING TARGETS

16.3.1 Noncoherently Integrated Returns

The appropriate fluctuation PDF is used to calculate the probability of detection as a function of average SNR, probability of false alarm, and the number of pulses noncoherently integrated. As noted previously, this computation is made on the assumption that the target return lies within a single range-gate-doppler-filter. The detection threshold is equal to that estimated for the nonfluctuating-target case, since it is based on the assumed interference PDF rather than the target characteristics. Following the previous development, the detection performance of noncoherent pulse integration will be defined in terms of its relation to single-pulse performance.

The target fluctuation corresponds to broadening the PDF of the signal-plus-noise receiver output. Hence, as illustrated in Figure 16.3, the SNR must be increased over the nonfluctuating target case to achieve the specified probability of detection. This increment in SNR is termed the *fluctuation loss*. As demonstrated in Figure 16.4, fluctuation loss varies with probability of detection. Detection is actually enhanced over the nonfluctuating case for detection probabilities less than approximately 0.3. Fluctuation loss is relatively insensitive to detection threshold variation as a function of probability of false alarm.

The performance of a noncoherently detected N-pulse train corresponds to the output of a given range-gate-doppler-filter noncoherently summed across N integration intervals. The PDF of a fluctuating target, f_{fl}, can be defined in terms of the conditional PDF of the voltage V at the output of an N-pulse noncoherent integrator as

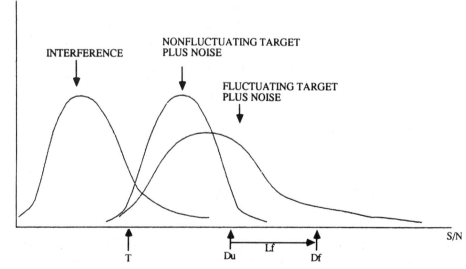

T   =   Detection threshold (fixed by P_{FA})

Du =   S/N required for P_{Do} on nonfluctuating target

Df =   S/N required for P_{Do} on fluctuating target

Lf   =   Fluctuation loss

Figure 16.3 Fluctuating target loss.

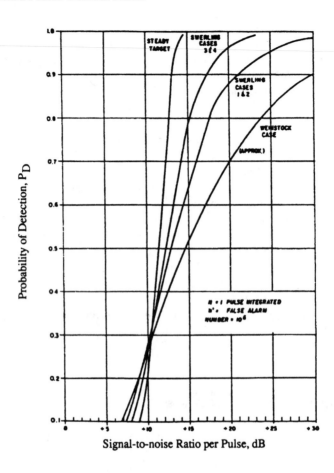

Figure 16.4 Single-pulse detection probabilities [6].

$$f_{nt}(V|y,|N) = \left(\frac{V}{Ny}\right)^{(N-1)/2} \exp[-(v + Ny))] I_{N-1}(\sqrt{NyV})\, U(V) \qquad (16.8)$$

where y corresponds to the IF SNR. $I_m(*)$ is the modified Bessel function of the first kind with order m. Although this expression was derived presupposing square-law detection, it is applicable with very small error to a linear detector output.

The corresponding probability of detection P_{dsf} of an arbitrary "slow" RCS PDF, $f_s(y)$, such as Swerling cases 1 and 3, can be defined as

$$P_s(V) = \int_0^\infty f_{nt}(V|y,|N)\, f_s(y)\; \mathrm{d}y \qquad (16.9)$$

where the RCS fluctuation PDF is defined in terms of the equivalent SNR fluctuation. This case corresponds to a constant target RCS over the integration period.

Definition of the probability of detection given a "fast" fluctuation model, $f_f(x)$, is more complex. The composite PDF is the N-fold convolution of the expression

$$\int_0^\infty f_{nt}(V|y,|i)\, f_f(y)\ \mathrm{d}y \tag{16.10}$$

This case corresponds to a target with interpulse fluctuation such as the Swerling 2 or 4. A brief survey of efficient computational techniques is conducted in [11]. A computer program has been published that estimates detection performance for the Swerling cases as a function of radar system parameters [12].

16.3.2 Fluctuation Loss

The effect of target RCS fluctuation on detection performance relative to the nonfluctuating case can also be estimated in terms of the corresponding fluctuation loss L_f. The fluctuation loss for the Swerling 1 and 3 cases can be approximated independently of the number of pulses noncoherently integrated. As noted by Barton [13], the fluctuation loss associated with the Swerling 2 and 4 cases can be approximated as

$$L_{f4} \approx L_{f3}(1)^{1/N} \tag{16.11}$$

$$L_{f2} \approx L_{f1}(1)^{1/N} \tag{16.12}$$

where $L_{fi}(N)$ is the fluctuation loss of N noncoherently integrated returns from a Swerling Case i target [12]. In the case of partially correlated returns, the effective number of independent samples is substituted for N in (16.11) and (16.12) to estimate fluctuation loss. These trends can be validated somewhat from Table 16.1, which lists fluctuation losses for $N = 1$, 10, and 100 noncoherently integrated pulses. The single-pulse SNR required detection probabilities of 0.5 and 0.9 as a function of the number of pulses noncoherently integrated for each of the Swerling cases are plotted in Figures 16.5 through 16.8.

In general, radar systems operating at a fixed frequency are assessed using the Swerling 1 fluctuation case as noted previously. Interpulse frequency agility can produce decorrelated returns so that Swerling 1 and 3 targets are effectively transformed into Swerling 2 and 4 targets. The resulting reduction in fluctuation loss is termed the *frequency diversity gain*. The minimum carrier frequency differential required for return decorrelation can be estimated as $c/(2L_t)$, where L_t is the range extent of the target.

Table 16.1

Fluctuation Loss Relative to Nonfluctuating Target (dB)

N	Swerling 1	Swerling 2	Swerling 3	Swerling 4
1	8.0	8.0	3.9	3.9
10	8.3	0.8	4.1	0.5
100	8.5	0.4	4.6	0.1
1,000	8.7	0.2	4.8	0

Note: Detection probability = 0.9; false alarm probability = 10^{-6}.
After: [11].

PDR operation and interpulse frequency agility are incompatible unless coherent frequency-jump burst processing is employed. (Since range resolution is inversely proportional to the coherent bandwidth, frequency jump waveforms can be used to generate high-resolution range profiles of targets to support identification of targets under track.) It may prove advantageous for an airborne radar to possess a noncoherent interpulse frequency agility mode for detection of long-range airborne and sea-surface targets.

16.4 CUMULATIVE DETECTION PERFORMANCE

As described in Chapter 14, operational detection requirements are often defined in terms of a cumulative detection range. Cumulative detection is inherent in any radar system not restricted by line-of-sight or maximum-range detection rejection. Maximum-range rejection logic may be implemented in the processing after detection to suppress detection reports of target returns outside some range of interest so as to mitigate data processor demands. Cumulative detection modeling is complicated in the case of pulse-doppler radar by the existence of blind zones and clutter distribution uncertainty.

Cumulative detection may be considered binary integration of target returns from multiple scans. Multiple-scan detection ensures independent observations of the radar RCS. Hence, there is an effective reduction in the fluctuation loss associated with single-pulse detection performance of Swerling 1 and 3 targets. More generally, the range enhancement afforded by cumulative detection is due to the nonzero probability of detection associated with each scan of the target prior to its attaining the specified minimum detection range.

Direct evaluation of the cumulative detection range for targets of arbitrary fluctuation model is laborious. Brookner [15] has condensed previous efforts into a series of universal cumulative-probability-of-detection curves normalized to the cumulative detection range to the radar reference range, which is defined as the range corresponding to unity SNR. Rusnak has further condensed cumulative range

(a)

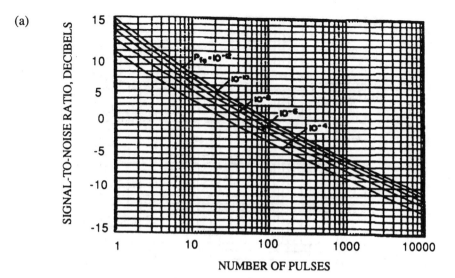

(b)

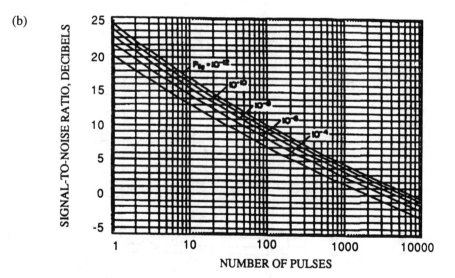

Figure 16.5 Required signal-to-noise ratio for detection with noncoherent integration of pulses; square-law detector, Swerling Case 1 fluctuation: (a) $P_d = 0.50$ and (b) $P_d = 0.90$ [14].

(a)

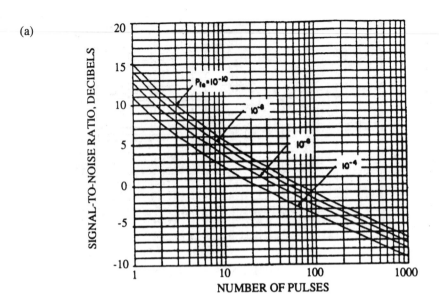

(b)

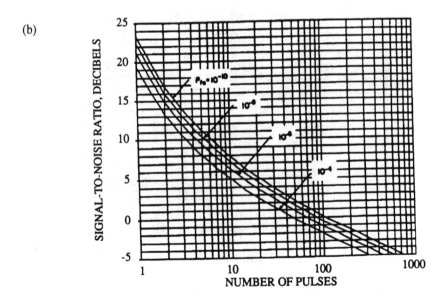

Figure 16.6 Required signal-to-noise ratio for detection with noncoherent integration of pulses; square-law detector, Swerling Case 2 fluctuation: (a) $P_d = 0.50$ and (b) $P_d = 0.90$ [14].

(a)

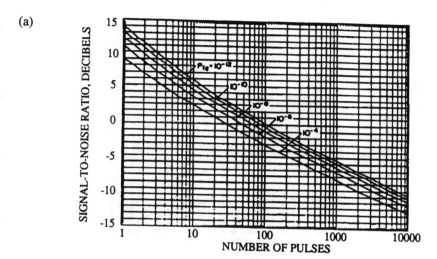

(b)

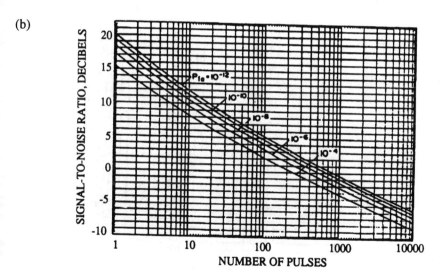

Figure 16.7 Required signal-to-noise ratio for detection with noncoherent integration of pulses; square-law detector, Swerling Case 3 fluctuation: (a) $P_d = 0.50$ and (b) $P_d = 0.90$ [14].

(a)

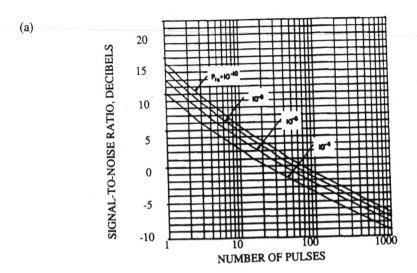

(b)

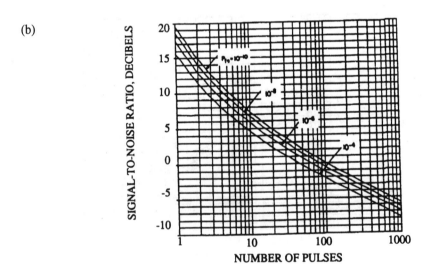

Figure 16.8 Required signal-to-noise ratio for detection with noncoherent integration of pulses; square-law detector, Swerling Case 4 fluctuation: (a) $P_d = 0.50$ and (b) $P_d = 0.90$ [14].

estimation by developing a family of normalized cumulative-probability-of-detection curves [16]. Both works consider the nonfluctuating, Swerling 1, and Swerling 3 cases. Swerling 2 and Swerling 4 cases are not directly considered, since one return per scan is assumed.

Figure 16.9 presents an example of the relative performance of cumulative detection. The abscissa is normalized to the specified cumulative detection range R_c divided by the product of the target closing velocity and scan time δR. The ordinate is the single-pulse SNR at which a cumulative detection probability P_c of 0.9 is attained. Note that the required SNR decreases with increasing normalized detection range. Increasing the abscissa corresponds to more opportunities for detection so that the same cumulative detection probability can be attained with reduced individual single-pulse probabilities.

Brookner has included the effect of nonsynchronized sampling in these curves. Nonsynchronized sampling denotes that the range of the target at some nominal range R_n must be considered to be a random variable uniformly distributed over the extent $R_n \pm \Delta R/2$. Inclusion of this factor increases the SNR required for cumulative detection and causes required single-pulse detection SNR to vary with normalized range.

Consider the example of a required detection range of 250 km with a closing velocity of 500 m/s and a scan period of 5 sec corresponding to 20 dB on the abscissa. Single-pulse detection of a Swerling 1 target would require 22-dB SNR while the corresponding SNR for a cumulative probability of detection at this point would be approximately 9 dB. This 13-dB differential in SNR requirement corresponds to a 13-dB differential in the required radar power-aperture product imposed by the two detection requirements.

16.5 COMPOSITE DETECTION ASSESSMENT

Composite evaluation of airborne PDR radar performance must consider the inter-related effects of target fluctuation, clutter characteristics, radar parameters, and target-radar encounter geometry. A mixture of Monte Carlo and deterministic modeling techniques may be used in performing detection performance assessment. This subsection will provide a brief overview of this process, since the component process modules are described elsewhere in the text.

A simulation of adequate detail is useful in determining radar subsystem requirements and in interpreting measured field data as described in [10,17,18]. This type of evaluation is generally conducted while evaluating a radar design or predicting performance. In addition, real-time clutter and target modeling may be required to support hardware-in-the-loop simulation of pulse-doppler seekers.

Composite detection performance assessment typically requires a complex simulation effort as indicated in Figure 16.10. Modeling of the signal synthesis-processing chain must include consideration of the specific waveforms and PRFs

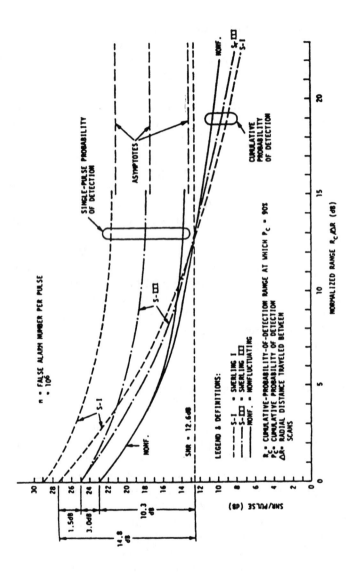

Figure 16.9 Cumulative detection performance [15].

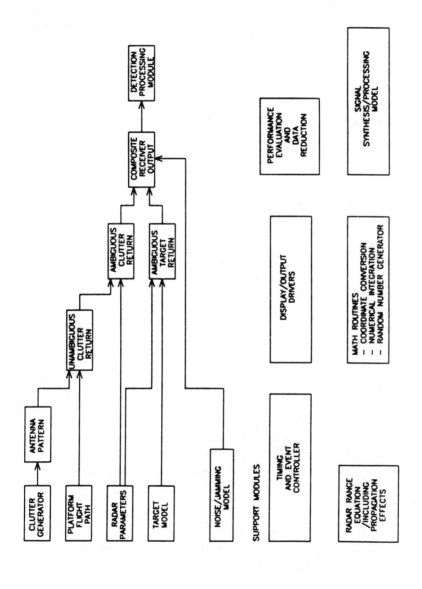

Figure 16.10 Detection performance modeling.

under consideration for the subject radar. In addition, spurious sideband characteristics should be modeled to the extent that initial analysis indicates it is of concern in meeting system-level subclutter visibility requirements.

The two-way antenna pattern weights the side-lobe clutter distribution processed by the radar. Its effect is of particular concern in medium-PRF operation due to the associated clutter foldover in range and doppler. High-PRF operation affords a nominally clear region for a target with closing velocities in excess of the radar platform velocity. A high-fidelity model should include provision for the main-lobe reflection associated with the pointed radomes of AI radars as well as other antenna pattern distortions induced by the airframe implementation.

The platform flight path model is listed as a separate entity to signify that the radar platform and target can potentially possess arbitrary flight paths with respect to one another. However, the complexity of the clutter model generally increases with the flexibility of the flight path model. A level flight path imposes reduced clutter computations than one with arbitrary dive angle.

As a minimum, the target model should consist of a simple fluctuation model such as the Swerling 1. An aspect-dependent model of median target RCS can significantly increase the simulation accuracy against specific targets. To provide increased target modeling fidelity, efficient multiple-scatterer models have been derived to simulate time-varying glint, RCS, and potentially doppler characteristics [19,20]. In a high-fidelity simulation involving detailed modeling of the detection processor, it may be necessary to incorporate JEM modeling.

Design of the clutter model generally forces a trade between desired overall model fidelity and computational complexity. In addition, the clutter model fidelity may be ultimately constrained by the availability of accurate measurement data and the uncertainty in clutter characteristics to be encountered by the subject radar.

The noise module provides additive thermal noise and potentially jamming signals. The range-doppler ambiguous return modules map the unambiguous target and clutter model outputs into the appropriate range-gate-doppler-filter outputs determined by the radar waveform parameters.

Finally, the detection simulation module provides an estimate of target detection as a function of the composite input which simulates that associated with the radar receiver output. The detection simulation module must reflect the actual detection processor design in terms of multiple-PRF detection processing and decision threshold estimation.

A typical performance measure is depicted in Figure 16.11 for a hypothetical example of radar detection performance derived from high and medium PRFs. The detection range is depicted as a function of target aspect relative to the interceptor radar. This range assessment presupposes that the two aircraft are flying a lead-collision course such that target aspect and closing velocity are constant.

The example illustrates that high-PRF operation can provide good performance in forward quarter target encounters but that medium-PRF operation can provide superior all-aspect performance. Pulse-doppler performance is generally

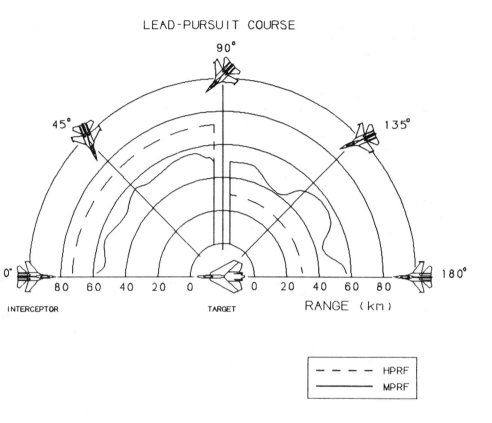

Figure 16.11 Example of high-PRF and medium-PRF performance.

poor relative to beam-aspect targets, since the target doppler is insufficient to separate its return from the main-lobe clutter.

References

[1] Farill, T. G., et al., "RCS Reduction Course Notes," Georgia Institute of Technology Engineering Extension Services, 1988.

[2] Holm, W. A., "Polarimetric Fundamentals and Techniques," *Principles of Modern Radar*, J. L. Eaves and E. K. Reedy, eds., New York: Van Nostrand Reinhold, 1987.

[3] Yaw, D. F., "Antenna Radar Cross Section," *Microwave Journal*, Vol. 27, No. 9, September 1984, p. 197.

[4] Papoulis, A., *Probability, Random Variables, and Stochastic Processes*, New York: McGraw-Hill, 1965.

[5] Swerling, P., "Probability of Detection for Fluctuating Targets," *RAND Report RM-1217*, March 1954, reprinted in IRE Trans. Information Theory, Vol. IT-6, No. 2, April 1960.

[6] Nathanson, F. E., *Radar Design Principles: Signal Processing and the Environment*, New York: McGraw-Hill, 1969.

[7] Gardner, R. E., "Doppler Spectral Characteristics of Aircraft Radar Targets at S-Band," *NRL Report 5656*, Naval Research Laboratory, Washington, D.C., 3 August 1961.

[8] Hynes, R., and R. E. Gardner, "Doppler Spectra of S-Band and X-Band Signals," *IEEE Trans. Supplement*, Vol. AES-3, No. 6, November 1967, pp. 356–365.

[9] Nelson, N., "Aircraft Tracking Problems From Range Rate 'Turbine Modulation,'" *NAECON 1977 Record*, pp. 679–682.

[10] Ringel, M. B., D. H. Mooney, and W. H. Long, "F-16 Pulse Doppler Radar (AN/APG-66) Performance," *IEEE Trans.*, Vol. AES-19, No. 1, January 1983, pp. 147–158.

[11] Barrett, C. R., "Target Models," *Principles of Modern Radar*, J. L. Eaves and E. K. Reedy, eds., New York: Van Nostrand Reinhold, 1987.

[12] Fielding, J. E., and G. D. Reynolds, *RGCALC: Radar Range Detection Software and Users Manual*, Norwood, MA: Artech House, 1987.

[13] Barton, D. K., "Detection and Measurement," *Radar Technology*, E. Brookner, ed., Norwood, MA: Artech House, 1977.

[14] Blake, L. V., *Radar Range-Performance Analysis*, Norwood, MA: Artech House, 1986.

[15] Brookner, E., "Cumulative Probability of Detection," *Radar Technology*, E. Brookner, ed., Norwood, MA: Artech House, 1977.

[16] Rusnak, I., "Search Radar Evaluation by the Normalized Cumulative Probability of Detection Curves," *IEEE Trans.*, Vol. AES-22, No. 4, July 1986, pp. 461–465.

[17] Arnoff, E., and N. M. Greenblatt, "Medium PRF Radar Design and Performance," *20th Tri-Service Radar Symposium, 1974, reprinted in Radars Volume 7: CW and Doppler Radar*, D. K. Barton, ed., Dedham, MA: Artech House, 1978.

[18] Holbourn, P. E. and A. M. Kinghorn, "Performance Analysis of Airborne Pulse Doppler Radar," *Record of the IEEE 1985 International Radar Conference*, Arlington, VA, 6–9 May 1985.

[19] Borden, B. H., and M. L. Mumford, "A Statistical Glint/Radar Cross Section Target Model," *IEEE Trans.*, Vol. AES-19, No. 5, September 1983, pp. 781–785.

[20] Sandhu, G. S., and A. V. Saylor, "A Real-time Statistical Radar Target Model," *IEEE Trans.*, Vol. AES-21, No. 4, July 1985, pp. 490–507.

CHAPTER 17

Automatic Detection

Melvin L. Belcher, Jr.

17.1 AUTOMATIC DETECTION

17.1.1 Automatic Detection Requirements

Detection processing may be characterized as the making of binary decisions for each range-gate-doppler-filter cell at the receiver output. Each decision is between two hypotheses, H_0 and H_1, signifying the absence and presence of a target return, respectively. As noted in Chapter 14, automatic detection schemes are designed to provide CFAR performance so as to maintain a manageable detection-report generation rate. An optimal detection scheme may be defined as one that provides the maximum probability of detection P_d for a given S/I while maintaining the specified probability of false alarm P_{fa}.

The temporal and spatial variation of interference encountered in airborne pulse-doppler applications results in suboptimal performance relative to that attained in the presence of homogenous interference having associated statistical characteristics that can be precisely estimated. Algorithms devised to accommodate this interference variation generally require a higher S/I to achieve a specified P_d than required by an optimal detection algorithm operating in homogeneous interference. This so-called CFAR loss is a general performance measure used to evaluate CFAR schemes. Effectively, it imposes an additional SNR margin requirement to that indicated by the P_d curves of the preceding chapters. CFAR loss varies markedly with the background interference characteristics.

17.1.2 Automatic Detection Implementation

As noted previously, realizable detection schemes are evaluated in terms of their effective CFAR loss relative to an optimal detection process. Optimal radar detection performance is customarily parameterized with respect to the Neyman-Pearson decision procedure [1].

A cell output composed solely of interference can be defined as

$$z_0 = (x_i^2 + y_i^2) \tag{17.1}$$

where x_i and y_i are the composite in-phase and quadrature cell components due to the sum of thermal noise, jamming, and clutter returns. In general, the cell contains foldover components due to ambiguous operation in range or doppler. Similarly, a cell containing a target return can be given as

$$z_1 = (x_i^2 + x_s^2) + (y_i^2 + y_s^2) \tag{17.2}$$

where x_s and y_s denote the return signal. In general, the target-return signal is represented as a random variable due to target fluctuations.

A likelihood ratio, L, can be formed as

$$L = P\{H_1(Z)\}/P\{H_2(Z)\} \tag{17.3}$$

where $P\{H_i(Z)\}$ is the probability that the magnitude Z from a specific output cell satisfies hypothesis i. The likelihood-ratio computed for the test cell is compared with a threshold L_t such that if $L > L_t$, a target is declared present, and if $L < L_t$, no target is declared. The likelihood-ratio threshold is a function of the desired probability of false alarm.

Assuming that the interference can be characterized as possessing the statistical characteristics of thermal noise, then the PDF of the interference and signal-plus-interference cells are given by (14.4) and (14.5), respectively. Under this definition, the threshold test can be reformulated as

$$D(Z) > K \hat{Q}$$

which implies presence of return, and

$$D(Z) < K \hat{Q}$$

which implies absence of return, where $\hat{Q}$ is the estimated mean interference power in the cell, $D(*)$ represents the detector law, and K is the CFAR constant calculated as a function of the specified P_{fa} and the algorithm used to estimate $\hat{Q}$. This procedure provides near optimal performance as $\hat{Q}$ approaches Q, the true interference power.

In theory, optimal performance is attained if the cell output is subjected to a detector law dependent upon S/I. The optimal detector law corresponds to a log function at high S/I and approximately to a square-law detector at low SNR. The variation in performance with detector law selection is actually insignificant [2]. A square-law detector is commonly presupposed in radar detection analyses.

The CFAR multiplier is typically precomputed and stored in a look-up table as a function of radar mode and desired P_{fa}. The nominal value of the CFAR multiplier(s) can be estimated under the assumption of uniform Rayleigh-distributed interference using tabulated data [3] or an approximation formula [4,5]. An adaptive detection processor may be designed so as to select a higher value of CFAR multiplier in response to the actual P_{fa} exceeding some predefined boundary. The data processor can estimate the false alarm rate by monitoring the number of detection reports that are not confirmed via multiple-PRF detection-correlation processing or do not lead to track file formation. This technique provides resilient performance against deviation of the interference PDF from that assumed in calculating the CFAR multiplier. Without this adaptive multiplier capability, it is necessary to establish a priori the statistical error variation anticipated in estimating the test cell interference. The CFAR multiplier is then increased by an appropriate amount to provide a threshold margin against a specified level of interference underestimation so as to prevent the false alarm rate from exceeding some maximum value. The average false alarm rate will generally be significantly lower than the design value and the CFAR loss increased over that achievable with adaptive processing.

In general, P_{fa} is sensitive to small changes in detection threshold and hence to CFAR multiplier selection. Barrett has computed P_{fa} as a function of the CFAR multiplier for Rayleigh (noiselike) and log-normal interference PDFs, as illustrated in Figure 17.1, for a reference window of 160 samples. It is readily apparent that correlation among the reference window samples or deviation of the interference from the assumed PDF can cause P_{fa} to increase multiple orders of magnitude beyond the design value. The nominal clutter characteristics and relatively small reference window of pulse-doppler radars should typically impose less variation in Pfa than implied by the preceding figure.

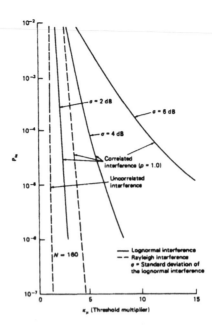

Figure 17.1 P_{fa} versus CFAR multiplier [6].

Estimation of the test cell interference tends to be the most critical real-time function in determining the performance of the detection processor. Stationary radar systems often employ clutter map techniques where the interference associated with a given test cell is listed in a look-up table as a function of range-angle position [7]. The test cell interference is estimated as a temporal average over a number of scans. This technique may be of limited utility in AEW radar detection, but is generally inapplicable to airborne radars due to the associated temporal variation in clutter return interference.

As described in Chapter 16, the range-doppler distribution of clutter return magnitude is characterized by two regions of strong magnitude: main-lobe clutter and altitude return. The ambiguous range-doppler extent of these clutter returns can be predicted as a function of the radar-platform flight path and antenna scan angle. In contrast, weather and terrain may produce localized clutter return areas yielding strong returns that cannot be predicted a priori. A medium-PRF receiver output will typically possess individual and groups of range-doppler cells that contain relatively strong clutter returns due to folded-over side-lobe returns. The magnitude and relative range-doppler position of ambiguous clutter returns will vary with PRF.

The clutter distribution present at the pulse-doppler receiver output is characterized by a mixture of diffuse clutter, which may have a large variance over the range-doppler map, and randomly positioned, strong point clutter returns. Hence, it is necessary to estimate the interference power in the test cell by examining a local reference window of M cells adjacent to the test cell. In low-PRF and high-PRF radars, this reference window is nominally one-dimensional in range and doppler, respectively. The reference window of a medium-PRF radar should be two-dimensional as indicated in Figure 17.2.

The outputs of the test cell and the guard cells immediately adjacent to it are not used in generating the local interference estimate. The adjacent guard cells are not used, so as to mitigate self-interference from a target return straddling contiguous range-doppler cells.

The abrupt increase in interference magnitude associated with the main-lobe clutter and altitude return range-doppler region generally results in severe degradation of CFAR performance [8]. It is customary to blank out the cells calculated to include these clutter returns so as to not incorporate them in the reference windows of adjoining cells.

An important robustness requirement of a CFAR interference estimation algorithm is its ability to deal with a reference window containing a clutter region boundary such that the interference samples are a mixture of strong and weak clutter returns. The resulting interference estimate should neither induce false detections at test cells on the clutter boundary nor result in a excessively high threshold magnitude of a test cell in the clear region. In addition, the CFAR interference estimator must accommodate multiple target returns distributed across the reference window extent. A strong target return in a reference cell can poten-

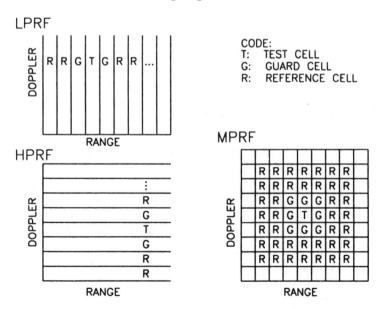

Figure 17.2 Examples of CFAR reference windows.

tially cause the detection threshold of the test cell to be overestimated, resulting in significant CFAR loss. Hence, two target returns of comparable magnitude and high SNR closely spaced in ambiguous range doppler could preclude each other from being detected. The presence of significant JEM lines inside the doppler extent of the reference window also induces this target suppression effect. The overestimation of the decision threshold due to either clutter boundaries or point interference is called *masking*.

CFAR performance is generally constrained by the accuracy of the test cell interference estimate. As noted above, statistical variation due to the limited number of samples in the reference window and heterogeneous interference distribution impose error in any CFAR estimator. In addition, the CFAR algorithm itself may introduce a bias error into the interference estimate. The interaction of these three sources of interference estimator error is computationally complex. CFAR bias is amenable to closed-form calculation only if the interference is of the same PDF and independent from cell to cell across the reference window. Simulation techniques are generally necessary to evaluate composite performance [8,9].

Estimates of interference from adjacent beam positions could potentially enhance CFAR performance. A mean-level detector has been suggested to control the receiver automatic gain control (AGC) setting so as to produce constant output power on the basis of the time-integrated receiver output [10]. The integration time constant may be chosen so as to effectively average received interference over an angular extent larger than the antenna beamwidth. More generally, a CFAR reference window could conceivably be generated across angle or time. The computational complexity of such an approach generally precludes its attractiveness.

The principal types of CFAR detection processing applicable to airborne pulse-doppler processing will be described in the following sections.

17.1.3 *M:N* Processing

The previous description of CFAR requirements has emphasized the processing of a single test cell. The associated input data represent either the output of a given range-gate-doppler-filter cell after a single coherent integration interval or a specific cell's output noncoherently integrated over several coherent integration periods. Automatic detection is a single-stage process under these conditions.

In contrast, medium-PRF radars employing *M:N* detection require a second level of detection processing such that a detection is declared to the data processor only if at least *M* out of *N* possible cell sets indicate a return at a specific unambiguous range-doppler coordinate. Each cell set corresponds to the composite range-gate-doppler-filter output derived from a specific PRF at the end of a coherent integration interval. Ambiguity resolution is an inherent component of this second stage of detection processing.

It follows that the composite probability of detection, P_{ds}, and composite probability of false alarm, P_{fas}, are related to the nominal single cell probabilities by

$$P_{ds} = \sum_{m=M}^{N} \frac{N!}{m!(N-m!)}(P_d)^m(1-P_d)^{N-m} \qquad (17.4)$$

$$P_{fas} = \sum_{m=M}^{N} \frac{N!}{m!(N-m)!}(P_{fa})^m(1-P_{fa})^{N-m} \qquad (17.5)$$

Hence, the P_d and P_{fa} required to meet system level specifications must be computed by iterative solution of (17.4) and (17.5).

A number of authors have investigated *M:N* detection performance under noise-limited conditions. This technique is sometimes referred to as coincident detection or binary integration. Analysis of the optimum value of *M*, M_{opt}, for a given *N* has indicated that this value is generally a broad maximum. Walker has estimated that M_{opt} should be approximately $0.44N$ for Swerling 1 targets and $0.34N$ for Swerling 2 [11]. The resulting performance of *M:N* detection is compared to noncoherent integration in Figure 17.3 for several target models [12]. A processing

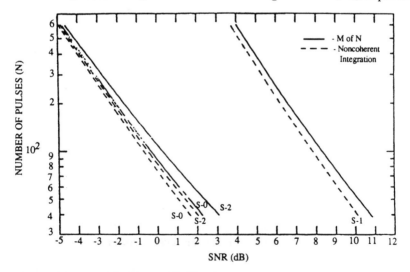

S-0 = Nonfluctuating target
S-1 = Swerling I fluctuating target
S-2 = Swerling III fluctuating target

Figure 17.3 *M:N* noise-limited detection performance [11].

loss of less than 1 dB relative to noncoherent integration results from $M{:}N$ processing of Swerling 1 target returns across the typical medium-PRF set sizes.

Usage of noise-limited performance expression and curves for medium-PRF detection processing analysis is a gross approximation, since the ambiguous range-doppler clutter return distribution will vary with PRF. Cells containing main-lobe clutter will be blanked so that a given unambiguous range-doppler coordinate may be obscured at specific PRFs. However, the N-PRF set should be chosen so as to provide a minimum of M "visible" PRFs for any given range-doppler coordinate. The resulting cumulative detection performance must generally be assessed using radar-clutter-target simulation, as described in the previous chapter.

The $M{:}N$ detection thresholding criteria may vary with radar mode. Specifically, detection processing for search and track operations may differ significantly. TWS radar operation requires detections to be correlated with tracks in progress so as to update existing track files, suppress clutter return detections, and ascertain the presence of previously undetected targets and initiate track files [13].

17.2 CELL-AVERAGING CFAR

Cell-averaging CFAR (CA-CFAR) processing estimates the test cell interference by simply computing the average of the power in the M reference cells as given by

$$\hat{Q}_{ca} = \frac{1}{M} \sum_{i=1}^{M} Z_i \tag{17.6}$$

The resulting CFAR detection process is illustrated in Figure 17.4 for the one-dimensional case. Note that the actual extent of the reference window is $M + M_b$, where M_b is the number of guard cells plus one (the test cell). Generally, M_b is 3 for a one-dimensional reference window and 9 for a two-dimensional window.

In the event of homogeneous interference, the performance of CA-CFAR is optimum as M increases to infinity. Specifically, $\hat{Q}_{ca}$ is an unbiased, minimum-variance estimate of the interference, assuming uniform interference across the reference window. The estimate variance will be decreased by a factor of M relative to that based on a single cell sample, assuming each reference cell contains an independent sample. Hence, $\hat{Q}_{ca}$ approaches Q with increasing M.

The resulting CFAR loss decreases with increasing M, as indicated in Figures 17.5 and 17.6 for Swerling 1 and 3 targets, respectively. A reference window on the order of 20 samples results in a CFAR loss of approximately 1.5 dB for the case of a Swerling 1 target with a homogeneous interference background [14].

Unfortunately, CA-CFAR performs poorly in the nonhomogeneous interference environment often encountered by pulse-doppler radar systems. A clutter boundary in the reference window can increase the test cell P_{fa} several orders of magnitude above the design value [15]. Several modifications of the CA-CFAR

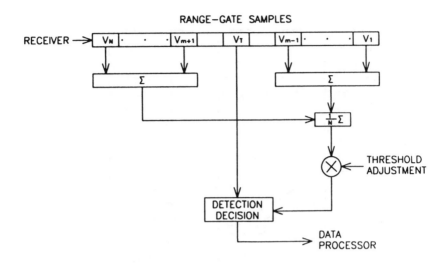

Figure 17.4 Cell-averaging CFAR (CA-CFAR).

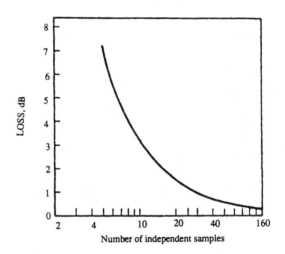

Figure 17.5 CFAR loss, Swerling case 1, unity postdetection integration. Probability of detection = 0.5 to 0.9; probability of false alarm = 10^{-6} [4].

technique have been developed to provide enhanced resilience to heterogeneous interference.

The greater-of-cell-average CFAR (GO-CFAR) technique computes the mean interference separately for each half of the reference window and selects the highest value as the interference estimate. The resulting interference estimate, Q_{go}, is then defined as

(a)

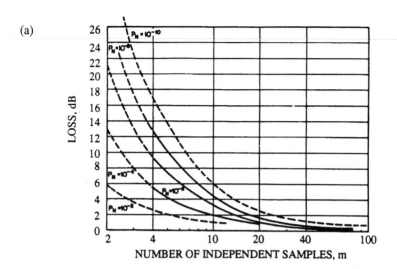

(b)

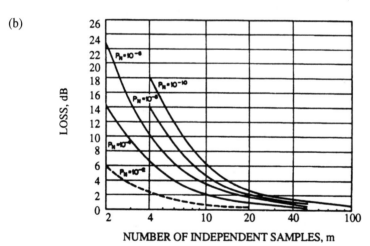

Figure 17.6 Adaptive detection of CFAR. Loss versus number of independent noise samples, m, to set threshold with P_N = false-alarm probability: (a) $P_D = 0.5$ and (b) $P_D = 0.9$. Fluctuating (Swerling Case 3) and steady targets [4].

$$\hat{Q}_{go} = \text{MAX}\left(\frac{2}{M}\sum_{i=1}^{M/2} Z_i, \frac{2}{M}\sum_{i=M/2+1}^{M} Z_i\right) \tag{17.7}$$

The operator MAX{*,*} denotes taking the maximum of its two arguments. As would be expected, the threshold computed from the GO-CFAR interference estimate tracks abrupt variation in interference magnitude significantly better than the nominal CA-CFAR. Less than an order of magnitude increase in P_{fa} is encountered at clutter boundaries. The number of reference cells used to estimate the interference power is decreased by a factor of two, so CFAR loss increases by 0.1 to 0.3 dB relative to CA-CFAR. The incremental CFAR loss imposed by GO-CFAR increases with decreasing P_{fa} and decreases with increasing reference window size. It is relatively insensitive to P_d [16].

However, GO-CFAR is vulnerable to capture from a strong interference source. Sensitivity to masking increases with decreasing window size. Hence, point interference and multiple target returns within the GO-CFAR result in increased CFAR loss over that imposed by CA-CFAR, since there are effectively fewer cells with which to estimate an average. Weiss has argued for employment of a smaller-of-cell-average CFAR processing to mitigate closely spaced target return masking by choosing the lesser of the split-reference-window interference estimates [17]. However, this technique results in dramatically increased P_{fa} in the presence of clutter boundaries across the reference window.

Deletion of the M_p largest magnitude reference window samples from the interference estimation significantly reduces CFAR loss in a multiple-target environment. The k smallest cell magnitudes from an M-sample window are employed to compute the interference in the test cell. Ritcey has evaluated design trades and performance sensitivity to the choice of M_p [18]. This technique results in increased P_{fa} when the number of censored cells is larger than the number of interference spikes, as illustrated in Figure 17.7. The associated CFAR loss due to incorrect estimation of the number of interference spikes is shown in Figure 17.8. Note that in the absence of censoring ($M_p = 0$), the CFAR loss is on the order of 1 to 2 dB for one to four interference spikes in this example.

In addition to the aforementioned difficulties, peak-cell censoring does not cope well with clutter boundaries. Weiss has suggested combining GO-CFAR with a censoring rule to provide adequate robustness [17].

Other techniques for minimizing the CFAR loss associated with test cells near clutter boundaries include gradient processing and heterogeneous-clutter-estimating CFAR (HGE-CFAR). Gradient CFAR computes the gradient of the interference power from that estimated in each half of the reference window [8]. The resulting average is used as the estimated test cell interference. HGE-CFAR is a computationally intensive procedure that estimates the statistical parameters of adjoining clutter areas and the associated transition point and produces a detection threshold on the basis of these computations [19].

A general caution on CA-CFAR techniques is that the maximum size of the reference window size should be on the order of the minimum interference extent anticipated. Performance degrades significantly as the window extent becomes larger than typical strong clutter regions.

17.3 ORDERED-STATISTIC CFAR PROCESSING

Ordered-statistic CFAR processing (OS-CFAR) has received significant attention since its introduction by Rohling [20,21]. This procedure appears resilient to clutter edge and masking effects while imposing acceptable CFAR loss.

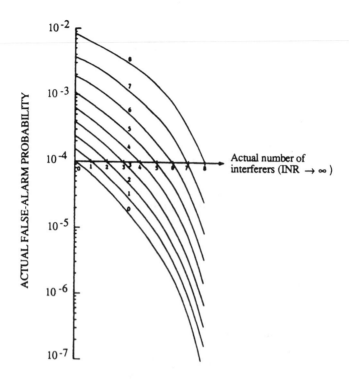

Actual probability of false-alarm for the CMLD with n = 32,
k = 24, c = 9, versus the actual number of INR → ∞
Swerling II interfering targets. Curves are indexed with
l_d , the number of interfering targets used in design to
achieve Q_0 = 10^{-4} .

Figure 17.7 False alarm performance of censored CA-CFAR [18].

To implement this scheme, the reference window cell magnitudes are rank-ordered such that

$$Z_{(1)}\langle Z_{(2)}\langle \ldots \langle Z_{(M)} \qquad (17.8)$$

where (i) denotes rank order i. The cell magnitude of rank order k_{os}, which lies within the range of 1 to M, is used as the interference estimate and is multiplied by the OS-CFAR multiplier. This process is depicted in Figure 17.9. As noted by Rohling, this approach to estimating background interference may be interpreted as a generalization of median filtering techniques employed in image processing.

The resulting P_{fa} is calculated in the same manner as in the CA-CFAR techniques. The interference PDF is integrated from the detection threshold defined

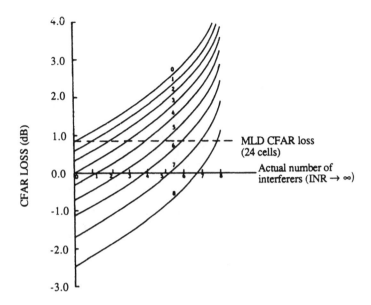

Actual SNR loss relative to the optimum Neyman-Pearson
detector to yield detection probability Q_d = 0.9 for the CMLD
with n = 32, k = 24, c = 9, versus the actual number of INR
→ ∞ Swerling II interfering targets. Curves are indexed
with l_d, the number of interfering targets used in design to
achieve Q_0 = 10^{-4}.

Figure 17.8 CFAR loss of censored CA-CFAR [18].

by the product of the CFAR multiplier and the interference estimate to infinity as
defined in (14.7). The OS-CFAR procedure is not distribution-free, since the
assumed interference PDF is employed in computing the CFAR multiplier. As
derived by Rohling, this procedure provides CFAR detection performance, since
the multiplier is calculated independently of the background interference power.

The performance of OS-CFAR processing in a clutter boundary and masking
environment is determined by the selection of the rank order cell for threshold
calculation. In order to avoid clutter edge detections, k should be greater than
$M/2$. To prevent masking degradation, $M - k$ should be greater than the number
of cells containing interference spikes. Rohling suggests that k_{os} = $3M/4$ is a good
value for typical radar applications. Selection of k_{os} to minimize CFAR loss is generally a broad minimum.

In comparison to CA-CFAR, OS-CFAR processing imposes additional CFAR
losses, ranging from 0.5 to 1 dB for equivalent reference window sizes ranging from
16 to 32 samples, and a design of 10^{-6}. In contrast to CA-CFAR techniques, OS-
CFAR adaptivity to local interference is relatively insensitive to reference window

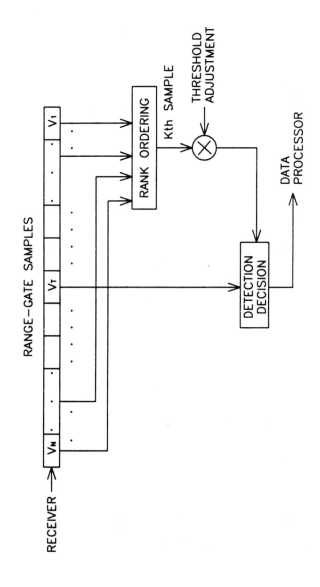

Figure 17.9 Ordered-statistic CFAR.

size. Rohling asserts that the CFAR loss relative to that of CA-CFAR can be reduced by expanding the reference window size over that feasible with CA-CFAR techniques.

A comparison of threshold calculation between CA-CFAR, GO-CFAR, and OS-CFAR is presented in Figures 17.10 and 17.11 for a two-target masking detection scenario. A reference window of M samples is employed; the OS-CFAR threshold is based on the cell rank k_{os}. In contrast to the CA-CFAR techniques, the OS-CFAR procedure readily detects each of the target returns. The insensitivity of the OS-CFAR detection threshold to capture by interference spikes is also apparent.

Weber and Haykin have developed a two-parameter implementation of OS-CFAR to accommodate interference PDFs such as log-normal [22]. The CFAR loss is reportedly greater than that associated with single-parameter OS-CFAR.

17.4 OTHER CFAR TECHNIQUES

The following detection procedures are not commonly encountered in airborne pulse-doppler radar applications but should be noted.

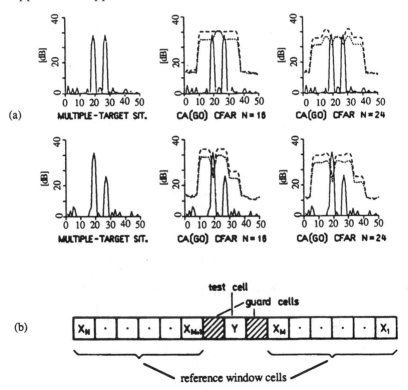

Figure 17.10 CA-CFAR multiple-target thresholding [20] with (a) two equal-amplitude closely spaced targets and (b) closely spaced large and small target.

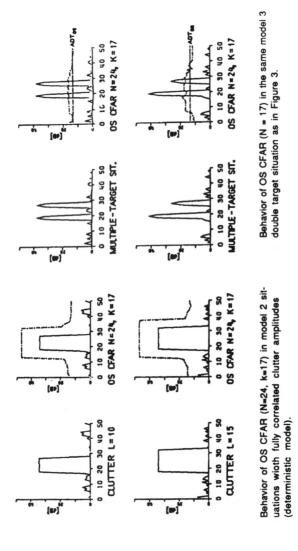

Figure 17.11 OS-CFAR multiple-target thresholding [20].

17.4.1 Distribution-Free CFAR

Distribution-free CFAR (DF-CFAR) procedures function without an assumption of interference PDF. More specifically, a detection procedure is termed *distribution-free* over a class of interference PDFs if the associated P_{fa} is constant over that class. DF-CFAR techniques have been developed that are applicable to the continuous interference PDFs encountered in radar detection.

However, DF-CFAR has not been widely applied in operational radar systems. Since DF-CFAR techniques are insensitive to the background interference PDF, their performance is suboptimal, typically imposing CFAR losses on the order of several decibels relative to optimal processes. In addition, DF-CFAR procedures typically degrade markedly against multiple-target masking, strong-signal capture, and reference window sample correlation.

A number of DF-CFAR techniques have been suggested that operate on the basis of ranking the cell magnitudes in the reference window and designating the kth rank as the threshold [23,24]. This technique contrasts with OS-CFAR, which computes the threshold using a CFAR multiplier that is a function of the assumed interference PDF. Dillard and Rickard have proposed a DF-CFAR based on doppler frequency [25]. Recent efforts in this area have included development of efficient and robust suboptimum ranking procedures [26] and evaluation of surface radar detection performance using DF-CFAR processing [27].

17.4.2 Statistically Adaptive CFAR Processing

Various decision procedures have been suggested to enhance performance via increased adaptability. Two-parameter schemes have been developed to employ the mean and variance of the reference window samples in estimating the threshold for the test cells. Farina, Russo, and various colleagues have suggested adaptive processor architectures that consider the statistical properties of the target as well as the interference [28–31]. These techniques have not been widely implemented to date due to the associated computational complexity.

17.5 SUMMARY

Variations of CA-CFAR are generally used in airborne pulse-doppler radar applications. As described previously, it is generally necessary to modify the nominal CA-CFAR scheme in order to achieve adequate performance against clutter boundaries and point interference. OS-CFAR processing appears to offer promising effectiveness against heterogeneous interference such as that associated with closely spaced targets, clutter, and some ECMs. The implementation of both classes of CFAR procedures are anticipated to benefit from continuing advances in digital signal processing technology.

Weber, Haykin, and Gray have demonstrated via simulation that both CA-CFAR and OS-CFAR techniques benefit significantly from blanking of strong-clutter cells [8]. These regions are associated with main-lobe clutter and altitude line returns so that the ambiguous range-doppler regions to be blanked can be calculated deterministically.

References

[1] Schwartz, M., "Statistical Decision Theory and Detection of Signals in Noise," in *Modern Radar: Analysis, Evaluation, and System Design*, R. S. Berkowitz, ed., New York: John Wiley & Sons, 1965.

[2] Blake, L. V., *Radar Range-Performance Analysis*, Dedham, MA: Artech House, 1986.

[3] Pachares, J., "A Table of Bias Levels Useful in Radar Detection Problems," *IRE Trans. Information Theory*, March 1958, pp. 38–45.

[4] Hansen, V. G., "Simple Expressions for Determining Radar Detection Thresholds," *IEEE Trans.*, Vol. AES-18, No. 4, July 1982, pp. 510–512.

[5] Urkowitz, H., "Corrections to and Comments on 'Simple Expressions for Determining Radar Detection Thresholds,' " *IEEE Trans.*, Vol. AES-21, No. 4, July 1985, pp. 583–584.

[6] Barrett, C. R., "Target Models," in *Principles of Modern Radar*, J. L. Eaves and E. K. Reedy, eds., New York: Van Nostrand Reinhold, 1987.

[7] Nitzberg, R., "Clutter Map CFAR Analysis," *IEEE Trans.*, Vol. AES-22, No. 4, July 1986, pp. 419–421.

[8] Weber, P., S. Haykin, and R. Gray, "Airborne Pulse-Doppler Radar: False-Alarm Control," *IEE Proc.*, Vol. 134, Part F, No. 2, April 1987, pp. 127–134.

[9] Bird, J. S., "Calculating Detection Printabilities for Adaptive Thresholds," *IEEE Trans.*, Vol. AES-19, No. 4, July 1983, pp. 506–512.

[10] Hovanessian, S. V., *Radar System Design and Analysis*, Norwood, MA: Artech House, 1984.

[11] Walker, J. F., "Performance Data for a Double-Threshold Detection Radar," *IEEE Trans.*, Vol. AES-7, No. 1, January 1971, pp. 142–146.

[12] Hansen, V. G., "Comments on 'Performance Data for a Double-Threshold Detection Radar,' " *IEEE Trans.*, Vol. AES-7, No. 3, May 1971, p. 561.

[13] Blackman, S. A., *Multiple-Target Tracking With Radar Applications*, Dedham, MA: Artech House, 1986.

[14] Petts, G. E., III, "Radar Systems," in *Handbook of Electronic System Design*, C. A. Harper, ed., New York: McGraw-Hill, 1980.

[15] Moore, J. D., and N. B. Lawrence, "Comparison of Two CFAR Methods Used With Square Law Detection of Swerling 1 Targets," *1980 IEEE International Radar Conference*, pp. 403–409.

[16] Hansen, V. G., and J. H. Sawyers, "Detectability Loss Due to 'Greatest of' Selection in a Cell-Averaging CFAR," *IEEE Trans.*, Vol. AES-16, No. 1, January 1980, pp. 115–118.

[17] Weiss, M., "Analysis of Some Modified Cell-Averaging CFAR Processors in Multiple-Target Situations," *IEEE Trans.*, Vol. AES-18, No. 1, January 1982, pp. 102–114.

[18] Ritcey, J. A., "Performance Analysis of the Censored Mean-Level Detector," *IEEE Trans.*, Vol. AES-22, No. 4, July 1986, pp. 443–454.

[19] Finn, H. M., "A CFAR Window for Spanning Two Clutter Fields," *IEEE Trans.*, Vol. AES-22, No. 2, March 1986, pp. 155–169.

[20] Rohling, H., "Radar CFAR Thresholding in Clutter and Multiple Target Situations," *IEEE Trans.*, Vol. AES-19, No. 4, July 1983, pp. 608–621.

[21] Rohling, H., "New CFAR-Processor Based on an Ordered Statistic," *Record of IEEE 1985 International Radar Conference*, Arlington, VA, 6–9 May 1985, pp. 271–275.

[22] Weber, P., and S. Haykin, "Ordered Statistic CFAR Processing for Two-Parameter Distributions With Variable Skewness," *IEEE Trans.*, Vol. AES-21, No. 6, November 1985, pp. 819–821.

[23] Dillard, G. A., and R. A. Dillard, "Radar Automatic Detection," *Microwave Journal*, Vol. 28, No. 6, June 1985, pp. 125–130.

[24] Barrett, C. R., "Adaptive Threshold and Automatic Detection Techniques," in *Principles of Modern Radar*, J. L. Eaves and E. K. Reedy, eds., New York: Van Nostrand Reinhold, 1987.

[25] Dillard, G. M., and J. T. Rickard, "A Distribution-Free Doppler Processor," *IEEE Trans.*, Vol. AES-10, No. 4, July 1974, pp. 479–486.

[26] Sanz-Galzalez, J. L., and A. R. Figueiras-Vidal, "A Suboptimum Rank Test for Nonparametric Radar Detection," *IEEE Trans.*, Vol. AES-22, No. 6, November 1986, pp. 670–680.

[27] Reid, W. S., K. D. Tschetter, and R. M. Johnson, "Analysis of a Rank-Based Radar Detection System Operating on Real Data," *Record of the IEEE 1985 International Radar Conference*, Arlington, VA, 6–9 May 1985, pp. 435–441.

[28] Farina, A., and A. Russo, "Radar Detection of Correlated Targets in Clutter," *IEEE Trans.*, Vol. AES-22, No. 5, September 1986, pp. 513–532.

[29] Farina, A., A. Russo, F. A. Studer, "Advanced Models of Targets and Disturbances and Related Radar Signal Processors," *1985 IEEE International Radar Conference*, pp. 151–158.

[30] Farina, A., A. Russo, F. A. Studer, "Coherent Radar Detection in Log-Normal Clutter," *IEE Proc.*, Vol. 133, Part F, No. 1, February 1986, pp. 39–54.

[31] Farina, A., A. Russo, F. Scannapieco, and S. Barbarossa, "Theory of Radar Detection in Coherent Weibull Clutter," *IEE Proc.*, Vol. 134, Part F, No. 2, April 1987, pp. 174–190.

Suggested Reading

Finn, H. M., and R. S. Johnson, "Adaptive Detection Mode With Threshold Control as a Function of Spatially Sampled Clutter-Level Estimates," *RCA Review*, Vol. 29, No. 3, September 1968.

Farina, A., and F. A. Studer, "A Review of CFAR Detection Techniques in Radar Systems," *Microwave Journal*, Vol. 29, No. 9, September 1986, pp. 115–128.

Clarke, J., and E. B. Cowley, "Approach to Study of PRF Sensitivity in Airborne Pulse-Doppler Radar," *IEE Proc.*, Vol. 134, Part F, No. 4, July 1987, pp. 335–340.

McLane, P. J., P. H. Wittke, and C. K.-S. Ip, "Threshold Control for Automatic Detection in Radar Systems," *IEEE Trans.*, Vol. AES-18, No. 2, March 1982, pp. 244–248.

CHAPTER 18

Fundamentals of Electronic Counter-Countermeasures

Guy V. Morris

The largest users of airborne pulse-doppler radars are the armed forces of the nations of the world. The radars are sometimes required to detect and track targets that do not wish to be tracked and therefore may use some form of jamming. No book about airborne pulse-doppler radars would be complete without some mention of the ECCM incorporated in modern systems. The security classification usually associated with this topic prevents discussion of individual ECCM techniques incorporated in specific current systems. However, the fundamentals and the considerations that should be addressed by the radar system engineer will be explained. The topics to be discussed include:

1. An explanation of the fundamental physics;
2. A definition of the terminology used by practitioners of the jammer and radar design community;
3. The methodology for analyzing the signal-to-interference ratio;
4. The fundamental basis of robust ECCM techniques;
5. A brief description of some of the ECCM techniques used in modern systems.

   The terminology recently adopted by the U.S. Department of Defense (DoD) shown below appears in some recent DoD documents and publications of the Association of Old Crows, the professional society of practitioners in electronic warfare. However, the former, more familiar terminology, ECCM, will be used in this chapter. A comparison of some of the new versus previous terminology is shown in Table 18.1. I will continue to use the terms *ECM* and *jamming* interchangeably.
   The purpose of this chapter is to serve as a guide for identifying ECCM technology available to enhance and maintain the capabilities of modern radar

---

Table 18.1
Jamming Terminology

Current DoD Term	Previous Term
Electronic protection (EP)	Electronic counter-countermeasures
Electronic attack (EA)	Electronic warfare (EW)
Electronic countermeasures	Electronic countermeasures

systems by reducing the EA vulnerability and increasing survivability on the future tactical battlefield. The goal of ECCM in this context is to achieve balanced radar protection over a wide variety of interference waveforms which may be deliberately introduced against the system. This requires an end-to-end assessment of each ECCM feature to each type of interference waveform under examination. Detailed knowledge of the degradation mechanisms and how they affect various radar system components is vital to the development of effective and efficient ECCM. The ECM techniques described are limited to those that are commonly used against coherent doppler radars.

Many ECCM techniques are very system-specific and are intended to reduce the vulnerability of a specific radar mechanization or to take advantage of the geometry (or some other aspect) of a particular scenario; these techniques are not described.

18.1 TRENDS IN ELECTRONIC PROTECTION

According to Morris and Kastle [1], the period 1985 to 1995 was a decade of transition for the radar community in its attitude toward EA. Prior to that time, the following attitudes were common:

- Little attention was paid to the initial design other than some basic features that could be incorporated without a major increase in cost.
- If more ECCM is ever needed (and many doubted that it would be), it can be added via engineering changes.
- ECCM was purely reactionary; that is, ECCM technology was developed in response to observed or postulated threats.
- ECCM technology development was ad hoc; that is, if the effect of an electronic countermeasure was observed in a specific system, a system-specific solution was developed.

The late 1990s should see a continuation of the following trends:

- ECCM requirements becoming an important part of the performance specifications for new system designs;
- Increased ECCM performance becoming a major factor in the planned upgrade programs for existing systems;
- All ECCM features, mode control, and waveform control being implemented in software so that counters to many new threats can be achieved without changing the hardware;
- A technology base of generic ECCM techniques being developed as far as possible prior to being applied to a specific system in order to reduce the reaction time necessary to counter a newly-observed threat;
- Emergence of the concept of "offensive ECCM," for example, inducing the enemy to use an ECM waveform that can be easily identified and tracked by our radar.

The technical impact of these trends on radar design includes the following:

1. Wide RF tuning bandwidths (e.g., more than an octave in the experimental electronically agile multimode radar (EMR) radar developed by the Air Force's Wright Laboratory);
2. More complex waveforms than current high-, low-, and medium-PRF waveforms that are interleaved in short bursts or even pulse by pulse;
3. Multiple receiver channels to implement functions such as four-channel monopulse, cancellation of side-lobe jammers, and guard channels for discrete (pulse) interference rejection;
4. Electronic scan antennas becoming economically feasible for more applications due to the cost reductions made possible by gallium arsenide monolithic microwave integrated circuits;
5. Implementation of more sophisticated ECCM algorithms made possible by the advances in digital signal processing power.

18.2 FUNDAMENTALS OF RADAR VULNERABILITY

The general effects of ECM on a radar are summarized before proceeding with a detailed description of ECM and ECCM techniques. The method used is as follows:

1. Basic terminology is defined.
2. The major classes of ECM techniques are described.
3. A generic radar is described.
4. The objectives and effects of various classes of ECM on the targeted functions of the radar are qualitatively described.
5. The calculation of jamming-to-signal ratio is treated.

6. A set of generic, robust ECCM features sometimes regarded as "good basic design" or often included in the radar for reasons other than ECCM are identified.

18.2.1 Terminology

Frequency bands: The ECCM designer must be bilingual in radar and ECM terminology. For example, the common airborne radar bands designated X and K_u correspond approximately to the ECM bands I and J. The relationships between the radar and ECM bands is shown in Table 18.2.

Robustness: Robustness implies that the ECCM technique is not tailored to address the weakness of a specific ECM system. The designer exploits basic principles or general limits of available technology, such as time delay through a repeater jammer. As applied to ECCM, robustness applies to whether the technique works when the jammer is either off or on. A technique that introduces a large performance penalty if engaged when no jammer signal is present would be nonrobust.

Perishability: The concept of perishability embodies two aspects. The first relates to how easy it is for the enemy to find out what ECCM is being employed. For example, the ability to rapidly change RF and PRF can be observed externally, but changes in tracking algorithms are difficult to assess. The second aspect is how easy it is for the enemy to counter even when changes are known.

Electronic Attack: EA consists of ESM, ECM, and ECCM. ESM includes intercepting, identifying, analyzing, and locating sources of enemy emissions. ECM includes

Table 18.2
Radar and ECM Frequency Bands

Radar Band	Frequency	ECM Band	Frequency
UHF	300 MHz–1 GHz	B	250–500 MHz
		C	500–1,000 GHz
L	1–2 GHz	D	1–2 GHz
S	2–4 GHz	E	2–3 GHz
		F	3–4 GHz
C	4–8 GHz	G	4–6 GHz
		H	6–8 GHz
X	8–12 GHz	I	8–10 GHz
K_u	12–18 GHz	J	10–20 GHz
K	18–27 GHz		
K_a	27–40 GHz	K	20–40 GHz
Millimeter	40–100 GHz	L	40–60 GHz
		M	60–100 GHz

jamming, disrupting, and deceiving enemy transmissions. ECCM refers to protecting our own systems from enemy EA.

18.2.2 Overview of ECM Techniques

The major classes of ECM techniques may be grouped by waveform and by operational utilization as shown in Table 18.3. A more detailed description of several important ECM techniques is contained in Appendix D. The operational utilizations are pictured in Figure 18.1.

Self-screening jamming is defined as jamming that the target employs for protecting itself.

Escort jamming is carried by an aircraft that is generally flying in formation with the target aircraft. The usual implication is that the jammer is also in the main beam of the radar antenna when the target is being illuminated.

Standoff jamming is generally performed by specialized aircraft that are at a greater distance from the radar than the target. Thus, standoff jammers are less vulnerable to direct attack. Standoff jamming can be in the radar main beam. However, standoff jamming is most often in the radar antenna side lobes due to the narrowness of the main beam. Standoff jamming is most often noise because the jammer has difficulty introducing repeatable, believable false targets into the radar via the side lobes.

Noise jamming is, as its name implies, a noise waveform that is used to obscure the target skin return and thus deny range and doppler.

Deceptive jamming, as employed against doppler radar, is usually repeater-based. The jammer receives the radar illumination, amplifies it, adds one or more types of modulation, and retransmits it to the radar as shown in Figure 18.2. The repeater jammer implementation maintains the pulse-to-pulse coherence necessary to generate a realistic false doppler target. The modulations can include false doppler shift, range delay, and angle deception.

Table 18.3
Types of ECM

Type	Self-Screening	Escort	Standoff
Noise	X	X	X
Deceptive	X	X	
Range			
Doppler			
Angle			
Passive		X	

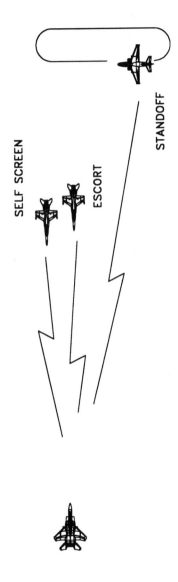

Figure 18.1 Jammer operational scenario.

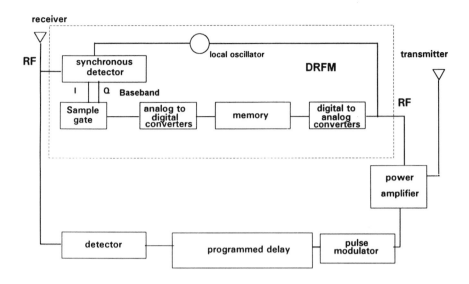

Figure 18.2 Repeater jammer.

Passive jamming is included for completeness but will not be discussed extensively. The most common form is chaff, metallized fibers that disburse to form a cloud of large RCS when dispensed from an aircraft. The chaff decelerates rapidly from the aircraft velocity and assumes the same apparent velocity as ground clutter. The processing that is provided to reject clutter also rejects chaff although some degradation results because the chaff has a broader spectral width than natural clutter and is often considerably greater in amplitude. Also, since a chaff cloud represents a large diffuse reflecting surface, it can be used to reflect active jamming waveforms.

18.2.2.1 Examples of Noise Jamming

The common types of noise jamming are described below and illustrated in Figure 18.3.

Spot noise (SN) refers to a jamming waveform that is intended to cover the current operating frequency of the radar. The jammer bandwidth is typically 10 to 20 MHz to ensure that the radar receiver instantaneous bandwidth is covered. Modern jammers measure the frequency of the radar rapidly and can set on the radar frequency for most or all of the coherent processing interval.

Barrage noise (BN) is used to cover the entire tuning band of the radar. It is used to simultaneously jam all radars in the band or to jam radars that use frequency agility or high-resolution waveforms. The jammer power spectral density, expressed

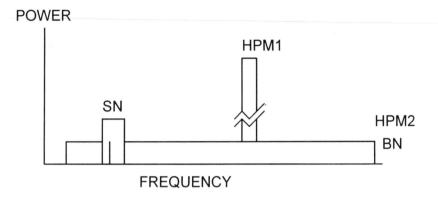

Figure 18.3 Types of noise jamming.

in watt/megahertz, of the barrage noise jammer is usually less than the spot noise jammer because the jammer is average power limited.

High-power microwave (HPM) jamming is a term that is used to describe two distinctly different techniques. One use of the term, depicted graphically as HPM1 in Figure 18.3, refers to a directional jammer that produces a sufficiently high level of jamming that it physically damages the sensitive radar receiver components. The second use of the term, depicted in Figure 18.3 as HPM2, refers to a standoff jammer that has an extremely large effective radiated power (ERP) and can produce effective levels of barrage noise jamming at much greater, and therefore safer, ranges.

18.2.2.2 Examples of Deceptive Jamming

Doppler deception is most often used against the single-target track mode of a doppler radar. The jammer begins by receiving and retransmitting the radar signal without modulation. The jamming signal is typically 10 to 20 dB greater than the skin return from the target. The jammer signal strength is sufficient to capture the doppler tracking gate. Then single-sideband modulation is applied to the received signal to alter the apparent doppler frequency. The objective of the jammer is to pull the tracking gate into the frequency region of clutter and cause the radar to transfer lock to the clutter or to cease jamming, which breaks track and forces the radar to reacquire. This jamming technique is usually called *velocity gate pull-off* (VGPO). The most common pull-off methods are linear and parabolic. A change in doppler corresponds to an acceleration. Parabolic pull-off is illustrated in Figure 18.4 for two values of acceleration.

Other doppler deception techniques include generating multiple false doppler targets. Variants of these techniques are used to overload or confuse the operator or the automatic signal processing in search and track.

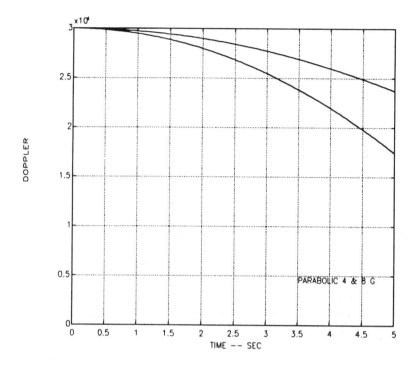

Figure 18.4 Parabolic velocity gate pull-off.

Range deception is analogous to doppler deception. The jammer receives the radar illumination, amplifies it, and retransmits with minimum time delay. The jammer signal is 10 to 20 dB greater than the target skin return. The objective of the jammer is to capture the range-tracking gate of the radar and pull it off of the target. Then the jammer introduces a progressively larger time delay from pulse to pulse to pull the radar range-tracking gate off the target. This jamming technique is usually called *range gate pull-off* (RGPO). The repeater-based jammer mechanization ensures that the false range target maintains the necessary pulse-to-pulse coherence. Variants of this technique can be used to generate multiple false range targets.

Until recently, analog delay devices, such as recirculating delay lines, were used to produce the time delay. Therefore, the jammer could only produce false targets that were at greater apparent ranges than the true target. Furthermore, analog delay devices could produce only relatively short delays while maintaining sufficient coherence and SNR to produce an effective false target. Recently, technology has made it possible to build jammers that use a delay device known as a *digital RF memory* (DRFM). With a DRFM it is possible to generate delays on the order of the PRI and thus create false targets with an apparent range that is less than the true target.

A range pull-off corresponds physically to a change in velocity. The most sophisticated jammers are capable of coordinated RGPO-VGPO. The false target has the proper apparent doppler that corresponds with the change in range and may be perceived as a second target separating from the original one being tracked.

Monopulse radar antennas are used to make it difficult for the jammer to perform *angle deception*. The radar mechanization assumes that a single planar electromagnetic wave, normally skin return from the target, impinges on the antenna. Angle deception against the radar requires that a second signal illuminate the radar antenna. This second signal can be produced by a jamming source that is physically displaced from the target or by a waveform emanating from the target that is capable of producing an erroneous angle-of-arrival measurement. Examples of physically displaced sources include two unresolved jamming aircraft, a jamming decoy that is towed behind the target, expendable jamming decoys that are ejected and fall away from the target, and jamming directed by the target toward the surface to create a false jammer location image.

The radar will tend to track the power centroid of two noise jammers that are unresolved in angle. If the two jamming sources are coherent and unresolved in both range and doppler, then the angle measured by the monopulse antenna may not accurately represent either source, and some measurements may be outside the physical boundaries of the two sources, as described in Chapter 13.

Cross-polarized jamming is an example of a waveform that may produce an improper angle tracking error measurement if the radar uses a monopulse antenna. The Δ/Σ ratio that results from a cross-polarized jamming signal at a certain angle may be very different from the true Δ/Σ ratio produced by the target. Also, it might have an incorrect sign that would, for example, cause the antenna to move left when it should go right. If the cross-polarized jamming signal can be made large enough to overcome the skin return, then erratic tracking and breaklock might occur. The gain of the radar antenna to the jammer signal with the orthogonal polarization is usually far less than the gain to copolarized target skin return. The jammer must overcome this deficiency in antenna gain and establish a 10 to 20 jamming-to-signal ratio (JSR). VGPO is often used with this waveform to separate the jammer signal from the target skin return and create an infinite JSR.

Masking refers to jamming techniques that are aimed at delaying or denying detection of the skin return. Some forms of masking are particularly insidious because they mask the target without radiating enough jamming energy to be readily detected. One example is narrowband doppler noise (NBDN). NBDN is produced by the repeater jammer by frequency modulating the received radar signal with a noise waveform having a bandwidth of 2 to 5 kHz, as depicted in Figure 18.5. Sliding-window cell-averaging CFAR will elevate the detection threshold when the jamming enters the reference window and may prevent detection. Other choices of modulation coefficients and frequencies can produce multiple discrete spectral lines that may produce a similar elevation of the detection threshold and loss of detection sensitivity.

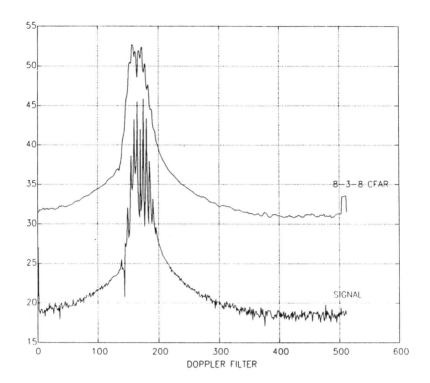

Figure 18.5 Example of masking by NBDN.

18.2.3 Generic Radar Description

The major functions of the generic radar are shown in Figure 18.6. Much of the intraradar signal flow has been omitted for simplicity. Other functions that are essential to a complete weapon system but not shown on the radar block diagram include controls and displays, a system computer, and communications links. The functions that might be included in each block of Figure 18.6 are described below.

18.2.3.1 Receiver Function

The receiver function includes low-noise RF amplification and down-conversion of the received signal to video. The receiver may contain several parallel reception channels (e.g., sum, azimuth difference, elevation difference, guard or blanking, and multiple side-lobe cancelers). AGC is often used to achieve the required dynamic range. The receiving function also generates low-power RF signals such as the transmit oscillator (TO) and the multiple LOs used for down-conversion.

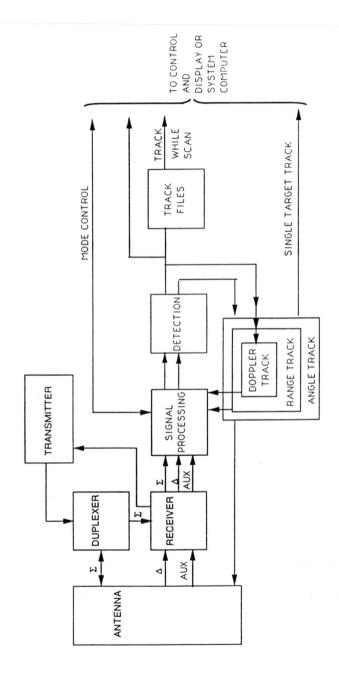

Figure 18.6 Generic radar block.

18.2.3.2 Transmitter Function

The transmitter function provides pulsed power amplification and contains high-voltage power supplies.

18.2.3.3 Duplexer

The duplexer permits transmission and reception through the same antenna.

18.2.3.4 Antenna

The antenna provides directive gain in the desired beam pointing direction. It may contain multiple channels corresponding to those listed for the receiver. The instantaneous pointing direction may scan a volume in space as directed by a scan controller or steer to point continuously at a single target under the control of the tracking function. In an active phased-array antenna system, the major portions of the transmitter, duplexer, antenna, and receiver functions may be contained in the antenna unit.

18.2.3.5 Signal-Processing Function

The signal-processing function includes clutter rejection and doppler filtering, often implemented as an FFT algorithm. Doppler filtering requirements vary substantially as a function of radar mode in a multimode radar. For example, a high-PRF mode might employ a single range gate and form 512 doppler filters. The medium-PRF mode might have 100 range gates containing 32 doppler filters each. A low-PRF mode might have 2,000 range gates with 8 doppler filters in each. Medium-PRF and low-PRF processing usually includes delay line cancelers before the FFT as a prefilter to suppress clutter, as explained in Chapter 8. The clutter rejection and doppler filtering functions are performed on each receiver channel, for example, sum, difference, and any auxiliary channels used for side-lobe blanking or side-lobe canceling.

If coherent side-lobe cancelers are implemented in the system, adaptive weights are developed from the correlation of signals received in the main and auxiliary channels. The adaptive weights are applied to the signals received through the auxiliary channels and subtracted from the sum and difference signals in the main channels to cancel noise jamming. The signals in the auxiliary channels are not processed further.

Digital pulse compression is sometimes implemented in the signal processor. The signal processor serves as the master synchronizer that generates the PRF and other timing signals. The signal processor responds to the mode control commands from the operator or system computer and controls the PRF, transmit frequency, and signal-processing algorithms.

18.2.3.6 Detection Function

The detection function consists of some type of CFAR circuit to provide the "raw" detections. Multiple detections in contiguous doppler filters are submitted to a centroiding algorithm to yield a single observation. Similar detection centroiding is performed in the range and angle dimensions also. If the waveform has range or doppler ambiguities, then ambiguity resolution will also be performed.

18.2.3.7 Search and Track-While-Scan Modes

In the search mode, the detections are supplied to the control and display subsystem of the system computer. The system usually displays the data with a relatively short-term display memory. The detections are displayed as symbols, and there is no scan-to-scan correlation. In the TWS mode, the system computer uses the detections of many scans to establish and maintain track files.

18.2.3.8 Single-Target Tracking

The radar mode most often referred to as *track* is more accurately called *single-target track*. The objective of single-target tracking is to provide continuous illumination of the target. The increased data rate relative to the TWS mode permits more accurate determination of the target angle, range, and doppler range rate. The increased accuracy is usually needed in the terminal phase of target attack used by missile guidance or fire control radars. Figure 18.6 depicts the single-target-tracking function as nested loops. A doppler track of the moving target is established first. The amplitudes of the target signal in both an early and a late range gate are compared to derive a range-tracking error and to close the range-tracking loop. After range tracking has been established, the amplitudes of target returns in the same range-doppler cell in the sum channel, elevation difference channel, and azimuth difference channel are compared to develop the angle error corrections.

The single-target track mode can be entered from either the search or TWS modes. The measurements of range, angle, and doppler are used to preposition the tracking gates to achieve lock-on. Often, the radar must search around the nominal values due to positioning errors or target maneuvers.

18.2.4 Objectives and Effects of Electronic Countermeasures

ECM techniques have been devised to attack each of the radar receiver and processor functions.

18.2.4.1 Antenna Function

All active jamming enters the system through the antenna and therefore may be considered as attacking the antenna. Some ECM techniques specifically exploit the undesirable characteristics of the antenna. The most significant characteristics are:

1. *Side lobes:* Noise jamming entering through the side lobes causes receiver desensitization.
2. *Main-lobe width:* Jamming sources not on the target can interfere with detection and tracking. Examples include standoff and escort jamming, surface bounce jamming, and decoys. Multiple signals within the antenna beamwidth that cannot be resolved in either range or doppler will not be angle-tracked properly.
3. *Cross-polarized response:* Jamming corrupts the difference-sum ratio and interferes with monopulse angle measurement.
4. *Conical scan and other sequential lobing techniques:* Moving a single beam in space and measuring the relative amplitude of the target return to determine angle-tracking error is susceptible to AM jamming.

18.2.4.2 Receiver Function

Jamming effects on the receiver include:

1. *Saturation:* Strong jamming attempts to prevent subsequent signal-processing stages from recovering the target return or properly angle tracking the jammer.
2. *Spurious signals:* Jamming may produce intermodulation products that result in false detections.
3. *AGC disturbance:* Since most receivers contain some type of AGC to control saturation, the jammers may use on-off modulation to capture or disturb the AGC.

18.2.4.3 Signal-Processing Function

Jamming may attack the signal processor in the following ways:

1. *Broadband noise:* Doppler processors may provide from 12 to 30 dB of processing gain, but noise that covers the IF bandwidth will fill all the doppler filters and may prevent detection of the target skin return.
2. *Cover pulse:* A cover pulse interferes with the pulse compression process and may prevent detection of the target skin return.

3. *Side-lobe jammer overload:* If there are more apparent side-lobe jammers, including strong reflections, than the number of auxiliary channels in the side-lobe canceler, then uncanceled jamming is passed to subsequent stages.

18.2.4.4 Detection Function

The detection function is a frequent target of jamming techniques. Some techniques are aimed at desensitizing the radar detector so that the target is not detected and no warning of jamming is presented to the operator. Other techniques create a large number of false "raw" detections that overload subsequent stages intended to remove false and redundant detections. Some examples of ECM against the detection function are:

1. *Anti-CFAR methods:* NBDN and false targets can elevate the estimate of receiver noise and cause the detection threshold to be erroneously raised, desensitizing the system.
2. *False range targets:* False targets can overload the range ambiguity resolution functions of medium PRF radars and cause ghosts.

18.2.4.5 Track-While-Scan Function

False targets from the detection function can overload the track association and updating logic.

18.2.4.6 Doppler-Tracking Function

Because the doppler-tracking function is the center of the single-target-tracking function, it is a leading object of jammer attack. VGPO may attempt to pull the tracking gate to clutter and cause a transfer of tracking to clutter. A second strategy is to simply turn off the jammer after pulling the tracking gate off the target skin return, forcing frequent reacquisitions and erratic tracking. A third strategy is to pull the gate off and increase the JSR so that angle deception techniques such as cross-polarization or cross-eye jamming are more effective.

18.2.4.7 Range-Tracking Function

RGPO has objectives similar to VGPO's. Older jammer systems could only introduce relatively short range delays and therefore could only pull the range gates to ranges greater than the jammer range. A jamming system using DRFM can introduce delays of a complete PRI and can therefore pull the range gate to a shorter range.

18.2.4.8 Angle-Tracking Function

Any interruption of the doppler- or range-tracking loops causes an interruption of angle tracking. Also, the unresolved signals of multiple targets and the antenna responses previously discussed can result in erroneous tracking.

18.3 ANALYZING THE EFFECTS OF ELECTRONIC COUNTERMEASURES

The steps include:

1. Calculating the power received from all signal sources, for example, target return, clutter return, and jamming;
2. Determining the effect of each of the steps in the radar signal processing on each of the signals.

18.3.1 Single-Pulse Jamming-to-Signal Ratio

A jamming technique will not generally be effective unless the jamming signal is 10 to 20 dB stronger than the desired signal at the point of attack within the radar. For example, if the jamming is intended to capture the velocity-tracking loop of the radar, then the JSR needs to be 10 dB at the input to the tracker with both the signal and jamming measured over the same bandwidth.

Calculating the jammer and signal powers impinging on the radar antenna is straightforward. The signal power S reflected from the target due to a single radar transmitter pulse is given by

$$S = \left(\frac{P_P G_{RT}}{4\pi R_T^2}\right)\sigma_T\left(\frac{1}{4\pi R_T^2}\right)A_{RR} \tag{18.1}$$

where:

P_P	=	peak radar power;
G_{RT}	=	radar transmit antenna gain;
R_{RT}	=	range from target to radar;
σ_T	=	target RCS;
A_{RR}	=	effective area of radar receive antenna.

Equation (18.1) is merely a common form of the radar range equation that has been factored to illustrate certain points relative to ECM. The first bracketed term on the right side of the equation indicates the power density at the target. The second term, σ_T, indicates the fraction of the power that is captured and reflected in the direction of the radar. The product of the first three terms represents the power density of the signal at the radar receive antenna. The signal power computed using (18.1) is the power appearing in the waveguide immediately behind the antenna. (The receive path ohmic losses are the same for the desired signal and the jammer and have been omitted in this discussion.)

The power received from a repeater jammer, J_R, is given by

$$J_R = \left(\frac{P_P G_{RT}}{4\pi R_J^2}\right) A_{JR} G_E G_{JT} \left(\frac{1}{4\pi R_J^2}\right) A_{RR} \qquad (18.2)$$

where:

G_E = electronic gain of the jammer;
A_{JR} = effective area of jammer receive antenna;
G_{JT} = jammer transmit antenna gain;
R_J = range from jammer to radar.

The illuminating power is captured by the jammer receive antenna, amplified, and retransmitted to the victim radar, usually with some modulation such as a false doppler. Comparing (18.1) and (18.2) clearly indicates that the repeater jammer may be understood as a device that augments the RCS. In fact, if it were not for the modulations that are intended to disrupt tracking, the jammer would present a signal that may be considerably easier to track than the normal skin return. Note also that the power received from the repeater jammer obeys the same $1/R^4$ relationship as the target power unless the jammer transmitter reaches saturation.

The power received at the radar, J, due to a nonrepeater jammer, is given by

$$J = \left(\frac{P_J G_{JT}}{4\pi R_J^2}\right) A_{RR} \qquad (18.3)$$

where P_J is the jammer transmit power.

The received nonrepeater jammer signal is independent of the strength of the illumination from the radar and therefore obeys a $1/R^2$ relationship.

The noise power N_R in the radar in the absence of jamming is given by

$$N_R = k T_E B_R \qquad (18.4)$$

where:

k = Boltzmann constant;
T_E = effective noise temperature of the radar receiver;
B_R = bandwidth of the radar receiver.

The effective capture area of an antenna is related to the gain by

$$A = \frac{G\lambda^2}{4\pi} \tag{18.5}$$

where λ is the radar wavelength.

Using (18.5), (18.2) may be rewritten into an alternate form:

$$J_R = \left[P_P G_{RT} \left(\frac{\lambda}{4\pi R_J} \right)^2 G_{JR} G_E G_{JT} \right] \left(\frac{\lambda}{4\pi R_J} \right)^2 G_{RR} \tag{18.6}$$

where:

$G_{RR}=$ radar receive antenna gain;

$G_{JR} =$ jammer receive antenna gain.

The term $(\lambda/4\pi R_J)^2$ is sometimes referred to as the *path loss*. The terms within the square brackets represent the ERP transmitted by the jammer. ERP is often expressed as

$$\mathrm{ERP} = P_J G_{JT} \tag{18.7}$$

As R_J approaches zero, the jammer transmitter will eventually saturate and will no longer be given by (18.6). The jammer power at the radar is then calculated using (18.3).

The preceding equations are general enough to be applicable to self-screening, escort, or standoff jamming by using the appropriate value of range and antenna gain in the direction of the jammer. For a self-screening jammer, the target and jammer are at the same range and enjoy the same radar antenna gains. The repeater JSR is given by

$$\frac{J_R}{S} = \left(\frac{\lambda^2}{4\pi\sigma_T} \right) G_{JR} G_E G_{JT} \tag{18.8}$$

18.3.2 Effect of Radar Signal Processing

The radar receiver and signal processor are designed to provide a matched filter to the transmitter waveform. The desired signal may enjoy signal-processing gain that the jamming signal may not. The JSR within the radar is calculated by starting with the impinging jammer and signal powers calculated using the preceding equations and applying the proper gain or loss individually to the jammer and signal powers. Typical potential sources of loss include:

1. *Jammer bandwidth:* Noise jammer waveforms frequently have greater bandwidth than the receiver IF bandwidth of the victim radar. The effective jammer power is reduced by the ratio (B_R/B_J).

2. *Antenna patterns:* Standoff jammers may be in the antenna side lobes. The antenna gain in the direction of the jammer should be used in the calculation. Similarly, the cross-polarized response of an antenna is typically much less than the copolarized response and constitutes an additional loss to cross-polarized jammers.

3. *Pulse compression:* The signal achieves a signal-to-noise improvement relative to a noise jammer but not a straight-through repeater.

4. *Doppler filtering:* Doppler filtering provides a signal-to-noise gain against a noise jammer but not a repeater. Doppler filtering may permit the interfering repeater signal to be identified and separated from the desired signal even if the JSR is adverse.

Figure 18.7 shows the effect of typical radar signal processing on the JSR for typical medium-PRF mode parameters.

18.4 GENERIC AND ROBUST ECCM TECHNIQUES

Some techniques that are incorporated in radars to improve performance in areas such as clutter rejection, reduction of multipath interference, range, and velocity resolution also make radars less vulnerable to ECM. These ECCM features are sometimes referred to as *generic* ECCM. *Robust* ECCM techniques are those that rely on fundamental physics rather that exploiting the unique characteristics of a specific jammer design.

18.4.1 Physical Basis for Robust ECCM Techniques

Protecting the radar from the effects of jamming may be subdivided into two processes, detecting that an interfering waveform is present and then either eliminating the undesired waveform or at least identifying the desired target return and ignoring the jamming.

18.4.1.1 Matched Filtering of the Expected Signal

The radar, of course, is designed to receive a delayed replica of the transmitted signal as modified by the reflection characteristics of the target. This matched filtering in a doppler radar can produce considerable enhancement of the target skin return relative to a noise jammer. Repeater-based jamming will generally enjoy the same processing gain as the target skin return, but may contain artifacts due to imperfections in the modulation that can be detected and used by the radar to discriminate against the jamming signal.

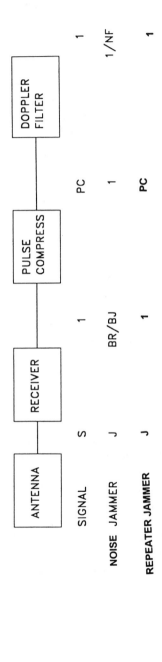

Figure 18.7 Effect on JSR of signal processing.

18.4.1.2 Time History of the Signal

The time history of the returned signal can contain considerable information about the onset of jamming and enhance discrimination of the jamming signal from the target return. One common weakness of some operational systems is that when they enter the single-target track mode, the situational awareness is greatly reduced. Specifically, the range-doppler region of interest is often narrowed greatly, and the ability to recognize the presence of two or more signals, for example, the skin return and the jammer, is lost. Newer systems have more signal-processing capability and can continue to process the entire range-doppler region in track. Other jamming indications that can be useful include sudden increases in signal strength and appearance of doppler sidebands.

18.4.1.3 Multiple Phase Center Antennas

If the composite return arriving at the radar antenna consists of multiple signals impinging from different angles, then sampling the signals from different points in space may provide the information needed to counter the jamming. One example is the use of auxiliary antennas in conjunction with the main antenna to provide side-lobe cancellation. Another example is the partitioning of the main antenna array into subapertures that can be used in conjunction with superresolution algorithms.

An antenna that provides dual simultaneous reception on two orthogonal polarizations or can be switched between the two polarizations can be used in the design of ECCM against cross-polarization jamming.

18.4.1.4 Preconceived Boundaries of Reasonable Target Characteristics

For each type of target of interest there is usually a set of characteristics, such as speed regime, acceleration, and target cross section, that are considered reasonable. Monitoring the measured characteristics versus the expectation may provide an alert of possible jamming.

18.4.2 Generic ECCM Techniques

The generic techniques are classified as robust because they provide capability against a wide variety of ECM techniques and require little a priori knowledge of the ECM. Techniques used by each component of the generic radar are listed below.

1. Antenna:
   a. Low side lobes;
   b. Low cross-polarized response.

2. Transmitter:
   a. Broad tuning range—Used to avoid or dilute jammer power density;
   b. Parameter diversity—The ability to rapidly change the transmit frequency, PRF, beam position, and pulse width can be used to avoid friendly and intentional interference;
   c. High transmit power.
3. Receiver:
   a. RF attenuator-receiver protection—prevents saturation;
   b. High LO frequency and first IF helps in increased image frequency rejection when a fixed RF bandpass filter is used. Tunable RF filters can also be used to reject the image frequency and out-of-band interference;
   c. Large instantaneous dynamic range.
4. Signal processor:
   a. Analog-to-digital converter—Uses the greatest number of bits achievable with the state of the art to exploit the receiver instantaneous dynamic range;
   b. Programmability to accommodate a variety of algorithms and provide the capability to respond readily to the changing threat;
   c. High range and doppler resolution to reduce vulnerability to decoys and repeater jammers;
   d. Signal-processing reserve capability in both memory and computing resource timeline.

18.5 DESCRIPTIONS OF SIGNIFICANT ECCM TECHNIQUES

This section will describe representative ECCM techniques that are applicable to doppler radars. The list is by no means exhaustive but is believed to contain the most important ones. Table 18.4 contains a list of the ECCM techniques cross-referenced to the ECM they are intended to counter. The ECCM techniques are described in the following subsections. Table 18.5 enables a user concerned with a specific ECM technique to identify ECCM techniques used to counter the jamming waveform.

18.5.1 Acceleration Limiting

The technique of acceleration limiting (AL) compares the measured target velocity, range, or angle accelerations with acceptable rates of change of range rate or angle rate. If these measured values exceed predetermined maximums representing actual target capabilities, the corresponding targets will not be tracked. A breaklock will be issued, causing the radar to enter the reacquisition mode. Velocity and range gate stealers that attempt high-acceleration pull-off will be defeated. Forcing the

Table 18.4
ECCM Techniques

ECCM Technique	Acronym	Used Against ECM Technique
Acceleration limiting	AL	VGPO, RGPO
Adaptive receive polarization	ARPOL	XPOL
Angle extent estimator	AEE	BSN
Bandwidth expansion	BE	VGPO
Beat frequency detector	BFD	VGPO
Censored (ordered-statistic CFAR)	C/OSCFAR	VGPO, MFDT, FRT
Cross-polarization cancellation	XPOLC	XPOL
Doppler display	DD	VGPOMFDT
Doppler/range rate comparison	D/RR	RGPO, VGPO
Frequency agility	FA	BN, BSN
Frequency diversity	FD	BN, BSN
Home-on-jam	HOJ	BN, BSN
Leading/trailing edge track	LET	RGPO, TB, TD
Narrowband doppler noise detector	NBDND	NBDN
Narrow pulse/pulse compression	NP	RGPO, BN, TB, RSN, BSN
Neural net	NN	MFDT, NBDN, XPOL, VGPO, RGPO
Off-boresight tracking	OBT	BSN, TB, TD
PRF jitter	PRFJ	RGPO
Side-lobe blanking	SLB	FRT, MFDT
Side-lobe canceler	SLC	BN, BSN
Sniff	SN	BN, BSN
Space-time adaptive processing	STAP	BN, BSN
Superresolution	SRES	BN, BSN
Transmit-receive polarization mismatch	POLMIS	XPOL
Velocity guard gates	VGG	VGPO
VGPO reset	VGPOR	VGPO
VGS ECCM—dual frequency	VGSDF	VGPO

jammer to use low-acceleration pull-off results in an increased time to achieve the desired error and less frequent opportunities to cause breaklock [2, p. 55].

18.5.2 Adaptive Receive Polarization

The ECCM technique of adaptive receive polarization (ARPOL) is used against main-beam self-screening or standoff jammers with a fixed polarization. This technique is specifically designed to counter nonadaptive, cross-polarization jammers. Upon detection of a cross-polarization jammer, the radar's receive antenna polarization is switched to match the enemy jammer transmit polarization. Thus, the enemy jammer becomes a beacon and may be tracked using passive techniques such as home-on-jam and digital beamforming [2, pp. 420–423].

Table 18.5
ECM Techniques

Technique Name	Acronym	ECCM Technique
Barrage noise	BN	FA, FD, HOJ, NP, SLC, SN, STAP, SRES, TET
Blinking spot noise	BSN	AEE, FA, FD, NP, SLC, SN, STAP, SRES
Cross-polarization jamming	XPOL	ARPOL, XPOLC, NN
Digital RF memory repeater	DRFM	IMPR, PRFJ
False range target	FRT	C/OSCFAR, NN, SLB
Multiple false doppler targets	MFDT	C/OSCFAR, DD, NN, SLB
Narrowband doppler noise	NBDN	NBDND, NN
Range gate pull-off	RGPO	AL, D/RR, LET, NN, NP, PRFJ
Terrain bounce	TB	LET, NP, OBT
Towed decoy	TD	LET, OBT
Velocity gate pull-off	VGPO	AL, BE, BFD, C/OSCFAR, DD, D/RR, NN, VGG, VGPOR, VGSDF

18.5.3 Angle Extent Estimator

The angle extent estimator (AEE) uses the received signals from a monopulse radar to provide an estimate of the angular extent of a target within the beam. The extent estimator measures the standard deviation of the monopulse angle measurement and may be used as an indicator of jamming. Figure 18.8 illustrates typical histograms of monopulse angle measurements of single and multiple target and jammer configurations [3, pp. 218–222].

Figure 18.9 illustrates monopulse angle histograms for one and two targets within the main beam. With two targets in the beam, the extent estimate will be large and will exhibit high fluctuations due to the target separation. If a single target is transmitting continuous noise jamming, the extent estimate will remain low with small fluctuations, indicating a strong point source. If a blinking noise waveform is being used by one or more jammers within the beam, the extent estimate will alternate between a small and large value, thus producing a bimodal histogram as the angular separation increases. A blinking noise jammer may be detected by monitoring the magnitude and the fluctuation of the monopulse angles.

Upon detection of two jammers within the beam, closed-form equations can be derived to estimate the angular location of each emitter. Mechanization of this process has been referred to as a *multi-in-beam detector*.

18.5.4 Bandwidth Expansion

The technique of bandwidth expansion (BE) shortens a radar's reacquisition time after a VGPO device has pulled off the radar velocity gate and has been turned

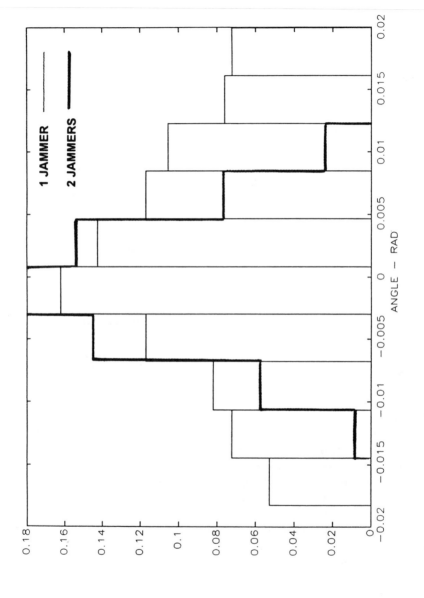

Figure 18.8 Monopulse angle histograms for various target configurations.

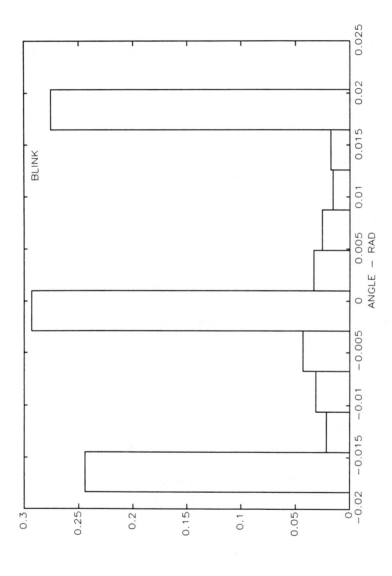

Figure 18.9 Monopulse angle error histograms.

off. A significant phase of a VGPO program is the off period during which the victim radar has no signal to track. During this phase, the jammer hopes to cause a breaklock and keep the doppler radar in reacquisition as long as possible, thus preventing detection of the jammer platform. Bandwidth expansion enables the victim radar to reacquire the true target much faster using digital techniques.

The rapid reacquisition of target signals in a doppler radar requires that the radar's doppler-tracking loop passband be positioned to the frequency of the desired signal. The maximum frequency search rate is inversely proportional to the square of the tracking bandwidth. Bandwidth expansion can speed up the search process. Increasing the effective bandwidth of the velocity gate and doppler spectrum increases the maximum tuning rate of the system and decreases reacquisition time. After the acquisition subsystem has located the true doppler signal, the velocity gate of the receiver can be reset to the correct bandwidth [2, pp. 95–99].

18.5.5 Beat Frequency Detector

The beat frequency detector (BFD) ECCM technique is used against velocity gate stealers. During a typical VGPO program, the repeater jammer attempts to introduce a signal that is at the same doppler frequency as the skin return but 10 or more dB stronger. The false doppler is then moved away from the target doppler. The tracking loop will tend to follow the jamming signal. Eventually, the target doppler is no longer in the tracking filter bandwidth. It is desirable to detect the presence of two signals simultaneously in the tracking filter passband before the track loop is pulled away from the true target doppler and an erroneous tracking rate (acceleration) is introduced.

If the target doppler and jammer signal are within the passband of the same doppler filter, the magnitude of the output of the doppler filter will be as shown in Figure 18.10. The large constant component of the output is the result of the jammer signal. The target doppler causes a modulation that can readily be detected by performing an FFT of the output magnitude to provide an alert of a VGPO. One use of the alert is to freeze the tracking gate position, which still contains the target doppler, and permit the jammer signal to move out of the passband and have no further effect [4].

18.5.6 Censored (Ordered-Statistic) CFAR

Censored cell and ordered statistic CFAR (C/OS CFAR) were originally developed to address performance degradation of conventional CA-CFAR in multitarget scenarios. CA-CFAR averages the amplitudes of a number of cells (range gates, doppler filters) to form an estimate of the local background interference. The interference or noise estimate is used to establish a detection threshold for the cell under test.

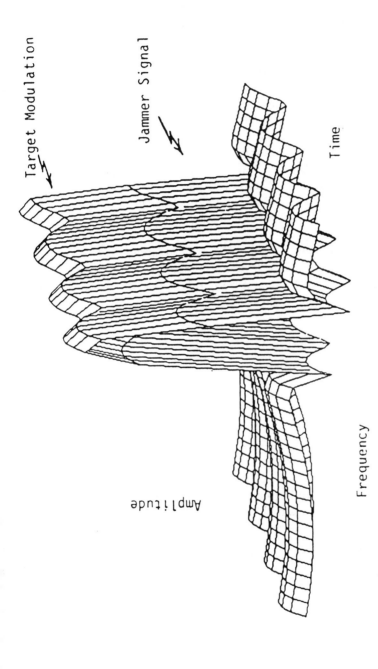

Figure 18.10 High-PRF spectrum containing a VGPO false target and the target skin return in a single doppler filter.

CA-CFAR is usually optimized for detecting a single target in receiver noise. If more than one signal is present, then it is possible that the second signal will contaminate the local noise estimation. The result is an erroneously high noise estimate and desensitization of the system. An example where the presence of two signals results in detection of neither is shown in Figure 17.11. As the sliding window is moved over the doppler spectrum, either one or the other of the two signals lies in the early or late window, artificially raising the noise estimation and denying detection of both signals.

Censored CFAR (C-CFAR) and ordered-statistic CFAR (OS-CFAR) are two approaches for preventing desensitization. In C-CFAR, a prespecified number of the largest signals occurring within the CFAR window are discarded from the average noise computation. In OS-CFAR, the samples are rank ordered in magnitude, and the kth largest sample is used as the noise estimate. Other processing techniques, such as spacing periodicity and relative motion, are required to interpret multiple detections.

C-CFAR and OS-CFAR, as described in Figure 17.11, are applied to the output spectrum of a single range gate. These techniques could also be applied to a radar that performs CFAR in the range dimension to prevent desensitization due to the presence of multiple false range targets.

18.5.7 Cross-Polarization Cancellation

Cross-polarization cancellation (CPOLC) ECCM is used against cross-polarization jamming. A cross-polarized auxiliary antenna is slaved in pointing direction with the principal radar antenna. The signal received by the auxiliary antenna is processed in an auxiliary channel matched in phase to the main radar receiver channel. This auxiliary channel signal is used in a nulling feedback cancellation loop to cancel the main channel cross-polarized signal prior to the angle track circuits in the radar. The residual copolarized signal is used by the radar to perform tracking functions. Both the sum and difference channels of a monopulse tracking radar must be canceled. This canceler works on the main lobe rather than the side lobes of the radar [2, pp. 161–163].

18.5.8 Doppler-Range Rate Comparison

The RGPO and VGPO deceptive ECM techniques attempt to force a victim radar to break track by generating false targets that capture the radar's velocity-range gates and lead them away from the true target. Doppler-range rate (D/RR) comparison is used by range-gated doppler radars to combat the use of uncoordinated RGPO and VGPO.

Two independent measures of target closing rate are collected: (1) measured doppler frequency, which is proportional to radial velocity, and (2) measured range rate, which is the time rate of change of the target range measurement. Inconsistencies between the two measurements can be used to identify an uncoordinated VGPO or RGPO. A VGPO device operating alone will cause the velocity gate of the victim radar to move away from the true target, but the time rate of change of the range measurement of the false target will not change, and vice versa for an RGPO device operating alone. (A coordinated RGPO-VGPO changes the range pull-off in accordance with the false velocity being transmitted; this technique requires a much more sophisticated jammer.) [2, p. 477; 5, pp. 12–14]

18.5.9 Frequency Agility

Frequency agility (FA) refers to the ability of the victim radar to change transmit frequencies on a dwell-to-dwell basis for a coherent radar or on a pulse-to-pulse basis for noncoherent radar modes. Typically, the rapid RF changes will be large compared with the radar's instantaneous IF bandwidth. Rapidly changing the radar frequency over a wide bandwidth forces a barrage noise jammer to broaden its radiated power over a significantly increased RF bandwidth, thus decreasing its effective jamming power density at the victim radar. The jamming power is reduced by a factor equal to the ratio of the frequency agile bandwidth to the radar signal bandwidth. Typically, a random process is utilized to select new transmit frequencies. FA is helpful to reduce unintentional interference and to avoid obsolete, slow-tuning jammers but is not effective against modern fast set on jammers as doppler radars must transmit many pulses at the same frequency [2, p. 217; 5, pp. 12–3].

18.5.10 Home-on-Jam

Home-on-jam (HOJ) is used by a missile guidance receiver to develop steering information on a noise jammer. The technique uses the jamming signal radiated by a barrage or spot noise jammer to extract angle tracking information. Either monopulse or conical scan concepts may be used to extract the angle information. Target range and velocity information are usually not obtainable, thus limiting the operational range of this technique.

18.5.11 Leading/Trailing Edge Track

Leading edge track (LET) and trailing edge track (TET) are ECCM techniques used by pulsed range-tracking radars to track target range in the presence of repeater jamming. In pulse tracking radars, two contiguous range gates having a total width slightly greater than the radar transmitted pulse width are used to measure the centroid of the energy contained within the received pulse. The first of the two gates is called the *early gate*, and the second is called the *late gate*. A comparator is used to measure the energy in the two gates. If the gates are not centered on the receive pulse, an error signal is produced. The error signal is used to move the gates, thereby equalizing the energy in the two gates, before measuring the target range. The radar angle-tracking loop operates only on signals that coincide with the radar's range-tracking gate.

The objective of an ECM system designed to interfere with radar tracking operation is to move the range-tracking gate away from the true target position to a position at which only the jamming signal is present. This implies that the ECM signal must initially transmit jamming signals directly over the true reflected target signal before attempting to implement any deception. Because of small delays in the ECM system due to required detection and amplification of the intercepted signals, the leading edge of the least-delayed signal from the ECM system will arrive at the radar slightly after the leading edge of the true reflected signal.

The leading edge tracker will attempt to exploit this delay by allowing the radar to gate through only the leading edge of signals entering the radar antenna. Thus, the ECM signal is denied the capability of influencing the AGC, range-tracking, or angle-tracking circuits. The split gates are just large enough to encompass the leading edge of the pulse, and the tracking loops operate on only that part of the pulse. Performance can be compromised due to the much smaller energy level detected by using only a portion of the pulse. The presence of a jammer pulse much stronger than the target signal may also compromise the success of a leading edge tracker. The strength of the jammer pulse will dominate the AGC, forcing the gain to become so low that the smaller target signal may not be detectable.

TET operates in exactly the same manner as the LET but on the trailing edge of received signals. TET provides ECCM capability against a tail aspect target that deploys chaff or decoys [2, pp. 309–315].

18.5.12 Narrowband Doppler Noise Detector

The narrowband doppler noise detector (NBDND) is used by high-PRF radars with a large number of doppler filters to defeat high-PRF NBDN ECM elevating conventional CFAR thresholds and desensitizing the radar. First, a narrow sliding-window noise estimator is used to remove targets and narrow spikes from the input data. Second, a detection threshold is computed by averaging signals in the remaining doppler filters. Then the power spectrum is smoothed by averaging signals in a small number of adjacent filters, perhaps four, around each test cell. The smoothed spectrum is compared to the noise threshold. A string of consecutive detections determines the existence of NBDN ECM.

18.5.13 Narrow Pulse/Pulse Compression

High range resolution using narrow pulses (NP) is a powerful ECCM against a variety of repeater jammers. Repeater jammers that are not DRFM-based systems have an inherent delay that is typically on the order of 100 ns. In addition, when the jammer is functioning as a terrain bounce jammer, the indirect path via the ground may introduce over 100 ns of additional delay relative to the direct reflection path from the target. If the radar has sufficient range resolution such that the target return and jamming occur in separate range cells, then proper angle tracking of the target in the presence of the false target can be achieved. The required range resolution can be achieved by using a narrow transmit pulse, but a narrow pulse may not supply the average power required for detection due to limitations in peak transmitter power or PRF. Pulse compression is often used to achieve the desired resolution.

Pulse compression, as described in Chapter 9, is a technique that uses long coded pulses to increase the energy on target while retaining the increased range resolution of a short-pulse system. It can be viewed as an ECCM technique by its ability to increase burn-through range. Noise from a common jammer will not compress if the jamming source does not know the pulse compression reference code. An artifact of the compression process is the production of time side lobes. The time side lobes of the jamming signal may corrupt the target return. The often-used 13-bit Barker code has time side lobes that are only 23.3 dB below the compressed pulse peak. Additional filtering to reduce the time side lobes will improve the ECCM performance.

18.5.14 Neural Net

The use of a neural net (NN) as an ECCM technique is currently being explored. The concept is based on the hypothesis that if a radar expert has available sufficient instrumentation data such as a time history of range-doppler maps and monopulse angle measurements, then it is likely that he could reliably determine: (1) that the radar is being jammed, (2) the type of jamming being used, and (3) the best ECCM technique available to the radar to counter the technique. Current ECCM algorithm implementations are often limited by the amount of processing time available and the amount of data that can be stored to support the algorithms.

A neural net offers the possibility that it could be trained to recognize a wide variety of ECCM techniques. Many jamming techniques have distinctive signatures in the range-doppler map. A neural net is a highly parallel processor; that is, it simultaneously tests the data to recognize all the ECMs that it has been trained to recognize. Research is being performed on digital computer simulations that emulate a neural net. The final implementation is envisioned as a custom integrated circuit that operates in conjunction with the conventional radar signal processors.

18.5.15 Off-Boresight Tracking

Off-boresight tracking (OBT) is a technique employed to track a target in the presence of a second target that is separated in angle by less than an antenna beamwidth. The second target could be another aircraft, a towed decoy, or terrain bounce interference. The tracking loop maintains the antenna pointing so that the target is at a fixed angular offset instead of the usual difference pattern null, as shown in Figure 18.11. The pointing direction is in the opposite direction from the interference. The antenna gain, both sum and difference, is greatly reduced in the direction of the interference, thereby minimizing the corruption of tracking [5, pp. 262–263].

Early detection of the presence of two targets unresolved in angle is helpful in transitioning from conventional tracking to OBT. If the targets or repeater jammers can be resolved in either range or doppler, then normal tracking can be used. Angle extent estimation can be used to detect noise jammers that are unresolved in angle.

18.5.16 PRF Jitter

Pulse repetition frequency jitter (PRFJ) has long been used in radars to decorrelate second-time-around echoes and to control doppler blind zones. This technique is also effective against repeater jammers such as RGPO, VGPO, multiple false doppler targets (MFDT), or false range target (FRT) that use a time delay just slightly less than the PRI to generate false targets at closer ranges than the true target. The

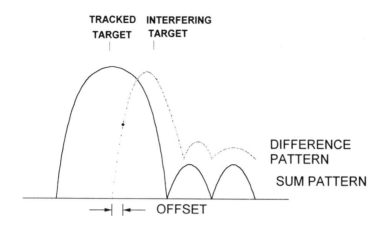

Figure 18.11 Off-boresight tracking.

radar PRF is automatically varied, either randomly or according to a predefined set of frequencies, between two or more frequencies.

False target generators assume that the measured PRF of the victim radar is constant from pulse to pulse. Thus, the ECM pulse is transmitted at the same PRF as the received pulse. ECM pulses that anticipate the true target return are by default ambiguous in range or doppler. Varying the PRF will cause the apparent range or doppler of these targets to vary, causing the A-scope jitter shown in Figure 18.12. Targets located behind the true target return may be created in an unambiguous manner such that their range or doppler remain constant regardless of a change in PRF. Variation in PRF is typically on the order of 5% during PRF jitter in order to minimize the spreading of ground clutter. PRFJ is normally used during low-PRF operation. The multiple-PRF range ambiguity resolution mechanization may accomplish similar benefits for the medium-PRF modes [2, pp. 306–308].

18.5.17 Side-Lobe Blanking

Side-lobe blanking (SLB) can be used as an ECCM technique that prevents pulsed ECM energy received through the radar antenna side lobes from adversely affecting radar performance. Both coherent and noncoherent radars may employ side-lobe blankers to defeat false target generators in either the doppler or range domain. An auxiliary antenna system is required Typically, an omnidirectional antenna with a gain slightly higher than the peak side lobe of the main channel antenna pattern is used. If the pulse signal in the auxiliary channel is greater in amplitude than the

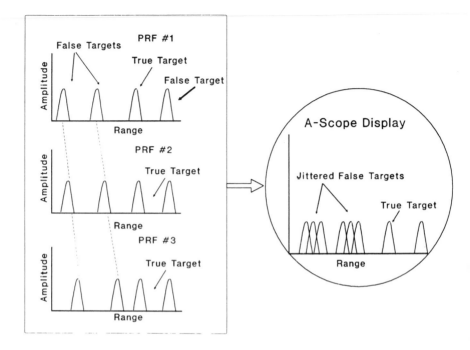

Figure 18.12 Apparent change in range for false ECM targets as a function of PRF.

signal in the main channel, the signal is identified as a side-lobe source, and the associated range-doppler cells are blanked and not processed further. Thus, false target generators operating in the side lobes of a victim antenna will not be effective. The auxiliary channel used to perform SLB is typically referred to as the *guard channel,* and the rejection criteria is referred to as the *main-to-guard ratio test* [2, pp. 544–546; 6, pp. 59–73].

18.5.18 Side-Lobe Canceler

The technique of adaptive side-lobe cancellation (SLC) is used to defeat self-screening or standoff noise jammers operating in the side lobes of a victim antenna. The technique uses auxiliary antenna elements to observe and cancel the jamming signal in the main antenna channel. The cancellation is performed by combining the weighted auxiliary signal and main channel signal, as shown in Figure 18.13. The auxiliary weights are chosen such that the power in the output signal is minimized. The combined signal from the auxiliaries is equal in amplitude and opposite in phase to the jamming component of the signal in the main channel. When

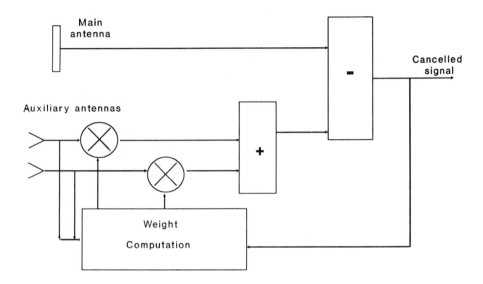

Figure 18.13 Hardware implementation of an adaptive side-lobe canceler.

combined with the main channel signal, the auxiliary signals cancel the main channel jamming components.

The success of the technique is primarily due to the observed path length difference between the auxiliary elements and the main channel of the received jamming waveform. The auxiliary channels are processed similarly to the main channel prior to cancellation and are discarded after cancellation. Figures 18.14 and 18.15 illustrate the processed range-doppler spectrum for a side-lobe canceler with two auxiliary antennas against a single side-lobe jammer before and after cancellation [6, pp. 95–31].

18.5.19 Sniff

Sniff is an ECCM technique for use on tunable radar systems. The radar's RF bandwidth of available transmit frequencies are monitored for jamming. If noise jamming is detected over this bandwidth, the radar receiver will search elsewhere in its tuning range. The transmitter is tuned to the bandwidth where jamming is a minimum. The receiver is tuned separately from the transmitter during a period

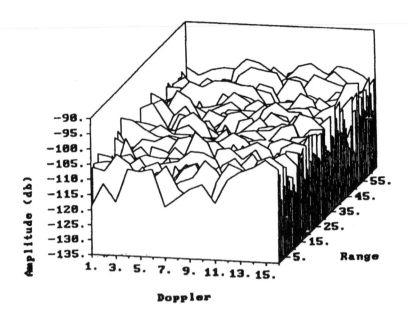

Figure 18.14 Medium-PRF range-doppler spectrum before cancellation.

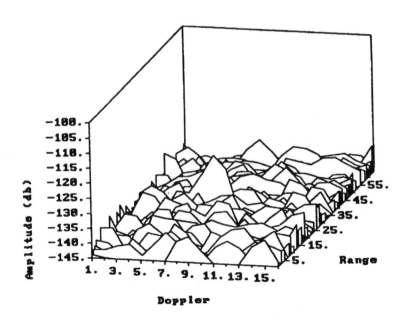

Figure 18.15 Medium-PRF range-doppler spectrum after cancellation.

when no signals are expected. This technique is similar to the ECCM technique of automatic radar tuning (ART); however, an intelligent selection of the radar frequency is determined prior to transmission. ART implies that the transmitter and receiver are always tuned together, whereas during sniff the transmitter will continue transmission while the receiver searches for a clear frequency [2, p. 565].

18.5.20 Space Time Adaptive Processing

Space-time adaptive processing (STAP) is a method for simultaneously canceling interference in angle and time (doppler). The objective is to maximize the S/I at positions that span the multidimensional, that is, angle-doppler, space. The interference sources may be a combination of noise sources, high-duty-cycle coherent sources, and main-lobe clutter.

There are basically two methods for implementing STAP: (1) factored and (2) combined. Factored STAP uses conventional adaptive array techniques such as fully adaptive arrays or side-lobe canceling followed by an adaptive doppler processor. The two adaptive functions are done separately. Combined STAP implies that angle and doppler adaptation are done jointly. Factored STAP is much less complicated than combined STAP but lacks the performance capability of combined STAP when multiple sources of interference are present having different temporal, spatial, and frequency characteristics. The major benefit of combined STAP in this combined interference is sometimes referred to as *supervisibility*. Supervisibility is achieved because all clutter and interference are resolved in angle and doppler, thus allowing the detection of targets that are resolved in *either* angle or doppler from the interference.

Main-lobe clutter in an airborne radar is one example of interference that can be resolved in angle. The clutter is spread in doppler frequency due to the platform motion by an amount proportional to the antenna beamwidth. The entire doppler region occupied by clutter is lost in conventional doppler processing. Combined STAP effectively resolves the clutter within the beamwidth into angle-doppler cells. Targets can be detected in cells that are not occupied by clutter. Thus, the system will exhibit a better low-velocity target detection capability.

The schematic of a combined STAP system is shown in Figure 18.16. The radar transmits a coherent pulse train to allow doppler filtering of the received signals. Multiple receiving antenna outputs are obtained from either individual array elements, overlapping beams, or subarrays. Several consecutive PRIs of received signals from the multiple antenna ports are processed. The antenna outputs are used for computing the required adaptive weights. These weights are then applied to a new set of inputs, or better, applied to the same data set from which the weights were derived. Application of the weights to the same data avoids losses due to a changing environment, such as antenna rotation or intermittent interference. A set of STAP outputs, covering the target doppler band, is obtained for each

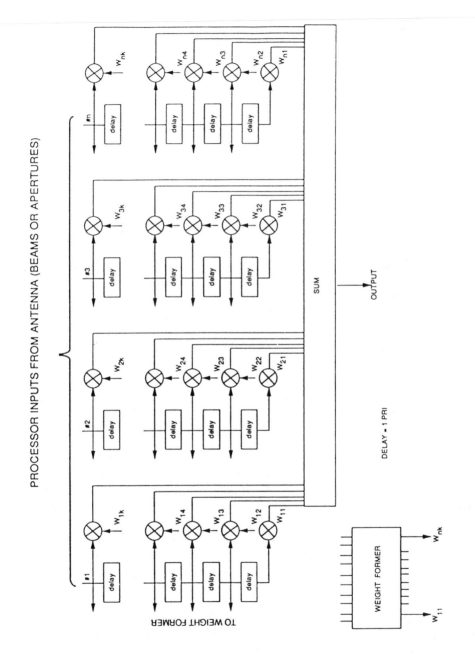

Figure 18.16 Combined space-time adaptive processor.

range cell by summing the weighted antenna outputs over multiple PRIs. The weights calculated by the adaptive processor suppress clutter and jamming.

The general method for determining the adaptive weights is as follows. The outputs from the N antennas for a set of K consecutive pulses are stored for each range gate. The following computations are performed for each range gate. An ordered vector $\mathbf{X}$ of the KN outputs is formed. Next, a sample covariance matrix M is formed by averaging the outer products of the range cell vectors

$$M = \overline{\mathbf{X}^*\mathbf{X}} \tag{18.9}$$

where * denotes the conjugate transpose of the vector, and the overbar denotes the averaging process. Since M must be an average, adjacent range cells or additional time samples must be used to form the M matrix, which has a dimensionality of $KN \cdot KN$. Sets of KN optimum weights, W_d, are computed using $W_d = M^{-1}S_d$, where S_d are the signal vectors. A separate signal vector and therefore a separate set of weights are computed for each doppler cell within the range cell being processed. The output for the range cell is thus a series of doppler filter outputs computed by applying each individual weight vector to the stored antenna output using $W_d^*\mathbf{X}$. The process is repeated for each range cell. The result of the total process is a range-doppler map that can be processed by a CFAR algorithm for target detection.

The size of the required digital processor is an important consideration. The size is proportional to the complex multiplication rate. For combined STAP, the rate, R_c, is proportional to the product of the square of the number of space-time adaptive degrees of freedom (NK) and the number of range gates to be processed, N_{rg}. Consequently, there is a strong motivation to reduce, if possible, the number of space-time adaptive degrees of freedom. In addition to the obvious method of reducing the number of antenna channels, research is being conducted concerning the trade-offs of reducing the temporal degrees of freedom. Finn et al. report [7] that the simplified signal-processing architecture of Figure 18.17 using two or three temporal degrees of freedom (one or two delay elements) may perform nearly as well for some applications as full temporal adaptation.

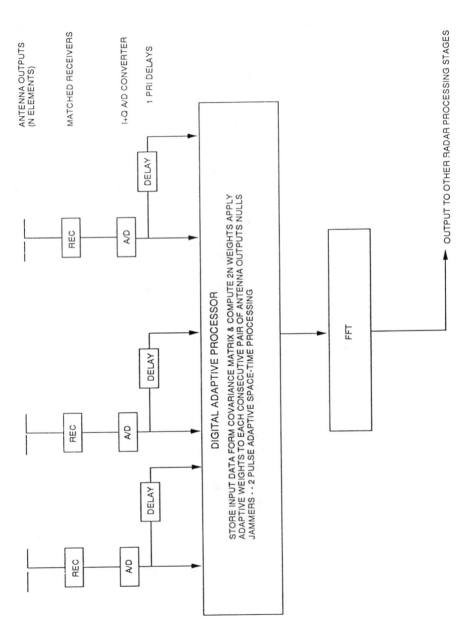

Figure 18.17 Simplified adaptive processor—STAP followed by an FFT.

18.5.21 Superresolution

Superresolution (SRES) techniques refer to a class of algorithms developed to measure the true angle of arrival of each jammer signal in a multiple-jammer engagement. These methods have been proposed to alleviate the inherent limitations of conventional beamforming techniques which use the FFT to estimate the spatial distribution of radiating sources. The most prominent limitations of the conventional FFT spectral estimation technique are that of spatial resolution and spectral "leakage" due to windowing. Superresolution techniques provide improved resolution at high SNRs versus FFT-based spectral estimation, which performs better at low SNRs. Superresolution methods typically require significantly higher computational power than conventional spectral estimation.

The maximum likelihood spectral estimate (MLSE) represents an example of a superresolution technique. The MLSE is defined as a filter designed to pass the power in a narrow band about the signal frequency of interest and to minimize or reject all other frequency components in an optimal manner. The MLSE is considered a member of a family of inverse filter techniques. These methods may be viewed as "extending" the antenna array by predicting the values at imaginary antenna elements from the N available measured values. It is this concept that provides an increase in resolution due to the "extension" of the array aperture. Figure 18.18 shows a maximum likelihood spectral estimate superimposed on a conventional beamforming spectrum. Two radiating sources of equal power were placed at +1 degree and +3 degrees with respect to the antenna normal. The two sources could not be resolved with the conventional beamformer, while the superresolution technique using MLSE resolved both sources.

The limitations of superresolution techniques include: high computational power requiring long computation time, low resolution at low SNRs, and performance degradation against coherent sources. Alternate superresolution techniques include the multiple signal classification (MUSIC) method and the maximum entropy spectral analysis (MESA) method. Figure 18.19 illustrates a typical implementation of superresolution technology [8, pp. 324–338; 9, pp. 431–433].

18.5.22 Velocity Guard Gates

Velocity guard gates (VGG) are used by automatic velocity tracking doppler radars to defeat the technique of VGPO. VGG uses two additional guard gates, a low-frequency and a high-frequency gate, to monitor the action of a velocity gate stealer.

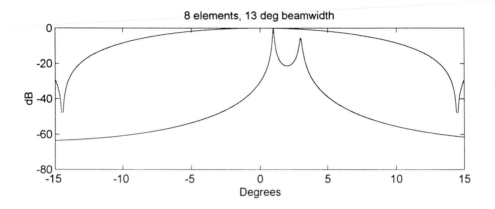

Figure 18.18 A comparison of MLSE and conventional beamforming.

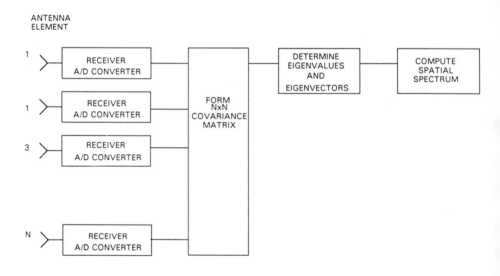

Figure 18.19 Block diagram of superresolution implementation.

If either gate senses the presence of a large signal, a feedback signal moves the tracking gate back in the direction of the true signal. The guard gate (which senses the VGPO) is blanked, allowing the false target signal to move away from the target skin return. Multiple closely spaced targets may adversely affect this ECCM technique by causing the tracker to blank true targets [2, pp. 252–254].

18.5.23 VGPO Reset

VGPO reset (VGPOR) is an ECCM technique that permits the velocity gate of a velocity tracking radar to be quickly reset to the true target doppler frequency during VGPO. This technique depends upon early detection of the presence of a velocity gate stealer using a technique such as the beat frequency detector.

Upon suspicion of the presence of VGPO, the radar receiver puts the control voltage into memory for future use. After the two signals are sufficiently far apart for the radar to determine that it is definitely under the influence of a VGS, the radar recalls the control voltage from memory and moves the tracking gate to the corresponding frequency. The radar begins a search mode in the reverse direction of the walk-off to reacquire the true target signal. The reset and reacquisition minimizes the amount of time the victim radar is without a tracking signal [2, pp. 464–467].

18.5.24 VGS ECCM—Dual Frequency

The dual-frequency velocity gate stealer (VGS) (VGSDF) ECCM technique uses two simultaneous transmit frequencies. (VGS is an alternate often used name for VGPO.) False velocity targets may be distinguished from the true target return by comparing the ratios of the two transmitted frequencies (18.9) with the ratio of the received signals (18.10). The success of this technique is based upon the assumption that the jammer will apply the same fixed doppler to each of the two received frequencies rather than dopplers that are proportional to each of the transmitter frequencies.

$$f_{d1}/f_{d2} = f_1/f_2 \text{ (no jamming)} \qquad (18.10)$$

$$(f_{d1} + f_J)/(f_{d2} + f_J) \text{ (with jamming)} \qquad (18.11)$$

where:

f_1 = transmit frequency 1;
f_2 = transmit frequency 2;
f_{d1} = two-way doppler frequency at $f_1 = 2 f_1 V/c$;
f_{d2} = two-way doppler frequency at $f_2 = 2 f_2 V/c$;
f_J = false target doppler offset;
c = speed of light;
V = radial speed of target.

As soon as the received signal ratio deviates from the transmitted frequency ratio, a VGPO is declared, and the velocity gate is put into reacquisition mode [2, pp. 591–595].

References

[1] Morris, G. V., and T. A. Kastle, "Trends in Electronic Counter-countermeasures," *J. Electronic Defense*, Vol. 15, No. 12, December 1992, pp. 49–54.

[2] Van Brunt, L., ed., *Applied ECM Volume 2*, EW Engineering, Dunn Loring, VA, 1989.

[3] Sherman, S. M., *Monopulse Principles and Techniques*, Norwood, MA: Artech House, 1984.

[4] Van Brunt, L., ed., *Applied ECM Volume 1*, EW Engineering, Dunn Loring, VA, 1989.

[5] Barton, D., C. Cook, and P. Hamilton, *Radar Evaluation Handbook*, Norwood, MA: Artech House, 1991.

[6] Farina, A., *Antenna-Based Signal Processing Techniques*, Norwood, MA: Artech House, 1992.

[7] Finn, H. M., J. D. Mallett, and L. E. Brennan, "Contributing Sections to the ECCM Design Reference for the Multimode Radar System," Adaptive Sensors, Inc., report under Contract No. DAAL01-90-D-0037, September 1991.

[8] Haykin, S., *Modern Filters*, New York: Macmillan, 1989.

[9] Kay, S. M., *Modern Spectral Estimation: Theory and Applications*, Englewood Cliffs, NJ: Prentice Hall, 1987.

Suggested Reading

Chrzanowski, E., *Active Radar Electronic Countermeasures*, Norwood, MA: Artech House, 1990.

Lothes, R., M. Szymanski, and R. Wiley, *Radar Vulnerability to Jamming*, Norwood, MA: Artech House, 1990.

APPENDIX A

Clutter Map Computer Program

The listing of the FORTRAN computer program that was used to generate the clutter maps of Chapter 2 follows. I refer to it as a demonstration program because some of the dimensions, such as the number of range gates, were limited to values that would run conveniently on a personal computer. The program is straightforward, and it will be readily readable by those familiar with FORTRAN. Due to the relatively small scale of the program, no attention was devoted to clever schemes to increase the speed. The average time it took to generate each of the illustrations in this book was approximately five seconds on my personal computer.

The program could be extended to a "professional" version by increasing some dimensions and adding the altitude line. Some speed improvements, such as an adaptable integration step size in azimuth instead of the fixed step of 0.002 radians, might be desirable, depending on the size of the problems to be run and the computer used.

The range-Doppler map produced by an actual airborne radar as the result of a single coherent processing interval will contain the statistical fluctuations discussed in Chapter 15. The clutter map produced by this program is idealized and may be thought of as the average on an infinite number of coherent processing intervals. Other assumptions and program characteristics are as follows.

1. The relative power axis is normalized to represent watts of received clutter power per watt of transmitted power, not including transmit and receive losses.

2. The antenna pattern is a pencil beam whose gain is calculated from the user input of beamwidth. The gain function is contained is a separate porgram module that the user may modify to represent other antenna types.

3. The backscatter coefficient entered by the user is converted from decibels to linear units and multiplied by the sine of the elevation (depression) angle. This is often referred to as the constant gamma approximation.

The program asks whether the user wishes to run interactively or from an input data file. If the user chooses to run interactively, the program will prompt the user to input various parameters. The input parameters are stored in a file named INPUT.DAT. The clutter map data are stored in a file named CLUT.DAT. for plotting by another program, "Data Analysis Plotting Software," authored by J. M. Baden of the Georgia Tech Research Institute. The plot program is executed by typing "plot clut clut" at the DOS prompt. The first "clut" refers to a header file that labels the axes; the second "clut" is the data file to be plotted. A subset of the program sufficient to produce the clutter maps may be obtained free of charge, courtesy of Mr. Baden, via anonymous ftp "rsd1000.gtri.gatech.edu" by downloading the contents of directory "pub/gmorris/graphics".

The data can also be plotted easily using other mathematical analysis and plotting software. I used Matlab™ to generate the plots shown in Chapter 2. The file, "clutplot.m", is included.

If the program is run using an input file, the file should be named DAT_IN.DAT. The file is simply a text file that contains the answers to the same questions that are asked in the interactive mode. Examples of files DAT_IN.DAT and CLUTIN.DAT are included. Annotations are included in parentheses that do not (and must not) appear in the actual files. All the files listed in this appendix are available for download via anonymous ftp from "rsd1000.gtri.gatech.edu", directory "pub/gmorris".

The program prevents overwriting previously generated INPUT.DAT or CLUT.DAT files existing within the directory as a result of provious runs. Therefore, it is necessary to delete or rename these files before running the program.

```
(CLUTER_2.FOR)
PROGRAM CLUTER_2
REAL INTI,INTJ,INTK,LAMBDA
CHARACTER LABEL1*40,LABEL2*40
COMMON/IN/ACALT,ACVEL,ANTAZ,ANTEL,PRFK,NFILT,ANTBW,FREQ
&          ,RMAX,NRGMX,SIG0DB,LABEL1,LABEL2,DAT_IN
COMMON/PRE/PI,DEGTORD,BWR,ANTI,ANTJ,ANTK,C,PRF,RRES
&          ,RUNAMB
COMMON/OUT/RNGDOP(64,64)
C
        CALL INPUT
        CALL PRELIM
        LAMBDA=C/(FREQ*1E9)
        RK=LAMBDA**2/(4*PI)**3
        DOPSF=2*ACVEL/LAMBDA
        GAMMA=(10.**(SIG0DB/10.))*RMAX/ACALT
```

```
      FILTBW=PRF/NFILT
      GMAX=4*PI/(BWR*BWR)
      DO 37 II=1,64
          DO 38 JJ=1,64
              RNGDOP(II,JJ)=1E-20
   38         CONTINUE
   37     CONTINUE
C
C***** BEGIN RANGE LOOP
C
      NRNG= Int(ACALT/RRES + 0.5)
      RANGE=Float(NRNG)*RRES
   39 CONTINUE
      IF (NRNG>NRGMX) THEN
          NRNG=NRNG - NRGMX
          GO TO 39
      END IF
C-
C-*****ADD ALTITUDE LINE
C
      SIG0=GAMMA*ACALT/RANGE
      SIGCLU=PI*(2.0*ACALT+RRES)*RRES*SIG0
      DEL=ACOS(ANTK)
      GANT=GAIN(DEL,BWR)*GMAX
      CLUT=RK*(GANT**2)*SIGCLU/ACALT**4
      RNGDOP(NRNG, 1)=RNGDOP(NRNG, 1) + CLUT
      NRNG=NRNG + 1
      RANGE=RANGE + RRES
C
C*****RANGE LOOP NORMAL ITERATION
C
   10 CONTINUE
      EL=-ASIN(ACALT/RANGE)
      SIG0=GAMMA*ACALT/RANGE
      CEL=COS(EL)
      INTK=-SIN(EL)
      DOPLER=-FILTBW/2.0
      NDOP=1
       DOPMAX=DOPSF*CEL
C
C***** BEGIN AZIMUTH LOOP
C
      WRITE(*,1052)RANGE,NRNG
 1052 FORMAT(' RANGE= ',F10.0,5X,'RNG GATE= ',I3)
      AZ=ACOS(-FILTBW/(2.0*DOPMAX))
C
```

```
C
   20   CONTINUE
         DOPLER= DOPLER + FILTBW
         cos_arg=DOPLER/(DOPMAX)
         if( cos_arg > 1.0) then
             cos_arg=1.0
         end if
         DAZ=AZ - ACOS(cos_arg)
         SIGCLU=RRES*RANGE*DAZ*SIG0/CEL
         INTI=CEL*COS(AZ-DAZ/2.0)
         INTJ=CEL*SIN(AZ-DAZ/2.0)
         DEL=ACOS(INTI*ANTI+INTJ*ANTJ+INTK*ANTK)
         GANT=GAIN(DEL,BWR)*GMAX
         CLUT=RK*GANT**2*SIGCLU/RANGE**4
         RNGDOP(NRNG,NDOP)=RNGDOP(NRNG,NDOP)+CLUT
         NDOP=NDOP+1
         IF (NDOP > NFILT) NDOP=1
         AZ=AZ-DAZ
         IF(DOPLER.LE.DOPMAX)GO TO 20
C
C*****END AZ LOOP
C
   12   RANGE=RANGE+RRES
         NRNG=NRNG+1
C
C*****SKIP MAIN BANG
C
         IF (NRNG > NRGMX) THEN
           RANGE=RANGE+RRES
           NRNG=1
           END IF
         IF(RANGE.GE.RMAX) GO TO 30
C
C*****CONTINUE TO NEXT RANGE CELL
         GO TO 10
C
C
   30 CONTINUE
C
C*****END OF RANGE LOOP
C
C
C*****OUTPUT TO PLOT FILE
C
         OPEN(UNIT=28,FILE='CLUT.DAT',STATUS='NEW')
C*****FLAG TO PLOT PROGRAM
```

```
      WRITE(28,1000) -1
C
      WRITE(28,1060) LABEL1
      WRITE(28,1060) LABEL2
 1060 FORMAT(A40)
      WRITE(28,1000) NRGMX
      WRITE(28,1000) NFILT
 1000 FORMAT(I3)
      DO 50 I=NRGMX,1,-1
      DO 51 J=1,NFILT
      WRITE(28,1001) 10.*LOG10(RNGDOP(I,J))
 51   CONTINUE
 50   CONTINUE
 1001 FORMAT(1P,E12.4)
      CLOSE(28)
      END
C
C
C
      SUBROUTINE INPUT
      CHARACTER LABEL1*40,LABEL2*40,DAT_IN*12
      COMMON/IN/ACALT,ACVEL,ANTAZ,ANTEL,PRFK,NFILT,ANTBW,FREQ
     &          ,RMAX,NRGMX,SIG0DB,LABEL1,LABEL2,DAT_IN
      COMMON/PRE/PI,DEGTORD,BWR,ANTI,ANTJ,ANTK,C,PRF,RRES
     &          ,RUNAMB
      COMMON/OUT/RNGDOP(64,64)
      WRITE(*,1000)
 1000 FORMAT(///,   '        GEORGIA INSTITUTE OF TECHNOLOGY',
     &/,/         ,  '        PRINCIPLES OF PULSE DOPPLER RADAR',
     & ///,         '       MEDIUM PRF CLUTTER MAP DEMONSTRATION',
     &/,/      ,  '             VER 2.0, JAN 92',
     &/,         '             GUY V. MORRIS',
     &//,' YOU CAN RUN THE PROGRAM USING AN INPUT FILE',
     &// ,' YOU CAN RUN THE PROGRAM INTERACTIVELY. IF YOU DO',
     &/,' YOU WILL BE ASKED TO INPUT SEVERAL PARAMETERS.',
     &/,' IN BOTH CASES, THE INPUTS WILL BE SENT TO FILE
INPUT.DAT.',
     &/,  ' THE OUTPUTS WILL BE SENT TO A FILE CLUT.DAT',
     &/  ,  '        FOR PLOTTING BY A SEPARATE PROGRAM.',
     & ///,' PRESS ''1'' (RETURN) TO RUN FROM A FILE ',
     $'             ''2'' INTERACTIVELY.')
      READ(*,1041)IDUMMY
 1041 FORMAT(I1)
      IF(IDUMMY==2) THEN
      WRITE(*,1003)
 1003 FORMAT(' AIRCRAFT ALTITUDE, KFT ')
```

```
      READ (*,*)ACALTK
      WRITE(*,1004)
      READ(*,*) PRFK
      WRITE(*,1005)
      READ(*,*) ACVEL
      WRITE(*,1006)
      READ(*,*) ANTAZ
      WRITE(*,1007)
      READ(*,*) ANTEL
      WRITE(*,1008)
      READ(*,*) ANTBW
      WRITE(*,1009)
      READ(*,1002) NRGMX
      WRITE(*,1010)
      READ(*,1002) NFILT
      WRITE(*,1011)
      READ(*,*) RMAXNM
      WRITE(*,1012)
      READ(*,*) FREQ
      WRITE(*,1014)
      READ(*,*) SIG0DB
      WRITE(*,1015)
      READ(*,1016) LABEL1
      READ(*,1016) LABEL2
 1001 FORMAT(F10.3)
 1002 FORMAT(I3)
 1004 FORMAT('  PULSE REPETITION FREQUENCY, KHZ ')
 1005 FORMAT('  AIRCRAFT VELOCITY, FT/S ')
 1006 FORMAT('  ANTENNA AZIMUTH, DEG ')
 1007 FORMAT('  ANTENNA ELEVATION, DEG (DOWN IS NEGATIVE) ')
 1008 FORMAT('  ANTENNA BEAMWIDTH, DEG ')
 1009 FORMAT('  NUMBER OF RANGE GATES (64 OR LESS) ')
 1010 FORMAT('  NUMBER OF DOPPLER FILTERS (64 OR LESS) ')
 1011 FORMAT('  MAXIMUM RANGE OF INTEREST, NMI ')
 1012 FORMAT('  TRANSMITTER FREQUENCY, GHZ ')
 1014 FORMAT('  BACKSCATTER COEFFICIENT, DB ')
 1015 FORMAT('  ENTER TWO LINES (UP TO 40 CHARACTERS) TO BE',
     &/,'    ADDED TO THE PLOTS AND FILES',/)
 1016 FORMAT(A40)
      ELSE
         WRITE(*,1031)
 1031    FORMAT('  FILE NAME?')
         READ(*,1032) DAT_IN
 1032    FORMAT(A12)
         OPEN(UNIT=10,FILE='DAT_IN.DAT',STATUS='OLD')
         READ(10,1001) ACALTK
```

```
        READ(10,1001) PRFK
        READ(10,1001) ACVEL
        READ(10,1001) ANTAZ
        READ(10,1001) ANTEL
        READ(10,1001) ANTBW
        READ(10,1002) NRGMX
        READ(10,1002) NFILT
        READ(10,1001) RMAXNM
        READ(10,1001) FREQ
        READ(10,1001) SIG0DB
        READ(10,1016) LABEL1
        READ(10,1016) LABEL2
        CLOSE(10)
END IF
ACALT=ACALTK*1000.
RMAX=RMAXNM*6076.1
WRITE(*,1003)
WRITE(*,1001)ACALT
WRITE(*,1004)
WRITE(*,1001)PRFK
WRITE(*,1005)
WRITE(*,1001)ACVEL
WRITE(*,1006)
WRITE(*,1001)ANTAZ
WRITE(*,1007)
WRITE(*,1001)ANTEL
WRITE(*,1008)
WRITE(*,1001)ANTBW
WRITE(*,1009)
WRITE(*,1002)NRGMX
WRITE(*,1010)
WRITE(*,1002)NFILT
WRITE(*,1011)
WRITE(*,1001)RMAXNM
WRITE(*,1012)
WRITE(*,1001)FREQ
WRITE(*,1014)
WRITE(*,1001) SIG0DB
OPEN(UNIT=9,FILE='INPUT.DAT',STATUS='NEW')
WRITE(9,1016) LABEL1
WRITE(9,1016) LABEL2
WRITE(9,1003)
WRITE(9,1001)ACALT
WRITE(9,1004)
WRITE(9,1001)PRFK
WRITE(9,1005)
```

```
      WRITE(9',1001)ACVEL
      WRITE(9,1006)
      WRITE(9,1001)ANTAZ
      WRITE(9,1007)
      WRITE(9,1001)ANTEL
      WRITE(9,1008)
      WRITE(9,1001)ANTBW
      WRITE(9,1009)
      WRITE(9,1002)NRGMX
      WRITE(9,1010)
      WRITE(9,1002)NFILT
      WRITE(9,1011)
      WRITE(9,1001)RMAXNM
      WRITE(9,1012)
      WRITE(9,1001)FREQ
      WRITE(9,1014)
      WRITE(9,1001) SIG0DB
      CLOSE(9)
      END
C
C
C     SUBROUTINE PRELIM
      CHARACTER LABEL1*40,LABEL2*40
      COMMON/IN/ACALT,ACVEL,ANTAZ,ANTEL,PRFK,NFILT,ANTBW,FREQ
     &          ,RMAX,NRGMX,SIG0DB,LABEL1,LABEL2
      COMMON/PRE/PI,DEGTORD,BWR,ANTI,ANTJ,ANTK,C,PRF,RRES
     &          ,RUNAMB
      COMMON/OUT/RNGDOP(64,64)
      DO 100 I=1,64
      DO 101 J=1,64
      RNGDOP(I,J)=1E-20
  101 CONTINUE
  100 CONTINUE
      C=1E9
      PI=4.*ATAN(1.)
      DEGTORD=PI/180.
      BWR=ANTBW*DEGTORD
      ANTI=COS(ANTEL*DEGTORD)*COS(ANTAZ*DEGTORD)
      ANTJ=COS(ANTEL*DEGTORD)*SIN(ANTAZ*DEGTORD)
      ANTK=-SIN(ANTEL*DEGTORD)
      PRF=PRFK*1000.
      RUNAMB=C/(2.*PRF)
      RRES=RUNAMB/FLOAT(NRGMX + 1)
      END
C
C
```

```
C
        FUNCTION GAIN(DEL,BWR)
        PI=3.1416
        Z=2*DEL/BWR
        IF(Z.LE.1) THEN
            GAIN=COS(Z*PI/3)
        ELSE
            GAIN=.01/Z
        END IF
        END
```

(FILE DAT_IN.DAT)

10.0	(altitude, kft)
10.0	(PRF, kHz)
150.0	(velocity, f/s)
30.0	(antenna azimuth, deg, positive=right)
−15.0	(antenna elevation, deg, negative=down)
2.5	(antenna beamwidth, deg)
64	(number of range gates)
32	(number of doppler filters)
30.0	(maximum range of clutter integration, nmi)
10.0	(frequency, GHz)
−20.0	(sigma zero, dB)
slow illustration	(first line of text, 40 characters max)
text line 2	(second line of text, 40 characters max)

(CLUT.DAT)

−1	(flag to plot program
slow illustration	(line 1 of text)
	(line 2 of text)
64	(number of range gates)
32	(number of doppler filters)
−1.9520E+02	(doppler filter 1 of range gate 64)
−1.9494E+02	(doppler filter 2 of range gate 64)
.	
.	
.	
−2.0000E+02	(doppler filter 32 of range gate 1)

(MATLAB FILE clutplot.m)

```
% clutplot
% clutplot makes a meshplot from files produced by cluter_2.exe
% The data produced by cluter_2 are stored in an ascii text file
% clut.dat with the following file structure:
%    line 1 = -1 flag to plot program
%    lines 2-3 = character string text for plot labels
%    line 4 = number of range gates
%    line 5 = number of doppler filters
```

```
%     lines 6 and up = filter amplitudes, dB
% To use:Use text editor to delete lines 1-3;
%     Load file into matlab;
%     run clutplot
n_range=clut(1);
n_dopp= clut(2);
for ii=1:n_range
    clutdat(:,n_range+1-ii)=clut(n_dopp*(ii-1)+3:n_dopp*(ii-
1)+n_dopp+2);
end
mesh(clutdat)
view(142.5,30)
text(40,40,-200,'RANGE GATE')
text(100,15,-200,'DOPPLER FILTER')
zz = axis;
axis([0 n_range 0 n_dopp zz(5) zz(6)]);
zlabel('dB')
text(110,20,-210,'Figure 2.13 Range-doppler clutter map using a
narrow beamwidth, slow platform.')
```

Range-Doppler Blind Zone Map
Computer Program

File "mprfcomm" that contains the operating instructions for the computer program used to generate the blind zone maps in this book follows. Files "mprfcomm.txt", the executable program "mprf.exe", and a sample data set "set1.-dat" are available free of charge via anonymous ftp from "rsd1000.gtri.gatech.edu" in directory "pub/gmorris".

The program computes the percentage of the range-doppler space that is clear, that is, is neither a range or a doppler blind zone, for two conditions. The "off axis" calculation excludes the regions around true zero doppler and zero range. These are true blind regions and cannot be mitigated by selection of various PRFs.

The program is intended primarily for validation of a set of medium PRFs chosen using the M:N method of Chapter 12. However, it can also be used to validate medium-PRF sets selected using the major/minor method as described in Chapter 12.

(FILE MPRFCOMM)

by Stan West

Syntax: MPRF [datafile.ext]

If a filename is specified on the command line, the required data will be read from the file in the order listed below. Otherwise, the program will prompt the user for the data in an interactive manner.

The order of the data in the data file is as follows:

Required data:

NumPrf		integer
PRF [1]	(KHz)	real
PulseWid [1]	(microsec)	real
ClutNotchWid [1]	(KHz)	real
PRF [2]	(KHz)	real
PulseWid [2]	(microsec)	real
ClutNotchWid [2]	(KHz)	real
...	...	...
PRF [NumPrf]	(KHz)	real
PulseWid [NumPRF]	(microsec)	real
ClutNotchWid [NumPRF]	(KHz)	real
LoRange	(NM)	real
HiRange	(NM)	real
LoDopp	(KHz)	real
HiDopp	(KHz)	real
NumBlank		integer

Optional data:

HitsReq		integer

(HitsReq is required data if the print device, below, is to be specified.)

Device	(DOS dvc)	string

If HitsReq is omitted from the file, the program will display the map then wait for user input. If HitsReq is included, the hit filter will be activated for the specified number of required hits. If Device is omitted from the file, the program will enter interactive mode after activating the hit filter. If it is included, however, the program will initiate printing to the specified device and, upon completion of printing, terminate execution.

The keys ''F'', ''P'', ''Q'', the up arrow, and the down arrow are usable while viewing the map to perform various actions, described below.

''F'' activates a three-color filter based on the number of hits required for detection (HitsReq, above). All regions receiving fewer than HitsReq hits are displayed in black, those receiving exactly HitsReq hits are displayed in the same color used in the multi-colored map, and those receiving greater than HitsReq hits are displayed in the color used for HitsReq + 1 hits in the multi-colored map. Also, the percentage in the right margin indicates how much area receives HitsReq or greater hits. Pressing ''F'' again causes the filter to use only two colors. Pressing ''F'' a third time deactivates the filter.

The arrow keys are active only while the filter is on. The up arrow increments HitsReq by one, while the down arrow decrements it by one. When either arrow is pressed, the colors are remapped and the percentage indicator is updated. HitsReq loops around each end of the scale; i.e., if HitsReq would become less than one, it is set equal to the number of PRFs, and if it would become greater than the number of PRFs, it is set to one.

''P'' prints the current map to an Hewlett Packard LaserJet series II printer. The program asks whether to print directly to the printer (using the DOS device ''PRN'') or to a file. Printing can also be cancelled at this point, returning the user to the interactive map display. If printing to a file is chosen, the program asks for a filename. Note that DOS device names such as ''LPT2'' can be used to force printing to a particular port. Printing then begins, and upon completion the user is returned to the map display.

''Q'' terminates execution of the program.

Sea Clutter Model

Sea State	Polarization	Frequency Range, GHz								
		0.44 to 0.87	0.87 to 1.73	1.73 to 3.4	3.4 to 6.75	6.75 to 13.4	13.4 to 26.4	26.4 to 52	Δ	I
0.1° Grazing Angle										
0	V	94	87	80	75	67	61	55	−6.5	100.5
	H	103	96	89	84	76	70	64	−6.5	109.5
1	V	89	82	75	70	62	56	50	−6.5	95.5
	H	96	89	82	77	69	63	57	−6.5	100.5
2	V	92	83	74	65	56	47	37	−9.1	101.1
	H	96	87	78	69	59	50	51	−9.1	104.9
3	V	86	76	67	57	47	38	28	−9.6	95.6
	H	90	80	71	61	51	42	32	−9.6	99.4
4	V	71	65	59	53	47	41	35	−6.0	77
	H	72	66	60	54	48	42	36	−6.0	78
5	V	71	63	56	48	41	34	26	−7.4	78.4
	H	71	63	56	48	41	34	26	−7.4	77.9
0.3° Grazing Angle										
0	V	84	78	73	67	61	55	49	−5.9	90.3
	H	94	89	83	78	72	67	61	−5.5	99.7
1	V	76	71	66	61	56	51	46	−5	81

	H	89	83	76	70	63	57	50	−6.5	95.4
2	V	77	70	64	57	50	44	37	−6.6	83.4
	H	84	76	69	61	53	46	38	−7.6	91.4
3	V	70	64	57	50	44	37	30	−6.7	77.1
	H	76	69	61	53	45	38	30	−7.7	84
4	V	62	57	53	48	43	38	33	−4.8	67
	H	62	57	52	47	42	37	31	−5	66.7
5	V	58	53	48	43	38	34	29	−4.8	62.3
	H	62	56	49	42	36	29	23	−6.6	68.7
1° Grazing Angle										
0	V	75	69	64	59	53	48	43	−5.3	80.1
	H	25	80	75	70	65	60	55	−5	90
1	V	68	64	59	55	50	45	41	−7.2	75
	H	81	74	66	59	52	45	38	−7.2	88
2	V	62	58	53	49	45	41	37	−4.1	66
	H	74	67	60	54	47	40	33	−6.75	80.6
3	V	57	53	49	45	41	37	32	−4.1	61.14
	H	67	61	55	49	42	36	30	−6.2	73.4
4	V	51	47	44	40	37	34	30	−3.4	54.3
	H	53	49	45	41	36	32	28	−4.2	57.4
5	V	43	41	39	36	34	32	30	−2.3	4.6
	H	49	52	46	39	33	26	20	−6.5	65.5
3° Grazing Angle										
0	V	64	61	59	57	54	52	50	−2.3	66
	H	76	72	67	63	59	54	50	−4.3	80.1
1	V	58	55	52	49	46	43	40	−3	61
	H	68	63	59	54	50	45	41	−4.5	72.4
2	V	55	52	49	46	42	39	36	−3.2	58.4
	H	63	59	54	50	45	41	36	−4.5	67.7
3	V	45	43	41	40	38	36	35	−1.64	48.8
	H	58	54	50	45	41	37	33	−4.1	61.7
4	V	39	38	37	36	34	33	32	−1.2	40.4
	H	51	47	44	41	37	34	31	−3.4	54

5	V	38	37	35	34	32	31	29	−1.54	39.9
	H	50	45	41	36	32	27	23	−4.6	54.7

10° Grazing Angle

0	V	48	47	46	46	45	45	44	−0.61	48.1
	H	62	59	57	54	52	49	47	−2.5	64.4
1	V	46	45	44	43	42	41	39	−1.1	47
	H	58	56	53	50	48	45	42	−2.7	61.1
2	V	38	38	37	36	36	35	34	−71	39.1
	H	55	52	49	46	44	41	38	−2.75	57.4
3	V	34	34	33	33	32	32	32	−0.43	34.6
	H	51	48	44	41	37	34	30	−3.4	54.4
4	V	31	31	30	30	30	30	30	−0.14	30.9
	H	47	44	41	37	34	31	28	−3.2	50.3
5	V	27	27	27	27	27	26	26	−0.07	26.9
	H	44	41	38	35	31	28	25	−3.3	47.7

30° Grazing Angle

0	V	43	43	42	42	41	41	40	−0.46	43.6
	H	47	48	48	49	50	50	51	+0.6	47.1
1	V	39	38	38	38	38	37	37	−0.39	39.1
	H	45	45	45	45	45	45	45	+0.1	44.6
2	V	34	33	33	32	32	31	31	−0.5	34.5
	H	42	41	41	40	40	39	39	−0.5	42.3
3	V	29	29	28	28	28	27	27	−0.46	29.9
	H	40	39	38	36	35	33	32	−1.4	41.8
4	V	29	28	27	26	24	23	22	−1.14	30.1
	H	39	38	36	35	33	32	31	−1.4	40.4
5	V	21	21	20	20	20	20	19	−0.23	21.3
	H	39	35	32	29	26	23	20	−3.1	41.6

60° Grazing Angle

0	V	33	33	32	32	32	31	31	−0.42	33.7
	H	33	32	32	32	31	30	30	−0.78	34
1	V	24	25	25	25	25	25	26	+0.18	24.3
	H	23	24	25	26	27	28	29	+1.0	21.6
2	V	22	22	22	21	21	21	20	−0.35	22.7

	H	21	22	22	22	22	23	23	+0.3	21.2
3	V	20	19	18	18	17	16	16	−0.64	20.1
	H	20	20	20	20	20	20	20	0	20.2
4	V	17	16	15	14	13	12	11	−0.93	17.4
	H	17	16	16	16	15	15	15	−0.3	16.9
5	V	14	13	11	10	9	7	6	−1.4	15.7
	H	16	14	13	12	11	9	8	−1.2	16.8

90° Grazing Angle*

0	V	−18	−18	−18	−18	−18	−18	−18	0	17.5
	H									
1	V	−15	−15	−14	−14	−13	−13	−13	0.39	−15.42
	H									
2	V	−13	−12	−11	−10	−9	−9	−9	0.71	13.29
	H									
3	V	−10	−9	−7	−6	−4	−4	−4	1.11	−10.71
	H									
4	V	−8	−6	−4	−2	+1	+1	+1	1.64	−9.00
	H									
5	V	−5	−3	0	+2.5	+5	+5	+5	1.82	−5.93
H										

*Median.

Note: Clutter return is given in decibels below 1 m^2 for each square meter illuminated.

Appendix D

ECM Descriptions

D.1 BARRAGE NOISE

Barrage noise (BN) is an ECM technique utilized by self-screening, stand-off, or support jammers to transmit noise-like energy over a wide frequency band to prevent detection of received target signals by a victim radar. Typical bandwidths range from 100 MHz to an octave. The barrage jammer will attempt to jam multiple victim radars by spreading noise-like energy over a frequency band encompassing the entire agile bandwidth of the victim radars. The success of barrage noise is directly related to the jamming power density. Widening the jammer's bandwidth lowers the resultant jamming power density, thus lowering the effectiveness of the technique and causing possible target burnthrough. The barrage noise jammer can produce wide band noise-like jamming power in a variety of ways.

D.2 BLINKING SPOT NOISE

Spot noise jamming is an ECM technique utilized by self-screening or support jammers to transmit narrow band noise-like energy to prevent detection of received target signals by an enemy radar. Spot noise implies narrow frequency band coverage, producing high jamming power density (watt/megahertz) in the bandwidth of the victim radar. The bandwidth of the spot noise jammer is usually just wide enough to cover the victim radar bandwidth. Typically, spot noise jammers can only jam a single radar and are not effective against pulse-to-pulse frequency agile radars. Many radars will have a home-on-jam (HOJ) or track-on-jam mode to counter spot noise. Blinking spot noise (BSN) jamming is a variation of spot noise jamming which alternates a spot noise waveform on and off at about a 50% duty cycle. The blinking rates are intended to confuse either a manual or automatic track-on-jam feature of the victim radar.

BSN is typically implemented with two or more jammers located on multiple aircraft separated in angle within the victim radar antenna main beam. The jammer's blink rates can be synchronous or asynchronous in an attempt to confuse the single target track mode of the victim radar. As the jammers alternately turn on and off, the victim radar antenna will shift its tracking direction, possibly causing transients and instability in the tracking function. Establishing the interaircraft control links that synchronize the multiple jammer blinking rates is critical to the success of this technique. Typical blink rates range from 0.1 to 10 Hz.

D.3 BLINKING COOPERATIVE DOPPLER NOISE

The technique of blinking cooperative doppler noise (BCDN) requires two aircraft to maintain an idealized two-ship cell configuration while flying toward the target. Each aircraft contains a repeater which receives, amplifies, and retransmits the received radar signal with doppler noise modulation. The narrow band doppler noise transmitted by each jammer is centered on the target skin line. The two aircraft employing the cooperative countermeasure are synchronized at a 50% duty factor blink rate.

D.4 COVER PULSE

Cover pulse (CP) jamming is a technique utilized by self-screening jammers to cover their platform skin return with an ECM pulse and prevent platform detection by a range constant false alarm rate (CFAR) detector. The CP can interfere with the pulse compression process, causing high time sidelobes that obscure the true target return. The ECM cover pulse is normally wider than the actual target skin return. Depending upon the complexity of the jammer, the ECM pulse may anticipate the platform return by several microseconds to completely cover the platform return.

D.5 CROSS-POLARIZATION JAMMING

Cross-polarization (XPOL) jamming is utilized as a self-screening or support ECM technique to induce angular tracking errors in tracking radars. The ECM system utilizes two orthogonally polarized receiving antennas to determine the polarization of the interrogating signal from the radar transmitter. Next, the jammer transmits a signal that is orthogonally polarized to the principal polarization of the radar transmitting and receiving antennas. The cross-polarized signal has sufficient gain and power such that the radar will attempt to use the cross-polarized lobe structure for tracking purposes. The erratic and unpredictable XPOL response of most tracking radars will produce an improper delta/sum ratio and unpredictable tracking errors with a tendency to move off the true target angle. In addition, cross-polarized sidelobe jamming cannot be countered with normal sidelobe suppression. EP techniques utilizing copolarized auxiliary antennas. The technique can be effective against monopulse antennas.

Adaptive cross-polarized jammers continually measure the polarization of the radar and adjust their transmit polarization to be orthogonal to the radar polarizations. Nonadaptive jammers sweep the polarization around the orthogonal polarization. The technique is used in conjunction with RGPO or velocity gate pull-off (VGPO) to achieve a high JSR.

D.6 GATE STEALER/TERRAIN BOUNCE

The technique of gate stealer/terrain bounce (GSTB) provides both velocity and angle deception against tracking radars. The technique begins by transmitting a VGPO waveform to pull the velocity gate of the victim radar away from the actual target skin return. After the VGPO program has pulled the velocity gate a sufficient distance, the jammer begins transmitting the terrain bounce ECM. While the jammer is transmitting the terrain bounce technique, the false doppler target remains in the maximum frequency pull-off position. The entire program recycles to ensure success.

D.7 MULTIPLE FALSE DOPPLER TARGETS

The ECM technique of multiple false doppler targets (MFDT) is a self-screening jamming technique aimed at complicating target velocity and range resolution, delaying target acquisition, and degrading tracking. The technique generates false doppler targets at equal frequency increments located symmetrically around the target skin line. Appropriate spacing of these false doppler targets can desensitize the CFAR detectors and prevent target detection.

D.8 NARROW BAND DOPPLER NOISE

Narrow band doppler noise (NBDN) is a self-screening, repeater jamming technique used against doppler radars. The victim radar's signal is amplified and modulated in frequency and amplitude to produce a narrow band of noise centered about the radar carrier frequency. Conventional CFAR algorithms may become desensitized by this ECM, preventing detection of the jammer skin return.

D.9 RANGE GATE PULL-OFF

Range gate pull-off (RGPO) is a self-screening repeater jamming technique aimed at automatic range tracking radars. The technique consists of three phases:

(1) range gate capture, (2) range gate pull-off, and (3) off time. During phase 1, the victim radar's signal is received, amplified, and retransmitted to provide a strong signal to capture the range gate of the victim radar. The time delay of the repeated signal is then modified on a pulse-by-pulse basis so as to walk the false signal away from the jammer platform return. The ECM pulses must be synchronized with the victim radar's pulse train to ensure the pull-off. The pull-off function is typically parabolic or linear, with a positive or negative slope. If this technique is used in conjunction with VGPO, the time derivative of the RGPO function should equal the VGPO function. Upon reaching the end of the pull-off sequence, the ECM repeater is turned off. With no signal to track, the victim radar is forced to break track and initiate a reacquisition phase. If possible, the true target is reacquired by the victim radar, and the entire process is repeated.

D.10 TERRAIN BOUNCE

Terrain bounce (TB) jamming is a self-screening or support ECM repeater technique. The technique consists of aiming the ECM antenna towards the Earth's surface to jam the victim radar. The ECM energy is bounced off the terrain into the victim radar antenna. Depending upon its goal of deception or denial, the terrain bounce jammer may use several different ECM waveforms. In the deception role, the ECM energy is reflected off the Earth's surface. This causes the false signal to appear at a different angle than that of the real target, thus producing false angle strobing in the victim radar. The tracking radar is forced to track the ground-imaged signal rather than the true target. TB ECM jammers utilize relatively narrow beam and low sidelobe antennas to ensure that the effect of the direct path of the jammer sidelobes on the victim radar is less than the effect due to the TB path. In the detection denial role, spot noise or narrow band doppler noise may be utilized to prevent detection and confuse HOJ modes.

D.11 TOWED DECOY

Towed decoy (TD) ECM utilizes a decoy carrying an active repeater towed behind the enemy aircraft. The repeater typically employs a variety of ECM techniques to generate false targets. Operating the repeater at a location separated from the real target creates a multi-target scenario and causes tracking errors in the victim radar.

D.12 VELOCITY GATE PULL-OFF

VGPO is a self-screening repeater jamming technique aimed at automatic velocity tracking radars. The technique consists of three phases: (1) velocity gate capture, (2) velocity gate pull-off, and (3) off time. During phase 1, the victim radar's signal is received, amplified, and retransmitted to provide a strong signal to capture the velocity gate of the victim radar. The doppler frequency of the false signal is sequentially varied or walked away from the true target doppler frequency. The pull-off rate cannot exceed the victim radar's tracking capability or reasonable target velocity maneuvers. The pull-off function can assume many shapes; the false target is usually pulled toward main lobe clutter. If this technique is used in conjunction with range gate pull-off, the time derivative of the RGPO function should equal the VGPO function. Upon reaching the end of the pull-off sequence, the ECM repeater is turned off. With no signal to track, the victim radar is forced to break track and initiate a reacquisition phase. If possible, the true target is reacquired by the victim radar, and the entire process is repeated.

Glossary

AL	acceleration limiting
ARPOL	adaptive receive polarization
AEW	airborne early warning
AI	airborne intercept
AM	amplitude modulation
A/D	analog-to-digital
AEE	angle extent estimator
AGC	automatic gain control
BE	bandwidth expansion
BN	barrage noise
BFD	beat frequency detector
BIT	bit in test
CRT	cathode-ray tube
CFAR	cell-averaging CFAR (C
C/OS CFAR	censored cell and ordered-statistic CFAR
C-CFAR	censored CFAR
CLT	central limit theorem
COHO	coherent oscillator
CPI	coherent processing interval
C^3	command, control, and communication
CAT	computer-aided tomography
CFAR	constant false alarm rate
CW	continuous wave
RCS	radar cross section
CPOLC	cross-polarization cancellation

DoD	Department of Defense
DRFM	digital RF memory
DFT	discrete Fourier transform
DTFT	discrete time Fourier transform
DPCA	displaced phase center antenna
DF-CFAR	distribution-free CFAR
DBS	doppler beam sharpening
D/RR	doppler-range rate
VGSDF	dual-frequency VGS
ERP	effective radiated power
EA	electronic attack
ECCM	electronic counter-countermeasures
ECM	electronic countermeasures
ESM	electronic support measures
EW	electronic warfare
EMCON	emissions control
FR	false range target
FFT	fast Fourier transform
FOV	field of view
FIR	finite impulse response
FA	frequency agility
FM	frequency modulation
FMCW	frequency-modulated continuous wave
GPS	Global Positioning System
GO-CFAR	greater-of-cell-average CFAR
HCE-CFAR	heterogeneous-clutter-estimating CFAR
HPA	high-power amplifier
HPM	high-power microwave
HOJ	home-on-jam
IFF	identification-friend-or-foe
IMU	inertial measurement unit
INS	inertial navigation system
IIR	infinite impulse response
IRST	infrared search andtrack
ISL	integrated side-lobe level
IF	intermediate frequency
JSR	jamming-to-signal ratio
JEM	jet engine modulation
LET	leading edge track
LCM	least common multiple

LOS	line of sight
LFM	linear frequency modulation
LO	local oscillator
LPG	loss in processing gain
LNA	low-noise amplifier
MLC	main-lobe clutter
MESA	maximum entropy spectral analysis
MLSE	maximum likelihood spectral estimate
MTBCF	mean time between critical failures
MOPA	master oscillator power amplifier
MTI	moving-target indication
MFDT	multiple false doppler targets
MUSIC	multiple signal classification
MTT	multiple-target tracking
NP	narrow pulse
NBDN	narrowband doppler noise
NBDND	narrowband doppler noise detector
NN	neural net
NCTR	noncooperative target recognition
OBT	off-boresight tracking
OS-CFAR	ordered-statistic CFAR
PSL	peakside-lobe level
PM	phase modulation
PDF	probability density function
PRF	pulse repetition frequency
PRFJ	pulse repetition frequency jitter
PRI	pulse repetition interval
RDP	radar data processor
RF	radio frequency
RGPO	range gate pull-off
RWS	range-while-search
rms	root-mean-square
SAR	synthetic-aperture radar
STC	sensitivity time control
SLB	side-lobe blanking
SLC	side-lobe cancellation
S/I	signal-to-interference ratio
STT	single-target tracking
SSPA	solid-state phased array
STAP	space-time adaptive processing

SN	spot noise
STALO	stable local oscillator
SRES	superresolution
TDS	track-during-scan
TWS	track-while-scan
TET	trailing edge track
TO	transmit oscillator
T/R	transmit/receive
VRR	variance reduction ratio
VGPO	velocity gate pull-off
VGG	velocity guard gates
VS	velocity search
VGPOR	VGPO reset
VCO	voltage-controlled oscillator

About the Editors

Guy V. Morris is chief of the Radar Systems Division, Sensors Laboratory at the Georgia Tech Research (GTRI) Institute. He is currently conducting research in electronic counter-countermeasures for airborne pulsed doppler radar and teaches continuing education courses in these topics. In 1996, Mr. Morris was elected as a Fellow of the Institute of Electrical and Electronics Engineers. He is a contributing author to *Principles and Applications of Millimeter Wave Radar, Principles of Modern Radar,* and *Coherent Radar Performance Estimation.* Prior to joining GTRI in 1984, he gained 25 years of experience in the radar industry at Hughes Aircraft, Magnavox, and Motorola as a design engineer in signal processing, as the lead system engineer on major new radar development and modification programs, and as an engineering manager. He has three patents in signal processing and is a licensed professional engineer. Mr. Morris holds an M.S. in electrical engineering from the University of Southern California and a B.E.E. from the Georgia Institute of Technology. He lives in Acworth, Georgia, with his wife Barbara, to whom he dedicated the first edition.

Linda L. Harkness is a senior research scientist in the Radar Systems Division, Sensors & Electromagnetic Laboratory at GTRI. She has over 15 years experience in radar systems performance analysis. She is currently involved in evaluating hardware obsolescence issues associated with the AN/APG-63 radar system on the F-15 aircraft. Her previous research activities have included modeling and simulation of airborne radar sensors and their susceptibility to electronic countermeasures (ECM). Ms. Harkness is a member of the Institute of Electrical and Electronics Engineers. She holds an M.S. in applied mathematics from Clemson University and a B.S. in mathematics from Stetson University.

Index

The Artech House Radar Library

David K. Barton, *Series Editor*

Active Radar Electronic Countermeasures, Edward J. Chrzanowski

Airborne Early Warning System Concepts, Maurice Long, *et al.*

Airborne Pulsed Doppler Radar, Second Edition, Guy V. Morris and Linda Harkness, editors

Analog Automatic Control Loops in Radar and EW, Richard S. Hughes

CARPET (Computer-Aided Radar Performance Evaluation Tool): Radar Performance Analysis Software and User's Manual, developed at the TNO Physics and Electronic Laboratory by Albert G. Huizing and Arne Theil

Coherent Radar Performance Estimation, James A. Scheer and James L. Kurtz, editors

DETPROB: Probability of Detection Calculation Software and User's Manual, William A. Skillman

Digital Techniques for Wideband Receivers, James Tsui

Electronic Homing Systems, M. V. Maksimov and G. I. Gorgonov

Electronic Intelligence: The Analysis of Radar Signals, Second Edition, Richard G. Wiley

Electronic Intelligence: The Interception of Radar Signals, Richard Wiley

Estimation and Tracking: Principles, Techniques, and Software, Yaakov Bar-Shalom and Xiao-Rong-Li

High Resolution Radar, Second edition, Donald R. Wehner

Introduction to Electronic Defense Systems, Fillippo Neri

Introduction to Electronic Warfare, D. Curtis Schleher

Introduction to Multisensor Data Fusion: Multimedia Software and User's Guide, TECH REACH Inc.

IONOPROP: Ionospheric Propagation Assessment Software and Documentation, Herbert Hitney

Kalman-Bucy Filters, Karl Brammer